AF746233

SEMA SIMAI Springer Series

Volume 41

Editors-in-Chief

Carlos Castro, Departamento de Matemática e Informática, ETSI Caminos, Canales y Puertos, Universidad Politécnica de Madrid, Madrid, Spain

Luca Formaggia, MOX–Department of Mathematics, Politecnico di Milano, Milano, Italy

Series Editors

Maria Groppi, Dipartimento di Scienze Matematiche, Fisiche e Informatiche, Università di Parma, Parma, Italy

Mats G. Larson, Department of Mathematics, Umeå University, Umeå, Sweden

Maria Lopez Fernandez, Department of Mathematical Analysis, Statistics and O.R., and Applied Mathematics, Faculty of Sciences, University of Malaga, Malaga, Spain

Tomás Morales de Luna, Departamento de Análisis Matemático, Estad. e I.O., y Matemática Aplicada, Facultad de Ciencias, Universidad de Málaga, Málaga, Spain

Lorenzo Pareschi, Dipartimento di Matematica e Informatica, Università degli Studi di Ferrara, Ferrara, Italy

Elena Vázquez-Cendón, Departamento de Matemática Aplicada, Universidade de Santiago de Compostela, A Coruña, Spain

Paolo Zunino, Dipartimento di Matematica, Politecnico di Milano, Milano, Italy

As of 2013, the SIMAI Springer Series opens to SEMA in order to publish a joint series aiming to publish advanced textbooks, research-level monographs and collected works that focus on applications of mathematics to social and industrial problems, including biology, medicine, engineering, environment and finance. Mathematical and numerical modeling is playing a crucial role in the solution of the complex and interrelated problems faced nowadays not only by researchers operating in the field of basic sciences, but also in more directly applied and industrial sectors. This series is meant to host selected contributions focusing on the relevance of mathematics in real life applications and to provide useful reference material to students, academic and industrial researchers at an international level. Interdisciplinary contributions, showing a fruitful collaboration of mathematicians with researchers of other fields to address complex applications, are welcomed in this series.

THE SERIES IS INDEXED IN SCOPUS

María José Ibáñez-Pérez · Paola Lamberti ·
Sara Remogna · Driss Sbibih
Editors

Mathematical and Computational Modelling, Approximation and Simulation

MACMAS 2023

Springer

Editors
María José Ibáñez-Pérez
Department of Applied Mathematics
University of Granada
Granada, Spain

Paola Lamberti
Dipartimento di Matematica "Giuseppe Peano"
Università di Torino
Turin, Italy

Sara Remogna
Dipartimento di Matematica "Giuseppe Peano"
Università di Torino
Turin, Italy

Driss Sbibih
Department of Mathematics
University Mohammed First
Oujda, Morocco

ISSN 2199-3041 ISSN 2199-305X (electronic)
SEMA SIMAI Springer Series
ISBN 978-3-032-01113-8 ISBN 978-3-032-01114-5 (eBook)
https://doi.org/10.1007/978-3-032-01114-5

© The Editor(s) (if applicable) and The Author(s), under exclusive license to Springer Nature Switzerland AG 2025

This work is subject to copyright. All rights are solely and exclusively licensed by the Publisher, whether the whole or part of the material is concerned, specifically the rights of translation, reprinting, reuse of illustrations, recitation, broadcasting, reproduction on microfilms or in any other physical way, and transmission or information storage and retrieval, electronic adaptation, computer software, or by similar or dissimilar methodology now known or hereafter developed.
The use of general descriptive names, registered names, trademarks, service marks, etc. in this publication does not imply, even in the absence of a specific statement, that such names are exempt from the relevant protective laws and regulations and therefore free for general use.
The publisher, the authors and the editors are safe to assume that the advice and information in this book are believed to be true and accurate at the date of publication. Neither the publisher nor the authors or the editors give a warranty, expressed or implied, with respect to the material contained herein or for any errors or omissions that may have been made. The publisher remains neutral with regard to jurisdictional claims in published maps and institutional affiliations.

This Springer imprint is published by the registered company Springer Nature Switzerland AG
The registered company address is: Gewerbestrasse 11, 6330 Cham, Switzerland

If disposing of this product, please recycle the paper.

Preface

This volume is devoted to the proceedings of the Second International Conference on Mathematical and Computational Modelling, Approximation and Simulation (MACMAS 2023) which was held from May 29 to June 1, 2023 in Torino, Italy (https://www.macmas2023.unito.it/).

The previous MACMAS Conference took place in Granada in 2019 and the MACMAS series Conferences are jointly organized by the University of Granada (Spain), the Mohammed First University—Oujda (Morocco), and the University of Torino (Italy).

The following are among its goals:

- gather the researchers interested in the field of approximation, numerical modelling and their applications;
- reinforce the scientific interchange between researches from different countries, with special emphasis on the Mediterranean basin and neighbourhood;
- encourage young researchers to present the results of their scientific works;
- help new international research projects to arise by facilitating contacts between tentative partners.

The book is divided into two parts. The first part presents four papers based on the plenary lectures given by the invited speakers Mariantonia Cotronei, Rachid Jennane, Juan Manuel Peña and Tatyana Sorokina.

In particular, Cotronei's paper is a revisitation of the construction of wavelet filters: this paper aims to illustrate and summarize the research carried out in the last few years by the authors and collaborators in the direction of exploring new strategies for realizing novel wavelet filterbank systems, but also of unifying different existing approaches, both in the univariate and in the bivariate cases.

Jennane's paper concerns the application of fractal analysis to biomedical imaging. Indeed, among the fractal models, the fractional Brownian motion (fBm) is well suited for the characterization of images involving a varying range of gray levels and this paper focuses on the use of fBm for image analysis.

Peña's paper is devoted to recent advances on accuracy, stability and optimality of B-bases, that play an important role in Approximation Theory and Computer Aided

Geometric Design, focusing on the relationship of these fields in the framework of totally positive bases and matrices.

Finally, Sorokina's paper deals with intrinsic supersmoothness of piecewise functions, that is a consequence of the geometry of the underlying partition. In this paper, the author provides insights into this phenomenon, reviews and generalizes known results, and demonstrates how intrinsic supersmoothness affects other properties of spline spaces.

The second part is devoted to the submissions received from participants. They underwent a peer-review process and nine manuscripts, from the different topics covered at the conference, were retained for publication.

In particular, the paper by Aimi et al. is a preliminary work for the analysis of strategies to valuate weather derivatives using the complete Italian archive of average temperature data of the city of Parma. The paper by Aràndiga is concerned with the approximation order of ENO-SR piecewise polynomial interpolatory techniques for piecewise smooth functions with isolated singularities. The paper by Barrera et al. is about Powell-Sabin finite element point interpolation via the constrained mock-Chebyshev least squares operator and presents a novel technique for approximating derivatives using an interpolation-regression operator. The paper by Ben Amara et al. addresses the problem of existence, uniqueness and approximation of the solution of hybrid fractional integro-differential equations with nonlocal boundary value conditions. The paper by El Bouayadi et al. explores a modified Nyström approach based on the zeros of the Jacobi polynomials for solutions of Volterra integral equations with weakly singular kernels in weighted spaces, employing the interpolating operator in suitable weighted spaces and the associated product integration formulae. The paper by El Mokhtari presents a method based on C^1 quadratic polynomial splines for approximating solution of Love's integral equation. The paper by Lamberti et al. deals with the study of the interpolation of a bidirectional B-spline curve network by both biquadratic tensor product B-spline surfaces and quadratic B-spline surfaces on criss-cross triangulations with free parameters. The paper by Lamnii et al. treats the numerical solution of Sturm-Liouville-type differential equation with Dirichlet boundary conditions based on wavelet bases of Hermite C^1 quadratic splines. Finally, the paper by Sefti et al. concerns a deep network-based spline active contour method for medical image segmentation, proposing a combination of two convolutional neural networks and a deformable model using spline approximation functions.

We want to thank all the contributors who have coauthored the papers contained in this volume, as well as the anonymous referees who have revised the manuscripts, improving them with their comments and suggestions.

We would also like to express our gratitude to Francesca Bonadei from Springer for her patience, attention and support at every step in the editorial process.

Granada, Spain — María José Ibáñez-Pérez
Turin, Italy — Paola Lamberti
Turin, Italy — Sara Remogna
Oujda, Morocco — Driss Sbibih
February 2025

Contents

Contributors

Alessandra Aimi Department of Mathematical, Physical and Computer Sciences, University of Parma, Parma, Italy;
Member of the INdAM-GNCS Research Group, Torino, Italy

Chafik Allouch The Multidisciplinary Faculty of Nador, Team of Modeling and Scientific Computing, Nador, Morocco

Francesc Aràndiga Departament de Matemàtica Aplicada, Universitat de València, Burjassot, Spain

Domingo Barrera Department of Applied Mathematics and Institute of Mathematics (IMAG), University of Granada, Granada, Spain

Khaled Ben Amara Higher Institute of Computer Science and Multimedia of Gabès (ISIMG), University of Gabès, Gabès, Tunisia

María Isabel Berenguer Department of Applied Mathematics and Institute of Mathematics (IEMath-GR), E.T.S. de Ingenieria de Edificación, University of Granada, Granada, Spain

Ahmed Boujraf ENSAH, LSA Laboratory, University Abdelmalek Essaadi, Al-Hoceima, Morocco

Costanza Conti DIEF, Università di Firenze, Firenze, Italy

Mariantonia Cotronei DIIES, Università Mediterranea di Reggio Calabria, Reggio Calabria, Italy

Francesco Dell'Accio Department of Mathematics and Computer Science, University of Calabria, Rende, Italy;
Istituto per le Applicazioni del Calcolo "Mauro Picone", National Research Council of Italy, Napoli, Italy

Salah Eddargani Department of Applied Mathematics and Institute of Mathematics (IMAG), University of Granada, Granada, Spain

Abdelmajid El Bouayadi The Multidisciplinary Faculty of Nador, Team of Modeling and Scientific Computing, Nador, Morocco

Fadila El Mokhtari LANO Laboratory, FSO, University Mohammed First, Oujda, Morocco;
Research Center, High Studies of Engineering School, EHEI, Oujda, Morocco

Chiara Guardasoni Department of Mathematical, Physical and Computer Sciences, University of Parma, Parma, Italy;
Member of the INdAM-GNCS Research Group, Torino, Italy

Rachid Jennane IDP Laboratory, UMR CNRS 7013, University of Orleans, Orleans, France

Rachid Jennane IDP Institute UMR CNRS 7013, University of Orleans, Orleans, France

Paola Lamberti Department of Mathematics "G. Peano", University of Torino, Torino, Italy

Abdellah Lamnii Abdelmalek Essaadi University, ISI, ENSATe, Tetouan, Morocco

Mohamed Lamnii Laboratoire LANO, Faculty of Sciences Oujda, University Mohammed First, Oujda, Morocco

Chaimae Mouhoub Laboratoire LANO, Faculty of Sciences Oujda, University Mohammed First, Oujda, Morocco

Federico Nudo Department of Mathematics and Computer Science, University of Calabria, Rende, Italy

Luis Ortiz-Gracia Department d'Econometria, Estadística i Economia Aplicada, Universitat de Barcelona (UB), Barcelona, Spain;
RISKcenter, Institut de Recerca en Economia Aplicada (IREA), Universitat de Barcelona (UB), Barcelona, Spain

Juan Manuel Peña Dep. Applied Mathematics/IUMA, Universidad de Zaragoza, Zaragoza, Spain

Sara Remogna Department of Mathematics "G. Peano", University of Torino, Torino, Italy

Driss Sbibih Laboratoire LANO, Faculty of Sciences Oujda, University Mohammed First, Oujda, Morocco

Driss Sbibih LANO Laboratory, Faculty of Sciences, University Mohammed First, Oujda, Morocco

Rania Sefti LANO Laboratory, Faculty of Sciences, University Mohammed First, Oujda, Morocco

Tatyana Sorokina Towson University, Towson, MD, USA

Mohamed Tahrichi EST, ANAA Team, LANO Laboratory, University Mohammed First of Oujda, Oujda, Morocco

Francesco Zanni Department of Mathematical, Physical and Computer Sciences, University of Parma, Parma, Italy

Plenary Lectures

Construction of Wavelet Filters: A Revisitation

Costanza Conti and Mariantonia Cotronei

This paper is dedicated to Maria Charina, a friend and collaborator whose energy and kindness in doing research and exploring mathematics in a team we immensely miss.

Abstract Wavelets have been playing a prominent role in approximation theory, signal processing, and many other different contexts for several decades. A lot of effort has been spent on designing wavelet bases, in particular in the multivariate setting. Nevertheless, general procedures for their construction remain challenging tasks. This paper aims to illustrate and summarize the research carried out in the last few years by the authors and collaborators in the direction of exploring new strategies for realizing novel wavelet filterbank systems but also of unifying different existing approaches, both in the univariate and in the bivariate cases.

1 Introduction

Wavelets have been introduced by mathematicians and physicists in the early '80s as tools for the analysis of functions. Over the last few decades, they have found extensive application across various fields, particularly in extracting information from many kinds of data, including audio signals and images.

Compiling an exhaustive list of literature on this topic from the 1980s onwards is a formidable task. Searching the "wavelet" keyword in Scopus using the "Article title" field yields nearly 85,000 papers as a result! Therefore, we limit our referenced list here to a selection of seminal monographs [10, 11, 26, 34, 35, 41, 43, 47], deferring

C. Conti (✉)
DIEF, Università di Firenze, Firenze, Italy
e-mail: costanza.conti@unifi.it

M. Cotronei
DIIES, Università Mediterranea di Reggio Calabria, Reggio Calabria, Italy
e-mail: mariantonia.cotronei@unirc.it

© The Author(s), under exclusive license to Springer Nature Switzerland AG 2025
M. J. Ibáñez-Pérez et al. (eds.), *Mathematical And Computational Modelling, Approximation and Simulation*, SEMA SIMAI Springer Series 41,
https://doi.org/10.1007/978-3-032-01114-5_1

the citation of specific references containing results most closely related to ours to the next sections.

Since Daubechies' groundbreaking work [25], numerous orthogonal wavelet and filterbank systems have been proposed, each employing specific strategies rather than a general procedure for their construction. In contrast, in some recent studies [8, 9], we have explored the possibility of unifying these disparate methodologies, both in univariate and bivariate scenarios. This paper aims to systematically present and unify such findings, accompanied by a straightforward introduction to wavelets necessary for comprehension.

Specifically, we provide a characterization of compactly supported orthogonal wavelet and multi-wavelet filters in the univariate case, relying on classical results from system theory [8]. The bivariate setting poses a more intricate challenge. Here, we show how to express the inherent structure of two-band orthogonal filters through normal forms of the corresponding bivariate polyphase representations, achieved via appropriate unitary transformations [9]. We enrich this review with comments and examples.

2 Background

We start by recalling Daubechies' orthonormal *scaling functions* and *wavelets* of order $N \geq 1$. They are a family of function pairs $\phi, \psi \in L^2(\mathbb{R})$ respectively satisfying a *refinement equation*

$$\phi = 2 \sum_{k=0}^{2N-1} \phi(2 \cdot -k) p_k \,,$$

and the relation

$$\psi = 2 \sum_{k=0}^{2N-1} \phi(2 \cdot -k) q_k \,,$$

(sometimes called the *wavelet two-scale relation*) for some *filter sequences* $p = (p_k \in \mathbb{R} : k = 0 \ldots, 2N-1)$, $q = (q_k \in \mathbb{R} : k = 0 \ldots, 2N-1)$, also known as *masks*. The order parameter N identifies the family member. They pair ϕ, ψ, is supported on $[0, 2N-1]$, and satisfies the *orthonormality conditions*

$$\int \phi(x)\phi(x-k)dx = \int \psi(x)\psi(x-k)dx = \delta_{k0} \,,$$

$$\int \phi(x)\psi(x-k)dx = 0 \quad k \in \mathbb{Z} \,.$$

While the integral of the scaling function is 1, the wavelet possesses N *vanishing moments*, that is:

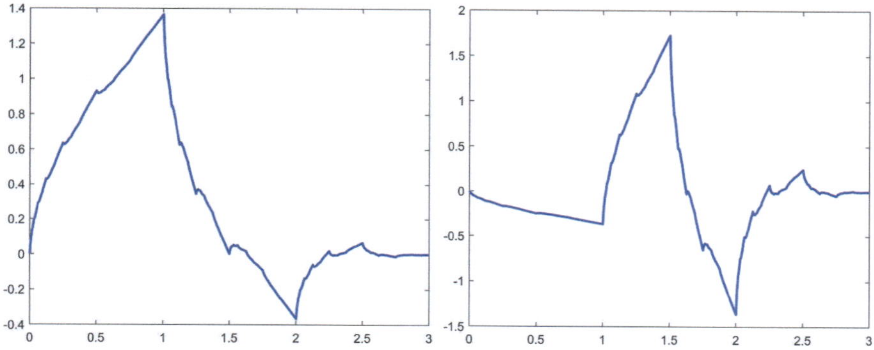

Fig. 1 Daubechies' scaling function (left) and wavelet (right) for $N = 2$. The respective filter sequences are $p = (\frac{1}{8}(1+\sqrt{3}), \frac{1}{8}(3+\sqrt{3}), \frac{1}{8}(3-\sqrt{3}), \frac{1}{8}(1-\sqrt{3}))$, $q = \frac{1}{8}(1-\sqrt{3}), -\frac{1}{8}(3-\sqrt{3}), \frac{1}{8}(3+\sqrt{3}), -\frac{1}{8}(1+\sqrt{3})$

$$\int \psi(x)dx = \int x\psi(x)dx = \cdots = \int x^k\psi(x)dx = 0, \quad k = 0, \ldots, N-1 .$$

The Daubechies' function pair of order 2 (2 vanishing moments and 4 mask elements or *taps*) is shown in Fig. 1.

Since the properties of the functions translate into properties on the filter sequences p, q, the elegant construction provided by Daubechies actually involves the manipulation of the *symbols*, i.e. of the Laurent polynomials:

$$p(z) = \sum_{j=0}^{n} p_j z^j, \quad q(z) = \sum_{j=0}^{n} q_j z^j ,$$

on the *unit circle*

$$\mathbb{T} = \{z \in \mathbb{C} \, : \, |z| = 1\}.$$

The orthogonality conditions on the functions imply the following orthogonality conditions on the symbols (on the unit circle):

$$\begin{array}{l} \bar{p}(z^{-1})\, p(z) + \bar{p}(-z^{-1})\, p(-z) = 2, \\ \bar{p}(z^{-1})\, q(z) + \bar{p}(-z^{-1})\, q(-z) = 0 \\ \bar{q}(z^{-1})\, q(z) + \bar{q}(-z^{-1})\, q(-z) = 2, \end{array} \quad , \quad z \in \mathbb{T}. \tag{1}$$

We mention that the *filterbank* pairs

$$(p, q) = (p_j, q_j : j \in \mathbb{Z}), \quad (p^*, q^*) = (\bar{p}_{-j}, \bar{q}_{-j} : j \in \mathbb{Z})$$

play a role in the *discrete wavelet transform* (DWT). If $c^{(1)}$ is a generic input signal, the generic step of the DWT can be described by the following flow:

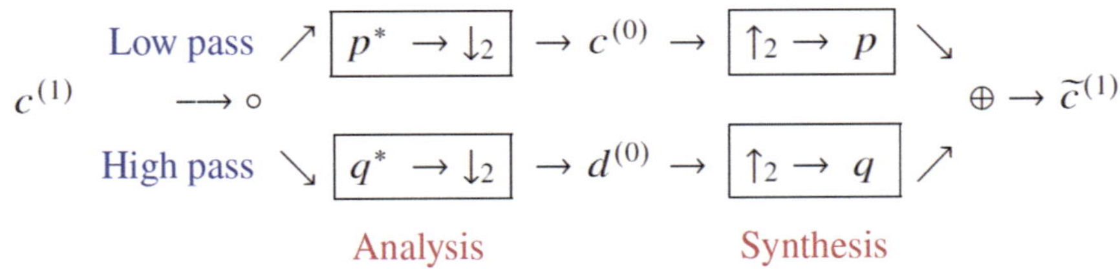

where the analysis is implemented in terms of convolution with p^*, q^* followed by downsampling, while the synthesis requires upsampling first and convolution with p, q afterward.

Perfect reconstruction is achieved whenever $c^{(1)} = \tilde{c}^{(1)}$ and this necessarily implies the following *perfect reconstruction conditions* on the symbols on the unit circle:

$$\begin{array}{l} p(z)\,\bar{p}(z^{-1}) + q(z)\,\bar{q}(z^{-1}) = 2, \\ p(-z)\,\bar{p}(-z^{-1}) + q(-z)\,\bar{q}(-z^{-1}) = 0, \end{array} \quad z \in \mathbb{T}. \tag{2}$$

It can be proved that such conditions are equivalent to the orthogonality conditions. On this assumption, according to the Daubechies' construction, the scaling filter p is found as a solution to the Bézout identity:

$$p^*(z)\,p(z) + p^*(-z)\,p(-z) = 2, \quad z \in \mathbb{T},$$

while the wavelet filter is found through the *alternating flip trick*:

$$q(z) = z^N p^*(-z)\,,$$

justifying the name of *quadrature mirror filter*. On the basis of this, with an abuse of terminology, we will later refer to (1) as QMF (quadrature mirror filter) conditions.

The question is how to generalize the construction to the following situations:

(a) Vector/matrix cases, where:

$$\phi, \psi : \mathbb{R} \to \mathbb{C}^{r\times m}, \quad r \le m,$$

and the two-scale relations are in matrix/vector form:

$$\phi = 2\sum_{j\in\mathbb{Z}} \phi(2\cdot - j)p_j, \quad \psi = 2\sum_{j\in\mathbb{Z}} \phi(2\cdot - j)q_j, \tag{3}$$

with

$$p_j, q_j \in \mathbb{C}^{r\times r}, \quad j \in \mathbb{Z};$$

(b) Multivariate case, where:

$$\phi, \psi : \mathbb{R}^s \to \mathbb{C}, \quad s > 1,$$

and, for a given *dilation* or *scaling matrix* matrix $M \in \mathbb{Z}^{s \times s}$ (required to be expansive, i.e. with spectral radius greater than one), with $d = |\det(M)|$, the following *d-band* relations must be satisfied:

$$\phi = d \sum_{j \in \mathbb{Z}^s} \phi(M \cdot -j) p_j, \quad \psi_\ell = d \sum_{j \in \mathbb{Z}^s} \phi(M \cdot -j) q_j^\ell, \quad \ell = 1, \ldots, d-1, \tag{4}$$

with

$$p_j, q_j^\ell \in \mathbb{C}, \quad j \in \mathbb{Z}^s.$$

In the next sections, we propose generalizations of Daubechies' construction in the 1-D case, including the vector/matrix situations, and in the 2-D case, with a specific dilation matrix M.

3 The 1-D Case: Wavelet Filters and System Theory

In this section, we specialize in the univariate setting and show that there are no intrinsic differences between the elegant construction of wavelets by Daubechies, in the scalar case, or any other construction of scalar/vector/matrix-valued wavelets, as provided, for example, in [4, 6, 12, 14, 15, 17, 18, 20–22, 27, 28, 32, 33, 40, 42, 44, 46, 49].

To this purpose, let us define the matrix polynomial map

$$F : \mathbb{C} \to \mathbb{C}^{2r \times 2r}, \quad F(z) = \sum_{j=0}^{N} F_j z^j, \quad z \in \mathbb{C}, \tag{5}$$

with matrix coefficients

$$F_j = \sqrt{2} \begin{pmatrix} p_{2j} & q_{2j} \\ p_{2j+1} & q_{2j+1} \end{pmatrix} \in \mathbb{C}^{2r \times 2r}, \quad j = 0, \ldots, N.$$

The entries of F are usually called the *polyphase components* of $p(z), q(z)$.

We now consider alternative ways to write the orthogonality and the perfect reconstruction conditions (1), (2). In particular, orthogonality is equivalent to

$$I - F^*(z) F(z) = 0, \quad z \in \mathbb{T},$$

which we call the *QMF (Quadrature Mirror Filter) condition*, while perfect reconstruction is equivalent to

$$I - F(z) F^*(z) = 0, \quad z \in \mathbb{T},$$

also known as *UEP (Unitary Extension Principle) condition* [45].

We remark that they are different equalities but represent equivalent conditions.

Let us now change perspective, and move from the unit circle $\mathbb{T}$ to the *unit disk*

$$\mathbb{D} = \{z \in \mathbb{C} \; : \; |z| < 1\}.$$

By the maximum principle, $I - F^*(z)F(z) = 0$ on the boundary of $\mathbb{D}$ implies

$$F(z) \quad \text{is contractive (i.e.} \|F(z)\| < 1) \text{ in } \mathbb{D}$$

so we can make use of the following classical result [1, 50].

Theorem 3.1 *Let $F : \mathbb{D} \to \mathbb{D}^{M\times M}$ be holomorphic. The following are equivalent:*

1. *F is contractive on $\mathbb{D}$;*
2. *there exists a unitary matrix $ABCD := \begin{pmatrix} A & B \\ C & D \end{pmatrix}$ such that*

$$F(z) = A + Bz(I - Dz)^{-1}C, \quad z \in \mathbb{D}.$$

Our goal is to characterize the polyphase matrix $F(z)$, and therefore the filters p and q, of any orthogonal wavelet system in terms of appropriate blocks A, B, C, D.

Remark 3.1 We observe that this incidentally establishes a connection to system theory. Specifically, $F(z)$ can be viewed as the "transfer function" of a continuous-time input/state/output (i/s/o) linear system, as defined by the system of equations:

$$\begin{cases} y(t) = A\,u(t) + B\,x(t) \\ \dot{x}(t) = C\,u(t) + D\,x(t) \end{cases}$$

where x is the state variable, u is the input variable and y is the output variable. □

3.1 Structure of A, B, C, D for $N = 1$

In order to provide a simple explanation of our arguments, we first restrict to the case of shortest support, that is $N = 1$ in (5).

We observe that on the boundary of $\mathbb{D}$ (i.e. for $z = e^{-iw}$) we have $\bar{z} = z^{-1}$, which implies, from the QMF:

$$I - \underbrace{(F_0^* + F_1^*\bar{z})}_{F^*(z)}\underbrace{(F_0 + F_1 z)}_{F(z)} = 0 \iff F_0^* F_1 = F_1^* F_0 = 0$$

while inside the unit disk $\mathbb{D}$:

$$I - (F_0^* + F_1^* \bar{z})(F_0 + F_1 z) = I - F_0^* F_0 - F_1^* F_1 - F_0^* F_1 z - F_1^* F_0 \bar{z}$$
$$+ F_1^* F_1 - F_1^* F_1 |z|^2 = (1 - |z|^2) F_1^* F_1.$$

Thus, additional information

$$I - F^*(z) F(z) = (1 - |z|^2) F_1^* F_1, \quad z \in \mathbb{D},$$

is available about $I - F^* F$ inside $\mathbb{D}$. It can be equivalently written as

$$F^*(z) F(z) + F_1^* F_1 = I + |z|^2 F_1^* F_1, \quad z \in \mathbb{D},$$

that is

$$(F^*(z) \;\; F_1^*) \begin{pmatrix} F(z) \\ F_1 \end{pmatrix} = (I \;\; \bar{z} F_1^*) \begin{pmatrix} I \\ z F_1 \end{pmatrix}, \quad z \in \mathbb{D}.$$

The last equality is interpreted as a norm-preserving condition. So

$$\begin{pmatrix} F(z) \\ F_1 \end{pmatrix} = \begin{pmatrix} A & B \\ C & D \end{pmatrix} \begin{pmatrix} I \\ z F_1 \end{pmatrix},$$

represents an isometry, that is the matrix $\begin{pmatrix} A & B \\ C & D \end{pmatrix}$ is unitary. This gives

$$\begin{cases} F_0 + F_1 z = A + B z F_1 \\ [8pt] F_1 = C + D z F_1 \\ [8pt] \begin{pmatrix} A & B \\ C & D \end{pmatrix} \begin{pmatrix} A^* & B^* \\ C^* & D^* \end{pmatrix} = \begin{pmatrix} I & 0 \\ 0 & I \end{pmatrix}, \end{cases}$$

which yield

$$F(z) = A + B z (I - D z)^{-1} C, \quad z \in \mathbb{D},$$

with

$$A = F_0, \; C = F_1, \; B = F_1 U, \; D = F_0 U \text{ and } U = F_0^* + F_1^*.$$

We have thus found a structure for the elements of the $ABCD$ matrix in the shortest support case.

3.2 Structure of A, B, C, D General Case

In [8] we have proved the following more general result.

Theorem 3.2 *$F(z)$ satisfies the QMF & UEP conditions on the unit circle iff*

$$F(z) = A + Bz(I - Dz)^{-1}C, \quad z \in \mathbb{D},$$

with

$$\left(\begin{array}{c|c} A & B \\ \hline C & D \end{array}\right) = \left(\begin{array}{c|cccc} F_0 & F_N & \dots & \dots & F_1 \\ \hline F_1 & F_0 & F_N & & F_2 \\ \vdots & F_1 & \ddots & \ddots & \vdots \\ \vdots & & \ddots & \ddots & F_N \\ F_N & F_{N-1} & \dots & F_1 & F_0 \end{array}\right) \left(\begin{array}{c|c} I & 0 \\ \hline 0 & U \end{array}\right),$$

where

$$U = \begin{pmatrix} F_0^* + F_N^* & F_1^* & \dots & F_{N-1}^* \\ F_{N-1}^* & \ddots & \ddots & \vdots \\ \vdots & \ddots & \ddots & F_1^* \\ F_1^* & \dots & F_{N-1}^* & F_0^* + F_N^* \end{pmatrix}.$$

Furthermore the matrices B, C, D satisfy:

$$B \begin{pmatrix} C \\ DC \\ D^2C \\ \vdots \\ D^{N-1}C \end{pmatrix} = C, \quad B\,D^N = 0.$$

In particular, this last statement guarantees that the $ABCD$ representation leads to a polynomial F in consideration that terms of order $\geq N$ in the Neumann series for $(I - Dz)^{-1}$ are annihilated.

The proof of the Theorem is found in [8]. The result is very general and interesting: We can unify wavelet and multiwavelet constructions - they all are of $ABCD$-nature.

In virtue of such a result, many well-known wavelet orthogonal filters, from scalar ones, i.e. Daubechies', to vector (e.g. [4, 12, 28, 46]) and matrix filters (e.g. [5, 15–18, 21, 22]) can be "re-discovered". Many examples can be found in [8].

In addition, the possibility of expressing the entire filterbank system (including both the scaling and the wavelet components) with an $ABCD$ representation, reveals particularly useful for deriving algorithms for the construction of new filters in the vector/matrix setting, where the non-commutativity constraints make it difficult even to associate a wavelet to a given scaling function. The characterization result offers the possibility to parametrize all orthogonal families of (multi) filters and to use the degrees of freedom to impose additional conditions, for example the so-called *full rank requirement* [16, 17], which fixes the unitary matrix U to be $U = \frac{\sqrt{2}}{2}\begin{pmatrix} I & W \\ I & -W \end{pmatrix}$, $W^*W = I$, and additional sum rules/vanishing moment conditions.

3.3 Remarks Concerning the Extension to the Bivariate Case

We conclude the section by mentioning that the core of our construction in the 1-D case, as explained in detail in [8], relies on the possibility of expressing:

$$I - F^*(\eta)F(\xi) = (1 - \bar{\eta}\xi)K(\eta, \xi), \quad \eta, \xi \in \mathbb{D},$$

with K positive-definite kernel on the disk $\mathbb{D}$. In order to generalize such a construction to the bivariate case, one needs the so-called *Schur-Agler decomposition* [36, 37]:

$$I - F^*(\eta)F(\xi) = (1 - \bar{\eta}_1\xi_1)K_1(\eta, \xi) + (1 - \bar{\eta}_2\xi_2)K_2(\eta, \xi),$$

with K_1, K_2 positive-definite kernels, on the polydisk

$$\mathbb{D}^2 = \{(\xi_1, \xi_2) \in \mathbb{C}^2 : |\xi_1|, |\xi_2| < 1\}.$$

As stated in [2], the existence of such a decomposition is in fact equivalent to the existence of the $ABCD$ representation. Nevertheless this decomposition is hard to find, except in a few particular cases.

For example, let us consider the scalar bivariate case represented by the polynomial map $F : \mathbb{C}^2 \to \mathbb{C}^{2\times 2}$ of bi-degree $(1, 1)$, i.e.

$$F(\xi) = F_{00} + F_{10}\xi_1 + F_{01}\xi_2 + F_{11}\xi_1\xi_2, \quad \xi \in \mathbb{D}^2,$$

that satisfies the QMF-condition on the bi-torus $\mathbb{T}^2 = \{(\xi_1, \xi_2) \in \mathbb{C}^2 : |\xi_1| = |\xi_2| = 1\}$. The only non-trivial case is when

$$\operatorname{rank}(F_{00}) = \operatorname{rank}(F_{10}) = \operatorname{rank}(F_{01}) = \operatorname{rank}(F_{11}) = 1.$$

It can be verified that the required Schur-Agler decomposition is given by

$$I - F(\eta)^*F(\xi) = (1 - \xi_1\bar{\eta}_1)\begin{pmatrix} 0 & 0 \\ 0 & 1 \end{pmatrix} + (1 - \xi_2\bar{\eta}_2)E_{01}(\xi_1)\,U\begin{pmatrix} 0 & 0 \\ 0 & 1 \end{pmatrix}U^*\,E_{01}(\eta_1)^*,$$

where

$$E_{01}(\xi_1) = \begin{pmatrix} 1 & 0 \\ 0 & \xi_1 \end{pmatrix}, \quad E_{10}(\eta_1) = \begin{pmatrix} \eta_1 & 0 \\ 0 & 1 \end{pmatrix},$$

and U is a unitary matrix. This allows us to employ the same arguments as in the univariate case and find an $ABCD$ representation of the polynomial map F.

Unfortunately, with the exception of few other particular cases, we could not find a general procedure for obtaining a proper decomposition of $I - F(\eta)^* F(\xi)$. This is indeed a challenging problem that deserves further exploration.

Nevertheless, while delving into the issue and aiming at obtaining a comprehensive characterization also in the 2-D realm, we devised an alternative approach. This approach enabled us to obtain such a parametrization for a very general class of filterbanks, as described in the next section.

4 The 2-D Case: Demystifying 2-Band Wavelets

Besides the several constructions proposed in the literature ([3, 13, 23, 24, 31, 38] just to mention a few), a complete characterization of bivariate, finitely supported orthogonal wavelet systems is a challenging task. We provide here such a characterization for the particular case where $M \in \mathbb{Z}^{2\times 2}$ is a *dilation matrix* with determinant 2 or -2, i.e. $d = 2$. The details can be found in [9].

We recall that all the possible dilation matrices with $\det(M) = \pm 2$ associated with wavelet systems are characterized in [39] and, up to an unimodular similarity, are given by:

$$\begin{pmatrix} 0 & 2 \\ \pm 1 & 0 \end{pmatrix}, \quad \pm\begin{pmatrix} 1 & 1 \\ -1 & 1 \end{pmatrix}, \quad \pm\begin{pmatrix} 0 & 2 \\ -1 & 1 \end{pmatrix}.$$

We also recall that the dilation matrix M defines a *sampling lattice* $\Gamma = M\mathbb{Z}^2$ where $2 = |\det(M)|$ is the number of *cosets*. The cosets are shifted version of Γ, i.e. they are expressed as $M\mathbb{Z}^2 + \mu_j$, $j = 0, 1$, in terms of the coset representatives:

$$\mu_0 = \begin{pmatrix} 0 \\ 0 \end{pmatrix}, \mu_1 = \begin{pmatrix} 1 \\ 0 \end{pmatrix}.$$

It is well known that

$$\mathbb{Z}^2 = M\mathbb{Z}^2 \cup (\mu_1 + M\mathbb{Z}^2).$$

We consider *2-band wavelet systems* generated by the function pair $\phi, \psi \in L^2(\mathbb{R}^2)$ satisfying

$$\phi = 2 \sum_{\alpha \in \mathbb{Z}^2} p_\alpha \phi(M \cdot - \alpha), \quad \psi = 2 \sum_{\alpha \in \mathbb{Z}^2} q_\alpha \phi(M \cdot - \alpha),$$

where now

$$p = (p_\alpha \in \mathbb{R} : \alpha \in \mathbb{Z}^2) \quad q = (q_\alpha \in \mathbb{R} : \alpha \in \mathbb{Z}^2)$$

are respectively the low- and high-pass bivariate filters. For these types of wavelets any specific constructions are available [7, 29, 38, 39, 48].

To get our complete and general characterization, it is convenient to consider the polyphase representation of the filters, expressed in terms of the bivariate matrix polynomial

$$F(z) = \sum_{\gamma \in \mathbb{Z}^2} F_\gamma \, z^\gamma, \quad z \in \mathbb{C}^2 \setminus \{(0,0)\},$$

with

$$F_\gamma = \sqrt{2} \begin{pmatrix} p_{M\gamma} & q_{M\gamma} \\ p_{M\gamma+\mu_1} & q_{M\gamma+\mu_1} \end{pmatrix},$$

where, in virtue of the orthogonality between ϕ, ψ, the following relation between the low-pass filter p and the high-pass filter q holds true:

$$q_\beta = (-1, -1)^{\beta - M\lceil M^{-1} \rceil} p_{\mu_1 - \beta}, \qquad \beta \in \mathbb{Z}^2.$$

The Quadrature Mirror Filter (QMF) and Unitary Extension Principle (UEP) conditions take the form:

$$\begin{aligned} F^*(z)F(z) &= I, \\ F(z)F^*(z) &= I, \end{aligned} \qquad z \in \mathbb{T}^2 = \{(\zeta_1, \zeta_2) \in \mathbb{C}^2 : |\zeta_1| = |\zeta_2| = 1\}.$$

It is worthwhile to mention that any wavelet decomposition is related to the approximation properties of the scaling function, which translate into the following *sum rules* (up to a certain order $L+1$, $L \in \mathbb{N}_0$) and *normalization* conditions:

$$\begin{pmatrix} 1 & -1 \end{pmatrix} \sum_{\gamma \in \mathbb{Z}^2} \begin{pmatrix} (M\gamma)^\beta & 0 \\ 0 & (M\gamma + \mu_1)^\beta \end{pmatrix} F_\gamma = \begin{pmatrix} 0 & \lambda_\beta \end{pmatrix}, \beta \in \mathbb{N}_0^2, \quad |\beta| \le L, \lambda_\beta \ne 0, \tag{6}$$

$$\sum_{\alpha \in \mathbb{Z}^2} p_\alpha = 1. \tag{7}$$

A crucial concept in our characterization is the so-called *normal form* of the polyphase representation, defined as follows.

We say that the polyphase matrix $F(z)$ possesses a *normal form* if

$$F(z) = U^T \, \widetilde{F}(z) \, V^T,$$

for some unitary transformations U and V, where:

1 every matrix coefficient in $\widetilde{F}(z) = \sum_{\gamma \in \mathbb{Z}^2} \widetilde{F}_\gamma z^\gamma$ possesses at most one nonzero element, except possibly for one matrix which possesses two nonzero elements on the diagonal or on the off-diagonal;
2 the nonzero elements of $\widetilde{F}(z)$ correspond to univariate QMF filters.

We show that the polyphase representation of the wavelet systems of the type described above always admits a normal form. To this purpose, we first discuss the simplest case, i.e. the $(1, 1)$ bi-degree case, we then state a more general result.

4.1 Characterization in Terms of Normal Forms: A Motivating Example

We consider the Belogay-Wang scaling filter [7], associated to the dilation matrix $M = \begin{pmatrix} 0 & 2 \\ 1 & 0 \end{pmatrix}$. The filter taps are explicitly given by

$$p_{0,0} = p_{1,0} = \frac{1}{4} - \frac{\sqrt{3}}{8}, \quad p_{2,0} = p_{3,0} = \frac{1}{4} + \frac{\sqrt{3}}{8},$$

$$p_{0,1} = -p_{1,1} = -p_{2,1} = p_{3,1} = -\frac{1}{8},$$

so that their polyphase representation

$$F(z) = F_{0,0} + F_{1,0}\, z_1 + F_{0,1}\, z_2 + F_{1,1}\, z_1 z_2\,,$$

is of of bi-degree $(1, 1)$ with

$$F_\gamma = \sqrt{2} \begin{pmatrix} p_{2\gamma_2,\gamma_1} & q_{2\gamma_2,\gamma_1} \\ p_{2\gamma_2+1,\gamma_1} & q_{2\gamma_2+1,\gamma_1} \end{pmatrix}, \quad \gamma = (\gamma_1, \gamma_2) \in \left\{ \begin{pmatrix} 0 \\ 0 \end{pmatrix}, \begin{pmatrix} 1 \\ 0 \end{pmatrix}, \begin{pmatrix} 0 \\ 1 \end{pmatrix}, \begin{pmatrix} 1 \\ 1 \end{pmatrix} \right\}.$$

In addition, they satisfy the sum-rule conditions (6) of order 2. The unitary transformations

$$U = -\frac{\sqrt{2}}{2} \begin{pmatrix} 1 & 1 \\ 1 & -1 \end{pmatrix}, \quad \text{and} \quad V = \frac{\sqrt{2}}{4} \begin{pmatrix} 1-\sqrt{3} & 1+\sqrt{3} \\ 1+\sqrt{3} & -1+\sqrt{3} \end{pmatrix},$$

lead to the normal form $\widetilde{F}(z) = U\, F(z)\, V$, with

$$\widetilde{F}_{0,0} = \frac{\sqrt{2}}{4}(-1+\sqrt{3}) \begin{pmatrix} 1 & 0 \\ 0 & 0 \end{pmatrix}, \quad \widetilde{F}_{0,1} = \frac{\sqrt{2}}{4}(-1-\sqrt{3}) \begin{pmatrix} 0 & 1 \\ 0 & 0 \end{pmatrix},$$

$$\widetilde{F}_{1,0} = \frac{\sqrt{2}}{4}(-1-\sqrt{3}) \begin{pmatrix} 0 & 0 \\ 1 & 0 \end{pmatrix}, \quad \widetilde{F}_{1,1} = \frac{\sqrt{2}}{4}(-1+\sqrt{3}) \begin{pmatrix} 0 & 0 \\ 0 & -1 \end{pmatrix}.$$

We note that the non-zero elements are one possible solution to the univariate quadrature mirror filters of Haar-type (i.e. masks of length 2) associated with the scaling factor 2, however $\widetilde{F}(z)$ does not even satisfy univariate sum rules of order 1.

Therefore, the role of U, V is to change the filter coefficients in the normal form so as to keep orthogonality and to satisfy the sum rule conditions of order 2.

This is true for every bivariate orthogonal filter pair: they always reduce to normal forms whose non-zero elements satisfy univariate QMF of dyadic Haar type, as proved in the following Theorem [9].

From now on, w.l.o.g. we fix $M = \begin{pmatrix} 1 & 1 \\ -1 & 1 \end{pmatrix}$.

Theorem 4.1 *Assume*

$$F(z) = U^T \widetilde{F}(z) V^T,\ \widetilde{F}(z) = \widetilde{F}_{0,0} + \widetilde{F}_{1,0}\, z_1 + \widetilde{F}_{0,1}\, z_2 + \widetilde{F}_{1,1}\, z_1 z_2,$$

with matrix coefficients $\tilde{F}_{i,j}$, $i, j = 0, 1$, *in the normal form*

$$\widetilde{F}_{0,0} = \sqrt{2}\begin{pmatrix} h_0 & 0 \\ 0 & 0 \end{pmatrix},\ \widetilde{F}_{1,0} = \sqrt{2}\begin{pmatrix} 0 & -h_1 \\ 0 & 0 \end{pmatrix},\ \widetilde{F}_{0,1} = \sqrt{2}\begin{pmatrix} 0 & 0 \\ h_1 & 0 \end{pmatrix},\ \widetilde{F}_{1,1} = \sqrt{2}\begin{pmatrix} 0 & 0 \\ 0 & h_0 \end{pmatrix}.$$

Then $F(z)$ *satisfies the QMF conditions iff* $h_0^2 + h_1^2 = \frac{1}{2}$. *Furthermore the sum rule of order* 2 *and the normalization condition hold iff the unitary transformations* U *and* V *are related by:*

1. $U = \begin{pmatrix} h_0 & -h_1 \\ h_1 & h_0 \end{pmatrix} V^T \begin{pmatrix} 1 & 1 \\ 1 & -1 \end{pmatrix}$,
2. *the rows* v_1^T, v_2^T *of* V *satisfy*

$$v_2^T \begin{pmatrix} -h_1^2 & h_0 h_1 \\ h_0 h_1 & h_1^2 \end{pmatrix} = \lambda_{10}\, v_2^T,$$

$$v_2^T \left((\frac{\sqrt{2}}{2} - \lambda_{01}) I + 2\sqrt{2} \begin{pmatrix} h_1^2 & h_0 h_1 \\ h_0 h_1 & \frac{1}{2} + h_0^2 \end{pmatrix} \right) = \frac{\sqrt{2}}{2}\, v_1^T,$$

for $\lambda_{10}, \lambda_{01} \in \mathbb{R} \setminus \{0\}$. □

We note that the displacement of the non-zero elements in the matrix elements of the normal form is arbitrary. We could for example consider the following configuration:

$$\widetilde{F}_{0,0} = \sqrt{2}\begin{pmatrix} h_0 & 0 \\ 0 & 0 \end{pmatrix},\ \widetilde{F}_{0,1} = \sqrt{2}\begin{pmatrix} 0 & h_1 \\ 0 & 0 \end{pmatrix},\ \widetilde{F}_{1,0} = \sqrt{2}\begin{pmatrix} 0 & 0 \\ h_1 & 0 \end{pmatrix},\ \widetilde{F}_{1,1} = \sqrt{2}\begin{pmatrix} 0 & 0 \\ 0 & -h_0 \end{pmatrix}.$$

Imposing univariate QMF and the sum rule conditions of order 2, we obtain $h_0 = -\frac{\sqrt{2}}{4}$, $h_1 = \sqrt{\frac{3}{8}}$ and the unitary transformations

$$U = \begin{pmatrix} h_0 & h_1 \\ h_1 & -h_0 \end{pmatrix} V^T \frac{\sqrt{2}}{2} \begin{pmatrix} 1 & 1 \\ 1 & -1 \end{pmatrix}, \quad \text{and} \quad V = \begin{pmatrix} -\cos\frac{\pi}{12} & -\sin\frac{\pi}{12} \\ \sin\frac{\pi}{12} & -\cos\frac{\pi}{12} \end{pmatrix},$$

which produce, depending on the chosen matrix M, the Belogay-Wang filters again, or a shifted version of the Kovačević-Vetterli filters from [38]. The latter, however, can be also associated to normal forms of bi-degree (2, 2), as illustrated in the following section.

4.2 Characterization of the (2, 2) Bi-degree Case

We reinterpret the wavelet construction proposed by Kovačević-Vetterli [38], where $M = \begin{pmatrix} 1 & 1 \\ 1 & -1 \end{pmatrix}$ and the non-zero low-pass p filter coefficients supported on $[1, 4] \times [-1, 1]$.

The filters satisfy the sum-rule conditions of order 2 and have the polyphase representation of bi-degree (2,2):

$$F(z) = F_{0,0} + F_{1,0}z_1 + F_{0,1}z_2 + F_{1,1}z_1z_2 + F_{2,1}z_1^2z_2 + F_{1,2}z_1z_2^2 + F_{2,2}z_1^2z_2^2. \quad (8)$$

It can be verified that $F(z)$ possesses a normal form, whose elements, up to the factor $\sqrt{2}$, are given by:

$$\tilde{F}_{1,2} = h_3e_2e_1^T, \quad \tilde{F}_{2,2} = g_3e_2e_2^T, \quad \tilde{F}_{0,1} = h_1e_1e_1^T, \quad \tilde{F}_{1,1} = (h_1e_2, g_2e_1),$$

$$\tilde{F}_{2,1} = g_1e_2e_2^T, \quad \tilde{F}_{0,0} = h_0e_1e_1^T, \quad \tilde{F}_{1,0} = g_0e_1e_2^T,$$

where we have used the canonical basis vectors $e_1 = (1, 0)^T$, $e_2 = (0, 2)^T$ and where the non-zero elements are

$$h_0 = -k\sqrt{w_2^2 + 1}, \quad h_1 = w_0h_0, \quad h_2 = -w_0w_1h_0, \quad h_3 = w_1h_0,$$

and

$$g_{3-j} = (-1)^{j+1}h_j, \quad j = 0, \ldots, 3,$$

with

$$w_0 = w_1 = \mp\sqrt{3}, \quad w_2 = 2 \pm \sqrt{3}, \quad k = \pm\frac{1}{\sqrt{2\left(w_0^2 + 1\right)\left(w_1^2 + 1\right)\left(w_2^2 + 1\right)}}.$$

The filters h and g are still orthogonal in the 1-D sense, however, no normalization or sum rule conditions are satisfied. In this case, the unitary transformations

$$U = \begin{pmatrix} 0 & 1 \\ -1 & 0 \end{pmatrix}, \quad \text{and} \quad V = \frac{1}{\sqrt{w_2^2 + 1}}\begin{pmatrix} w_2 & -1 \\ 1 & w_2 \end{pmatrix},$$

change the filter coefficients of the normal form in such a way sum rules of order 2 are satisfied.

Also in this case, a general result related to the existence of normal forms for the (2,2) bi-degree case can be proven [9].

Theorem 4.2 *$F(z)$ as in (8) satisfies the QMF conditions iff there exist unitary transformations $U, V \in \mathbb{R}^{2\times 2}$ such that*

1. *every matrix $\widetilde{F}_{i,j} = U\,F_{i,j}\,V = \begin{pmatrix} a_{ij} & b_{ij} \\ c_{ij} & d_{ij} \end{pmatrix}$, for $(i, j) \neq (1, 1)$, has at most one nonzero element;*
2. *$U\,F_{1,1}\,V = \begin{pmatrix} 0 & b_{11} \\ c_{11} & 0 \end{pmatrix}$ with $b_{11}, c_{11} \in \mathbb{R}$, $|b_{11}| = |c_{11}|$;*
3. *the non-zero coefficients satisfy univariate QMF conditions:*

$$
\begin{aligned}
&a_{00}^2 + a_{01}^2 + c_{12}^2 + c_{11}^2 = 1,\\
&a_{00}a_{01} + c_{12}c_{11} = 0,\\
&d_{21} = -a_{01}, \quad c_{11} = b_{11},\\
&c_{12} = b_{10}, \quad d_{22} = -a_{00}.
\end{aligned}
$$

Furthermore, $F(z)$ satisfies the normalization and sum rule conditions of order 2 iff:

4. *$U = \widetilde{F}(1)\,V^T\,\frac{\sqrt{2}}{2}\begin{pmatrix} 1 & 1 \\ 1 & -1 \end{pmatrix}$;*
5. *$v_2^T\,\sqrt{2}\,\widetilde{F}(1)^T \sum_{0\le|\gamma|\le 4} (\gamma_1 - \gamma_2)\widetilde{F}_\gamma = \lambda_{01}\,v_2^T$, for $\lambda_{01} \in \mathbb{R}\setminus\{0\}$;*
6. *$v_2^T \left(2\,\widetilde{F}(1)^T \sum_{0\le|\gamma|\le 4} (\gamma_1 + \gamma_2)\widetilde{F}_\gamma + (1 - \sqrt{2}\lambda_{10})I\right) = v_1^T$, for $\lambda_{10} \in \mathbb{R}\setminus\{0\}$.*

Remark 4.1 As for the previous 1-D situation, this characterization enables us to "re-discover" all known 2-band wavelet filters. However, it goes beyond mere rediscovery, allowing the construction of entirely new families of filters, as illustrated in the subsequent section, where a couple of novel examples are presented.

4.3 Some New Constructions

Let us start with two different configurations for the normal-form block matrix $\frac{\sqrt{2}}{2}\begin{pmatrix} 0 & \widetilde{F}_{1,2} & \widetilde{F}_{2,2} \\ \widetilde{F}_{0,1} & \widetilde{F}_{1,1} & \widetilde{F}_{2,1} \\ \widetilde{F}_{0,0} & \widetilde{F}_{1,0} & 0 \end{pmatrix}$ precisely

	d_{12}	d_{22}
c_{01}	b_{11} c_{11}	b_{21}
a_{00}	a_{10}	

(9)

associated to the dilation matrix $M = \begin{pmatrix} 0 & 2 \\ -1 & 1 \end{pmatrix}$, and

	c_{12}	d_{22}
a_{01}	b_{11} c_{11}	d_{21}
a_{00}	b_{10}	

(10)

with $M = \begin{pmatrix} 1 & 1 \\ -1 & 1 \end{pmatrix}$, where the absence of an element means the presence of a zero.

We impose the 1-D QMF conditions, letting a_{00}, b_{11} as free parameters in both cases. If we denote with $K = \frac{1}{2}\sqrt{\left(2\,a_{00}{}^2 + 2\,b_{11}{}^2\right)^{-1} - 1}$, we get

$$d_{12} = K b_{11},\ c_{01} = K a_{00},\ a_{10} = d_{12},\ b_{21} = -c_{01},\ c_{11} = -b_{11},\ d_{22} = a_{00}\,,$$

in the first case and

$$a_{01} = K b_{11},\ b_{10} = -K a_{00},\ c_{11} = -b_{11},\ c_{12} = b_{10},\ d_{21} = -a_{01},\ \ d_{22} = -a_{00}\,,$$

in the second case.

By parametrizing V as

$$\begin{pmatrix} \cos\alpha & -\sin\alpha \\ \sin\alpha & \cos\alpha \end{pmatrix}, \quad \alpha \in [0, 2\pi)\,,$$

and imposing the sum rules of order 2 and the normalization conditions, we get all possible filters coefficients in terms of α, giving rise to different scaling functions (existing at least in a distributional sense).

For the configuration (9), fixing, for example, $\alpha = \pi/2$, a possible choice for the pair (a_{00}, b_{11}) is $\approx (-0.1, -0.3)$. The corresponding scaling filter has support $[2, 5] \times [-1, 1]$ and the resulting scaling function is shown on the left of Fig. 2. We would like to emphasize that the smoothness of the functions is neither guaranteed nor required by our construction, which only ensures their existence in the distributional sense.

For the configuration (10), taking $\alpha = \pi/12$, we re-obtain the Kovačević-Vetterli filters, while, for example, the choice $\alpha = 2.24$ and $(a_{00}, b_{11}) \approx (0.3, 0.6)$ leads to a filter of support $[0, 5] \times [-1, 1]$, with scaling function shown on the right of Fig. 2.

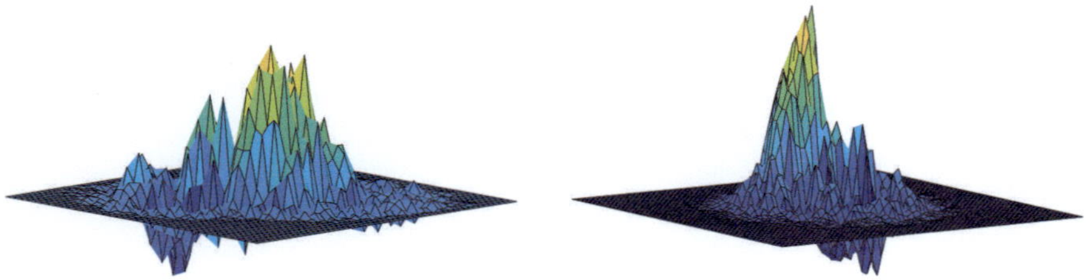

Fig. 2 The scaling functions constructed in Sect. 4.3

Remark 4.2 Our characterization easily extends to the cases of larger support, i.e. $N \geq 3$. However, several possible configurations of normal forms can be considered, for example, in the case $N = 3$,

$$\begin{pmatrix} \widetilde{F}_{0,3} & \widetilde{F}_{1,3} \\ \widetilde{F}_{0,2} & \widetilde{F}_{1,2} \\ \widetilde{F}_{0,1} & \widetilde{F}_{1,1} \\ \widetilde{F}_{0,0} & \widetilde{F}_{1,0} \end{pmatrix} \quad \text{or} \quad \begin{pmatrix} 0 & \widetilde{F}_{1,4} & \widetilde{F}_{2,4} \\ 0 & \widetilde{F}_{1,3} & \widetilde{F}_{2,3} \\ \widetilde{F}_{0,2} & \widetilde{F}_{1,2} & \widetilde{F}_{2,2} \\ \widetilde{F}_{0,1} & \widetilde{F}_{1,1} & 0 \\ \widetilde{F}_{0,0} & \widetilde{F}_{1,0} & 0 \end{pmatrix},$$

whose non-zero elements can be related to univariate QMF conditions.

While the arguments and the detailed computations given in [9] can be straightforwardly repeated, the practical use and implementation however require longer computations. □

5 Conclusions

In this paper, we have summarized and illustrated the research conducted over the past few years by the authors and collaborators aimed at characterizing orthogonal wavelet filterbanks and exploring new general strategies for defining novel wavelet systems. The univariate and bivariate cases are explored using different approaches: leveraging classical results from system theory in the former case, and employing normal forms of the corresponding bivariate polyphase representations in the latter case.

Nevertheless, a new example provided here shows that the use of system theory is worth exploring in the bivariate setting as well. This is certainly something we plan to investigate in the near future.

Another interesting aspect we plan to investigate is the potential extension of the "normal form" arguments to scaling matrices M beyond those with $\det = \pm 2$.

Acknowledgements The authors are members of RITA (Research Italian Network on Approximation), of UMI-TAA research group, and Gruppo Nazionale Calcolo Scientifico-Istituto Nazionale di Alta Matematica (GNCS-INdAM) which partially supported this work with Project CUP E53C24001950001. This work was also partially supported by the EU under the Italian National Recovery and Resilience Plan (NRRP) of NextGenerationEU, partnership on "Telecommunications of the Future" (PE00000001 - program "RESTART").

References

1. Agler, J.: Some interpolation theorems of Nevanlinna-Pick type, unpublished manuscript (1988)
2. Agler, J.: On the representation of certain holomorphic functions defined on a polydisc. In: Topics in Operator Theory: Ernst D. Hellinger Memorial Volume, in: Oper. Theory Adv. Appl. 48, Birkhäuser, Basel, pp. 47–66 (1990)
3. Ayache, A.: Construction of nonseparable dyadic compactly supported orthonormal wavelet bases for $L^2(\mathbb{R}^2)$ of arbitrarily high regularity. Rev. Mat. Iberoamericana **15**, 37–58 (1999)
4. Bacchelli, S., Cotronei, M., Lazzaro, D.: An algebraic construction of k-balanced multiwavelets via the lifting scheme. Numer. Algorithms **23**, 329–356 (2000)
5. Bacchelli, S., Cotronei, M., Sauer, T.: Wavelets for multichannel signals. Adv. Appl. Math. **29**, 581–598 (2002)
6. Bacchelli, S., Cotronei, M., Sauer, T.: Multifilters with and without prefilters. BIT **42**(2), 231–261 (2002)
7. Belogay, E., Wang, Y.: Arbitrarily smooth orthogonal nonseparable wavelets in $\mathbb{R}^2$. SIAM J. Math. Anal. **30**, 678–697 (1999)
8. Charina, M., Conti, C., Cotronei, M., Putinar, M.: System theory and orthogonal multi-wavelets. J. Approx. Theory **238**, 85–102 (2019)
9. Charina, M., Conti, C., Cotronei, M., Sauer, T.: Bivariate two-band wavelets demystified. Linear Algebra Appl. **608**, 13–36 (2021)
10. Chui, C.: An Introduction to Wavelets. Academic, London (1992)
11. Chui, C., De Villiers, J.: Wavelet Subdivision Methods: GEMS for Rendering Curves and Surfaces. CRC Press (2011)
12. Chui, C.K., Lian, J.: A study of orthonormal multi-wavelets. Appl. Num. Math. **20**(3), 273–298 (1996)
13. Cohen, A., Daubechies, I.: Nonseparable bidimensional wavelet bases. Rev. Mat. Iberoamericana **9**, 51–137 (1993)
14. Colella, D., Heil, C.: Characterizations of scaling functions: Continuous solutions. SIAM J. Matrix Anal. Appl. **15**, 469–518 (1994)
15. Conti, C., Cotronei, M.: Construction of multichannel wavelets from orthogonal full rank refinable function. In: Cohen, J., Zayed, A.I. (eds.) Wavelets and Multiscale Analysis. Birkhauser, pp. 109–130 (2011)
16. Conti, C., Cotronei, M., Sauer, T.: Full rank interpolatory vector subdivision schemes. In: Cohen, A., Merrien, J.L., Schumaker, L. (eds.) Curve and Surface Fitting: Avignon 2006, vol. 71, pp. 71–80 (2007)
17. Conti, C., Cotronei, M., Sauer, T.: Full rank positive matrix symbols: interpolation and orthogonality. BIT **48**, 5–27 (2008)
18. Conti, C., Cotronei, M., Sauer, T.: Full rank interpolatory subdivision schemes: Kronecker, filters and multiresolution. J. Comput. Appl. Math. **233**(7), 1649–1659 (2010)
19. Conti, C., Cotronei, M., Sauer, T.: Convergence of level-dependent Hermite subdivision schemes. Appl. Numer. Math. **116**, 119–128 (2017)
20. Cotronei, M., Lo Cascio, L., Sauer, T.: Multifilters and prefilters: Uniqueness and algorithmic aspects. J. Comput. Appl. Math. **221**, 346–354 (2008)

21. Cotronei, M., Holschneider, M.: Partial parameterization of orthogonal wavelet matrix filters. J. Comput. Appl. Math. **243**, 113–125 (2013)
22. Cotronei, M., Sauer, T.: Full rank filters and polynomial reproduction. Commun. Pure Appl. Anal. **6**, 667–687 (2007)
23. Cotronei, M., Rossini, M., Sauer, T., Volontè, E.: Filters for anisotropic wavelet decompositions. J. Comput. Appl. Math. **349**, 316–330 (2019)
24. Cotronei, M., Rüweler, D., Sauer, T.: Multiple filterbanks for image processing: implementation issues. Math. Comput. Simulation **176**, 147–159 (2020)
25. Daubechies, I.: Orthonormal bases of compactly supported wavelets. Commun. Pure Appl. Math. **41**(7), 909–996 (1988)
26. Daubechies, I.: Ten lectures on wavelets. CBMS Conf. Series in Appl. Math. **61** (1992)
27. Daubechies, I., Lagarias, J.C.: Sets of matrices all infinite products of which converge. Linear Algebra Appl. **162**, 227–263 (1992)
28. Geronimo, J.S., Hardin, D.P., Massopust, P.R.: Fractal functions and wavelet expansions based on several scaling functions. J. Approx. Theory **78**, 373–401 (1994)
29. Gundy, R.F., Jonsson, A.L.: Scaling functions on $\mathbb{R}^2$ for dilations of determinant ± 2. Appl. Comput. Harmon. Anal. **29**, 49–62 (2010)
30. Han, B.: Compactly supported tight wavelet frames and orthonormal wavelets of exponential decay with a general dilation matrix. J. Comput. Appl. Math. **155**, 43–67 (2003)
31. Han, B.: Symmetry property and construction of wavelets with a general dilation matrix. Linear Algebra Appl. **353**, 207–225 (2002)
32. Han, B.: Vector cascade algorithm and refinable function vectors in Sobolev spaces. J. Approx. Theory **124**, 44–88 (2003)
33. Heil, C., Colella, D.: Matrix refinement equations: existence and uniqueness. J. Fourier Anal. Appl. **2**, 363–377 (1996)
34. Hernandez, E., Weiss, G.: A First Course on Wavelets. CRC Press, Boca Raton, Stud. Adv. Math. (1996)
35. Keinert, F.: Wavelets and Multiwavelets. Chapman & Hall/CRC (2004)
36. Knese, G.: Rational inner functions in the Schur-Agler class of the polydisk. Publ. Mat. **55**(2), 343–357 (2011)
37. Knese, G.: Kummert's approach to realization on the bidisk. Indiana Univ. Math. J. **70**(6), 2369–2403 (2021)
38. Kovačević, J., Vetterli, M.: Nonseparable multidimensional perfect reconstruction filter-banks. IEEE Trans. Inform. Theory **38**, 533–555 (1992)
39. Lagarias, J.C., Wang, Y.: Haar-type orthonormal wavelet bases in $\mathbb{R}^2$. J. Fourier Anal. **2**, 1–14 (1995)
40. Lebrun, J., Vetterli, M.: High-order balanced multiwavelets: theory, factorization, and design. IEEE Trans. Signal Process. **49**(9), 1918–1930 (2001)
41. Mallat, S.: A Wavelet Tour of Signal Processing. Academic Press (2008)
42. Micchelli, C.A., Sauer, T.: Regularity of multiwavelets. Adv. Comput. Math. **7**, 455–545 (1997)
43. Novikov, I.Y., Protasov, V.Yu., Skopina, M.A.: Wavelets Theory. Translations Mathematical Monographs, vol. 239. AMS (2011)
44. Plonka, G., Strela, V.: Construction of multiscaling functions with approximation and symmetry. SIAM J. Math. Anal. **29**, 481–510 (1998)
45. Ron, A., Shen, Z.: Affine systems in $L_2(\mathbb{R}^d)$: the analysis of the analysis operator. J. Funct. Anal. **148**, 408–447 (1997)
46. Strang, G., Strela, V.: Orthogonal multiwavelets with vanishing moments. J. Optical Eng. **33**, 2104–2107 (1994)
47. Vetterli, M., Kovačević, J.: Wavelets and Subband Coding. Prentice Hall PTR (1995)
48. Villemoes, L.F.: Continuity of nonseparable quincunx wavelets. Appl. Comput. Harmon. Anal. **1**, 180–187 (1994)
49. Viscardi, A.: Semi-regular Dubuc-Deslauriers wavelet tight frames. J. Comput. Appl. Math. **349**, 548–562 (2019)
50. von Neumann, J.: Eine Spektraltheorie für allgemeine Operatoren eines unitären Raumes. Math. Nachr. **4**, 258–281 J. (1951)

Fractal Analysis. Application to Biomedical Imaging

Rachid Jennane

Abstract Fractal analysis for the characterization of complex structures is widely used in many fields. Among the fractal models, the fractional Brownian motion (fBm) is well suited for the characterization of images involving a varying range of gray levels. This chapter focuses on the use of the fBm for image analysis. Starting from the classical fBm of unique parameter H, different stochastic models are defined and discussed. The fractal anisotropy is considered and different methods emphasizing this property will be discussed. An application of these models to the field of biomedical imaging is also presented.

1 Introduction

The main attraction of fractal geometry stems from its ability to describe the irregular or fragmented shape of natural features as well as other complex objects that traditional Euclidean geometry fails to analyze. This phenomenon is often expressed by spatial or time-domain statistical scaling laws. This concept enables a simple geometrical interpretation and is frequently encountered in a variety of fields, such as geophysics, biology or fluid mechanics. To this end, Mandelbrot introduced the notion of fractal sets [24], which enables to take into account the degree of regularity of the organizational structure related to the physical system's behavior.

Texture is known to provide information about depth and surface orientation and, as such, describes the content of both natural and artificial images. One particular textural model which has received recent interest, especially for its ability to model "natural" shapes and forms is that of fractional Brownian motions (fBm) [27]. This class of mathematical functions have been used to characterize the geometrical properties of sets. They have been used to generate images [13, 24, 25], classify texture [23, 30, 31], to describe the surface area of chemical reactants [35], and the length of coastlines [26]. FBm is governed by the single Hurst parameter, $H (0 < H < 1)$.

R. Jennane (✉)
IDP Institute UMR CNRS 7013, University of Orleans, Orleans, France
e-mail: Rachid.Jennane@univ-orleans.fr

© The Author(s), under exclusive license to Springer Nature Switzerland AG 2025
M. J. Ibáñez-Pérez et al. (eds.), *Mathematical And Computational Modelling, Approximation and Simulation*, SEMA SIMAI Springer Series 41,
https://doi.org/10.1007/978-3-032-01114-5_2

When applied in image analysis, fractal geometry is often brought to the evaluation of the fractal dimension, D, from a given dataset. D is linked to the H parameter by: $H = E + 1 - D$, where E is the Euclidean dimension. It has been shown that H and the intuitive notion of roughness are closely linked: the lower H is, the higher is the roughness [31]. The interest of fBm is that it simply represents, with the H parameter, a wide class of $1/f^{\alpha}$ nonstationary signals with α in the range $]1, 3[$. Another property of fBm is that it is statistically self-similar meaning that as scale is changing, the process looks like identical. For this reason, it has been applied in image processing for modeling texture having multiscale patterns such as natural scenes [17, 31]. Another interest of these models, is that they exhibit long-term memory. This notion has been widely encountered in experimental data sets (e.g., in hydrology [16] and in communications [1, 6]).

These three properties ($1/f$ spectral behavior, self-similarity, and long-term memory) are strictly equivalent for fBm and make it convenient for many applications. However, in some experimental cases, the classical definition of fBm is not sufficient to properly describe real data.

As a first example, short- and long-term correlations could be different. A unique frequency regime or self-similar behavior is not efficient to entirely describe the observed data. For example, in [37], aggregates of silica particles showed an $1/f$ spectral behavior with two slopes: a low- and a high-frequency regime separated by a cut-off frequency corresponding to the inverse mean size of particles. This kind of dual self-similarity, i.e., different long- and short-term correlations [12], was also shown on bone radiographs [15] and on smectite clay images [14]. A possible solution has been proposed to overcome this difficulty [20].

FBm-based models that are able to capture these long- and short term correlations were proposed in [11, 21]. They are either based on a filtered version of fBm or on a generalization of fBm from its structure function [39]. However, these models have no simple analytic definition, and some difficulties arise in the choice of the filter or in the range of variation of the process parameters. Other studies of fractal models were performed by the authors in [7, 8] and, by others [4]. The change in the slope may be interpreted as a change in fractality according to the scale or the frequency, leading to the notion of a piecewise fractal process. Thus, it would be necessary to define a new continuous model based on fBm having this property. In [2], the general multifractional Brownian motion (gmBm) considered the parameter H as a continuous function of time. This extension leads to a more general model than that needed for the previous application. But this model does not lead to stationary increments nor to self-similar processes.

A second limitation of the model could be given. Values above 1 have been observed as, for example, in Nile River data ([5], p. 84) or in bone radiographs [22]. These two applications suggest the elaboration of a framework to take into account the fact that H could be greater than 1 for true data. The range for H has already been enlarged numerically [28] by incrementing/decrementing the modelized signal. On the other hand, a mathematical theory of generalized fBm has been developed in [3].

To give a new approach between the numerical approach and the purely theoretical approach, this work presents two extensions of the classical fBm. First, the piece-wise fractional Brownian motion model (pfBm) of parameters H_o, H_i and γ and which presents two spectral regimes. Second, the nth-order fractional Brownian motion (nfBm) which enables the H parameter to be within the range $]n-1, n[$, where n is any strictly positive integer. The properties of these processes and their increments are presented. Finally, using bone radiograph images, an application of these processes is investigated for the biomedical diagnosis of osteoporotic patients.

2 Fractional Brownian Motion

The fractional Brownian motion (fBm) model proposed by Mandelbrot and Van Ness [27] is a stochastic fractal process governed by a unique parameter $0 < H < 1$ and is defined as:

$$B^H(t) - B^H(0) = \frac{1}{\Gamma\left(H+\frac{1}{2}\right)}\left[\int_{-\infty}^{0}\left((t-s)^{H-1/2} - (-s)^{H-1/2}\right) dB(s) + \int_0^t (t-s)^{H-1/2}\, dB(s)\right], \tag{1}$$

where t represents time, Γ represents the Gamma function and $B(s)$ is the standard Brownian motion.

A spectral definition of the fBm has been proposed by Reed et al. in [36], for every $-\infty < t < +\infty$:

$$B^H(t) = \frac{1}{2\pi}\int_{-\infty}^{+\infty}\left(\frac{1}{i\omega}\right)^{H+\frac{1}{2}}\left(e^{it\omega}-1\right) d\beta(\omega). \tag{2}$$

where $\beta(\omega)$ is the complex Brownian motion in frequency.

The convergence of (2) in the sense of L^2 is ensured when $\omega \to 0$ for $H < 1$ and when $\omega \to \infty$ for $H > 0$. This process is real, Gaussian, and continuous.

The covariance function of fBm is:

$$E\left[B^H(t)\overline{B^H(s)}\right] = \frac{C_H}{2}\left[|t|^{2H} + |s|^{2H} - |t-s|^{2H}\right], \tag{3}$$

where $C_H = \Gamma(1-2H)(cos(\pi H))\pi H$ is the constant calculated in [36]. Note from this expression that fBm is not stationary. This is due to the special role that the kernel plays $e^{it\omega} - 1$, while a kernel based on $e^{it\omega}$ would have rendered a stationary process.

FBm being non-stationary, it is more convenient to study its increments, called fractional Gaussian noises (fGn), which turn out to be stationary. FGn's are defined as:

$$G_u^H(t) = B^H(t+u) - B^H(t), t \in \Re. \tag{4}$$

Their covariance function might be expressed both in time and in frequency as:

$$E\left[G_u^H(t)G_u^H(s)\right] = \frac{C_H}{2}\left[|\tau - u|^{2H} + |\tau + u|^{2H} - 2|\tau|^{2H}\right]. \tag{5}$$

It only depends on the lag, $\tau = |t - s|$ and therefore, shows that fGn is a stationary process.

It is also direct to see that its power spectral density (PSD) is:

$$F_u^H(\omega) = \frac{4sin^2(\frac{u\omega}{2})}{|\omega|^{2H+1}}. \tag{6}$$

The self-similarity [27] of fBm is expressed for $a \succ 0$ as:

$$\left\{B^H(t_0 + at) - B^H(t_0)\right\} \overset{law}{=} a^H \left\{B^H(t_0 + t) - B^H(t_0)\right\} \tag{7}$$

For time scales dilated by a positive factor a, the process looks identical to itself, with an amplitude factor a^H. All this means that whatever the observation scale and the time t_0 we start with, the process seems identical, i.e., presents the same roughness.

The next section defines a more general model that takes into account the limitations of the fBm model for true data, especially to capture long- and short-term correlations that could be different.

3 Piece-Wise Fractional Brownian Motion

Dual self-similarity or different long- and short-term correlation can be defined in the same model as follows.

For $H_o < 1$, $0 < H_i < 1$, $\gamma > 0$, and for every $-\infty < t < +\infty$, let $B^{H_o,H_i,\gamma}(t)$ be defined by the stochastical integral:

$$B^{H_o,H_i,\gamma}(t) = \frac{1}{2\pi}\int_{-\infty}^{+\infty} F^{H_o,H_i,\gamma}(\omega)\left(e^{it\omega} - 1\right) d\beta(\omega). \tag{8}$$

with

$$F^{H_o,H_i,\gamma}(\omega) = \frac{1_{[0,\gamma)}(|\omega|)}{(i\omega)^{H_o+(1/2)}} + \gamma^{H_i-H_o}\frac{1_{[\gamma,\infty)}(|\omega|)}{(i\omega)^{H_i+(1/2)}}, \tag{9}$$

where $\beta(\omega)$ is the complex brownian motion in frequency, and $1_I(w)$ is the indicator function over interval I (i.e., $1_I(\omega) = 1$ if $\omega \in I$, and $1_I(\omega) = 0$, otherwise). Then

$B^{H_o,H_i,\gamma}(t)$ is called the piece-wise fractional Brownian motion (pfBm) model of parameters H_o, H_i, and γ.

This integral is convergent in the L^2 sense when $\omega \to 0$ if $H_o < 1$ and when $\omega \to \infty$ if $H_i > 0$. The continuity of $F^{H_o,H_i,\gamma}(\omega)$ is ensured at the cut-off frequency γ by the factor $\gamma^{H_i - H_o}$.

From the definition of pfBm, the variance of the process increments, which are also called the structure function $f(s)$ [39], easily results in:

$$\begin{aligned} f(s) &= Var\left[B^{H_o,H_i,\gamma}(t+s) - B^{H_o,H_i,\gamma}(t)\right] \\ &= \frac{1}{\pi^2}\int_{-\infty}^{+\infty} sin^2\left(\frac{\omega s}{2}\right)\left|F^{H_o,H_i,\gamma}(\omega)\right|^2 d\omega. \end{aligned} \tag{10}$$

From the definition of $F^{H_o,H_i,\gamma}(\omega)$, and by making the change of variable $\omega \to \omega s$, it follows that $f(s)$ behaves as $|s|^{2H_i}$ for fine scales, i.e., as $s \to 0$, and as $|s|^{2H_o}$ for large scales, i.e., as $s \to +\infty$. The covariance of pfBm directly results and clearly indicates that this process is non-stationary.

Thus, pfBm is real, Gaussian, non-stationary, and can be seen to be continuous. Note that pfBm is the classical fBm of the parameter H when $H = H_o = H_i \in]0; 1[$.

The nonstationarity of the pfBm makes it difficult to study such processes; for this reason, it is more convenient to study piece-wise fractional Gaussian noises.

3.1 *Piece-Wise Fractional Gaussian Noises*

This section defines the increments of the pfBm process.

Let $B^{H_o,H_i,\gamma}(t)$ be a pfBm, and let its increments be defined by

$$G_u^{H_o,H_i,\gamma}(t) = B^{H_o,H_i,\gamma}(t+u) - B^{H_o,H_i,\gamma}(t), t \in \Re \tag{11}$$

where the lag $u \in \Re$ is fixed. The processes $\left\{G_u^{H_o,H_i,\gamma}(t)\right\}$ are called piecewise fractional Gaussian noises (pfGn) of the parameters H_o, H_i, and γ.

Their cross-correlation function is given by:

$$G_{u,v}^{H_o,H_i,\gamma}(t,s) = E\left[G^{H_o,H_i,\gamma}(t)\overline{G^{H_o,H_i,\gamma}(s)}\right]. \tag{12}$$

pfGn are stationarity processes. Their PSD is given by:

$$4sin^2\left(\frac{u\omega}{2}\right)\left|F^{H_o,H_i,\gamma}(\omega)\right|^2. \tag{13}$$

At this point, we know that pfBm is a continuous, Gaussian, and nonstationary process, having stationary increments termed pfGn. To complete the description of the pfBm process, its self-similarity will be assessed, but unlike fBm, the two frequency regimes cannot lead to a unique self-similarity property, but to two asymptotic self-similarity regimes.

3.2 Asymptotic Self-Similarity of PfBm

The process $B^{H_o,H_i,\gamma}(t)$ behaves like a fBm of parameter H_o for low frequencies and like a fBm of parameter H_i for high frequencies. More precisely, for any $t_0 \in \Re$, and $H_o, H_i \in]0, 1[$, we have the following:

$$\lim_{a\to 0}\left\{\frac{B^{H_o,H_i,\gamma}(t_0+at)-B^{H_o,H_i,\gamma}(t_0)}{a^{H_i}\gamma^{H_i-H_o}}\right\}_t \overset{law}{=} \left\{B^{H_i}(t_0+t)-B^{H_i}(t_0)\right\}_t, \qquad (14)$$

and

$$\lim_{a\to \infty}\left\{\frac{B^{H_o,H_i,\gamma}(t_0+at)-B^{H_o,H_i,\gamma}(t_0)}{a^{H_o}\gamma^{H_i-H_o}}\right\}_t \overset{law}{=} \left\{B^{H_o}(t_0+t)-B^{H_o}(t_0)\right\}_t. \qquad (15)$$

A complete proof of this results can be found in [33]. To conclude with self similarity, it can be pointed out that classical fBm/fGn are self-similar, whereas pfBm/pfGn presents an asymptotical self-similarity regime.

3.3 Synthesis of Piece-Wise Fractional Brownian Motion

We know that for a large scale of observation $a \to \infty$, the process behaves like a self-similar process with parameter H_o. For a closer scale of observation when $a \to 0$, the process is governed by the self-similar parameter H_i. This finding can be noticed on the realizations of pfBm on Figs. 1 and 2. These realizations were obtained using the Fourier synthesis method [24]. For an illustration purpose, this classical method is suitable. The applicability of recent development of fast and exact synthesis method to the pfBm [34] is still an open problem. Two realizations of the process are presented with $H_o = 0.2$ and $H_i = 0.8$ for the first one and with $H_o = 0.8$ and $H_i = 0.2$ for the second one. For both signals, $\gamma = 0.05$. Top drawings show the processes with a large scale, and it is clear that it behaves like H_o: rough for $H_o = 0.2$ and smooth for $H_o = 0.8$. On the opposite side, the bottom drawings show the behavior following H_i for a closer scale: smooth for $H_i = 0.8$ and rougher for $H_i = 0.2$.

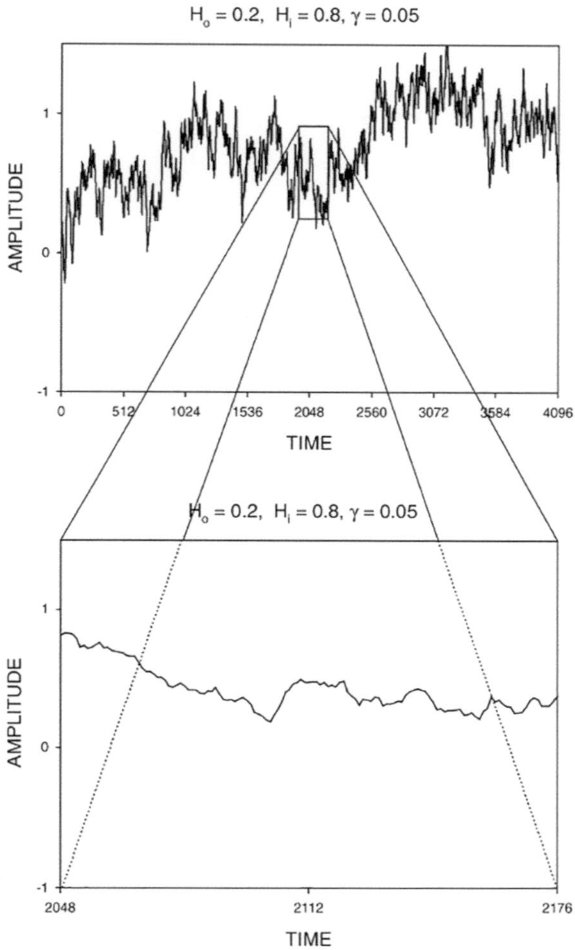

Fig. 1 A realization of pfBm with $H_o = 0.2$, $H_i = 0.8$, $\gamma = 0.05$ presented across two scales. Large scale (Top). Small scale (Bottom)

The aim of the following section is the elaboration of a framework to take into account the fact that the H parameter can be greater than 1 for true data. More precisely, the purpose is to propose a generalization of fBm with higher range of variation for H parameter.

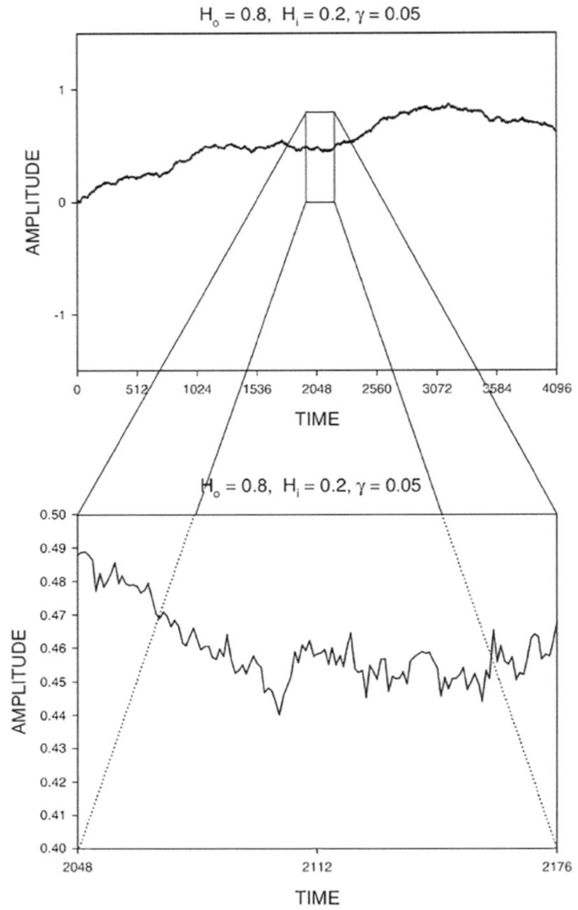

Fig. 2 A realization of pfBm with $H_o = 0.8$, $H_i = 0.2$, $\Gamma = 0.05$ presented across two scales. Large scale (Top). Small scale (Bottom)

4 *n*th-Order Fractional Brownian Motion

The nth-order fractional Brownian motion (n-fBm) [32], B_H^n is defined as:

$$B_H^n(t) = \frac{1}{\Gamma\left(H + \frac{1}{2}\right)} \left\{ \int_{-\infty}^{0} \left[(t-u)^{H-1/2} - (-u)^{H-1/2} - \dots \right. \right.$$
$$\left. - \left(H - \frac{1}{2}\right) \dots \left(H - \frac{2n-3}{2}\right) . (-u)^{H-n+1/2} \frac{t^{n-1}}{(n-1)!} \right] dB(u)$$
$$\left. + \int_0^t (t-u)^{H-1/2}\, dB(u) \right\} . \tag{16}$$

where n is any strictly positive integer, $B(t)$ is the standard Brownian motion. This definition allows H to be in the range $]n-1, n[$.

When $n = 1$, this definition corresponds to fBm taking value 0 at time 0.

Remark 1 Note that from (16), $B_H^n(0) = 0$ and that the derivative of $B_H^n(t)$ is $B_{H-1}^{n-1}(t)$. In addition, the nth-order fractional Brownian motion can be seen as the antiderivative of order n of B_{H-n+1}^1 having derivatives up to order $n-1$ vanishing at 0. In particular, B_H^n has continuous derivative up to order $n-1$, and the last one has the same smoothness as the fBm of order $H-n+1$. When $n = 1$, Eq. (16) corresponds to fBm taking value 0 at time 0.

The n-fBm can also be defined using the spectral representation as:

$$B_H^n(t) = \frac{1}{2\pi} \int_{-\infty}^{+\infty} \frac{1}{(i\omega)^{H+1/2}} \left[e^{it\omega} - \sum_{k=0}^{n-1} \frac{(it\omega)^k}{k!} \right] d\beta(\omega). \tag{17}$$

where β is the complex Brownian process in frequency.

As the classical fBm, the n-fBm is a zero mean, self-similar, and nonstationary process.

Using Remark 1, we know that:

$$B_H^n(t) = \int_0^t B_{H-1}^{n-1}(u)du. \tag{18}$$

From bilinearity, we get the covariance function, $G_{H,n}(t,s)$ of n-fBm:

$$G_{H,n}(t,s) = E\left[B_H^n(t) B_H^n(s)\right] = \int_0^t \int_0^s G_{H-1,n-1}(u,v)dudv. \tag{19}$$

Using the expression of $G_{H-n+1,1}$, we get:

$$G_{H,n}(t,s) = (-1)^n \frac{C_H^n}{2} \left\{ |t-s|^{2H} - \sum_{j=0}^{n-1} (-1)^j \binom{2H}{j} \left[\left(\frac{t}{s}\right)^j |s|^{2H} + \left(\frac{s}{t}\right)^j |t|^{2H} \right] \right\}, \tag{20}$$

where

$$\binom{\alpha}{j} = \frac{\alpha(\alpha-1)...(\alpha-(j-1))}{j!} \tag{21}$$

and for $n \geq 2$

$$C_H^n = \frac{C_{H-n+1}^1}{(2H)(2H-1)...(2H-(2n-3))} = \frac{1}{\Gamma(2H+1)\,|sin(\pi H)|}. \tag{22}$$

Equation (20) clearly proves that n-fBm is nonstationary.

Using the fact that

$$\sum_{j=0}^{n-1} \binom{2H}{j} (-1)^j = (-1)^{n-1} \binom{2H-1}{n-1} \tag{23}$$

for $t = s$, we get the variance, of B_H^n

$$V\left[B_H^n(t)\right] = C_H^n \binom{2H-1}{n-1} |t|^{2H}. \tag{24}$$

As for fBm, the variance of this nonstationary process follows a t^{2H} law.

From (20), with $t = at$, the self-similarity of n-fBm follows:

$$B_H^n(t) \stackrel{law}{=} a^{-H} B_H^n(at) \tag{25}$$

In the next section, we will be interested in nth-order fractional Gaussian noises.

4.1 n-th Order Fractional Gaussian Noises

In this section, we are interested in the kth-order increments of the n-fBm process. We will see for which k the increments become stationary, leading to the definition of nth-order fractional Gaussian noises (n-fGn).

4.1.1 k-th Order Increments of n-fBm

For a function g, we denote its increments by:

$$\Delta_l g(x) = g(x+l) - g(x) \tag{26}$$

where l is the lag.

Recursively, the kth-order increments are defined as follows:

$$\Delta_l^{(k)} g(x) = \Delta_l^{(k-1)} g(x+l) - \Delta_l^{(k-1)} g(x). \tag{27}$$

The kth-order increments can be obtained as:

$$\Delta_l^{(k)} g(x) = \sum_{j=0}^{k} (-1)^{k-j} \binom{k}{j} g(x+jl). \tag{28}$$

The covariance function of the kth-order increments of n-fBm is obtained by:

$$G_{H,n}^{(k)}(t,s) = E\left[\Delta_{l,t}^{(k)} B_H^n(t) \Delta_{l,s}^{(k)} B_H^n(s)\right]. \tag{29}$$

We start from (20), which corresponds to $k = 0$. We use linearity, which leads to

$$\begin{aligned} G_{H,n}^{(k)}(t,s) = \frac{C_H^n}{2}(-1)^n \Big\{ & \Delta_{l,t}^{(k)} \Delta_{-l,t}^{(k)} \left(|t-s|^{2H}\right) \\ & - \sum_{j=0}^{n-1} \Big[(-1)^j \binom{2H}{j} \Delta_l^{(k)}(t^j) \Delta_l^{(k)} \left(s^{-j}|s|^{2H}\right) \\ & + \Delta_l^{(k)}(s^j) \Delta_l^{(k)} \left(t^{-j}|t|^{2H}\right) \Big] \Big\}. \end{aligned} \tag{30}$$

4.1.2 *n*-th Order Stationary Increments of *n*-fBm

Let us now define the n-th stationary increments of n-fBm, using the following remark:

$$\Delta_l^{(k)}(t^j) = 0, k \geq j+1. \tag{31}$$

We observe that for $k \geq n$, the expression (30) simplifies, and increments appear to be stationary. The case $k = n$ corresponds to the first k value rendering stationarity of the increments, namely, n-fGn.

We get the covariance of n-fGn for $\tau = (t-s)$

$$\begin{aligned} G_{H,n}^{(k)}(\tau) &= E\left[\Delta_l^{(n)} B_H^n(t-\tau) \Delta_l^{(n)} B_H^n(t)\right] \\ &= (-1)^n \frac{C_H^n}{2} \Delta_l^{(n)} \Delta_{-l}^{(n)} \left(|\tau|^{2H}\right). \end{aligned} \tag{32}$$

A more explicit expression of the covariance function of n-fGn can be written as follows:

$$G_{H,n}^{(k)}(\tau) = \frac{C_H^n}{2}(-1)^n \sum_{j=-n}^{n} (-1)^j \binom{2n}{n+j} |\tau + jl|^{2H}. \tag{33}$$

Let us discuss the physical meaning of the expression, the long-term dependence properties of fractional Gaussian noises (increments of fBm) are determined by the shape of their covariance functions at infinity. Thus, the behavior of $G_{H,n}^{(n)}(\tau)$ for $\tau \rightarrow +\infty$ is of interest. For $H \neq n - 1/2$, it is equivalent to

$$\frac{C_H^n}{2}(2H)(2H-1)...(2H-2n+2)(2H-2n-1+1)\,|\tau|^{2H-2n}\,l^{2n}. \tag{34}$$

The sign of $G_{H,n}^{(n)}(\tau)$ is the one of $2H - 2n + 1$. For $n - 1 < H < n - 1/2$, the process is negatively correlated. For $n - 1/2 < H < n$, the process is positively correlated, and in addition, since it decays as a power function, it is said to be a long-term memory process [5]. The case when $H = n - 1/2$ is a half integer is of interest. Denoting $G_n(\tau) = G_{n-1/2,n}^{(n)}(\tau)$, it is seen that $G_n(\tau) = 0$ for $|\tau| > n\,|l|$.

Finally, note that for $n = 1$, n-fGn is the classical fGn.

4.2 Power Spectral Density of n-fGn

The power spectral density of n-fGn, noted $I_{H,n}(\omega)$ is defined as:

$$\begin{aligned} I_{H,n}(\omega) &= \int_{-\infty}^{+\infty} G_{H,n}^{(n)}(\tau)e^{-i\omega\tau}d\tau \\ &= l^{2n}\left(\frac{sin\frac{\omega l}{2}}{\frac{\omega l}{2}}\right)^{2n} |\omega|^{2n-1-2h}, \end{aligned} \tag{35}$$

when $\omega \rightarrow 0$, $I(\omega) \rightarrow 1/\omega^{2H-2n+1} = 1/w^{\alpha}$ with $-1 < \alpha < 1$. At low frequencies, n-fGn behaves like fGn.

Although nonstationary, it would be tempting to evaluate the spectral properties of n-fBm. Roughly speaking, it can be said that the spectral shape of n-fBm is equal to $1/\omega^{2H+1}$. With $n - 1 < H < n$, we get a $1/\omega^{\alpha}$ spectral shape with α in the range $]2n-1, 2n+1[$. These new models, with n being any strictly positive integer, allow a global range of variation for α in the range $]1, +\infty[$. It offers a larger framework than classical fBm for nonstationary $1/f^{\alpha}$ signals, where α can only vary in the range $]1, 3[$.

4.3 Synthesis of n-th Order Fractional Gaussian Noises

In this section, we propose a synthesis method for n-fGn. Without loss of generality, the case $n = 2$ is considered which means that H is considered in the range $]1, 2[$. This choice is motivated by two reasons. First, this case was the one that motivated the generalized model for values of H just above 1. Secondly, higher order n-fBm $(n > 2)$ will require much larger signal length than 2-fBm to illustrate their particular property of smoothness. Definition and properties of 2-fBm and 2-fGn are first briefly presented.

4.3.1 2-fBm and Its Second-Order Increments

Second-order fractional Brownian motion (2-fBm) is defined as:

$$B_H^2(t) = \frac{1}{\Gamma\left(H+\frac{1}{2}\right)}\left\{\int_{-\infty}^{0}\left[(t-u)^{H-1/2}-(-u)^{H-1/2}-\left(H-\frac{1}{2}\right)(-u)^{H-3/2}\right]dB(u) + \int_0^t (t-u)^{H-1/2}\,dB(u)\right\}. \tag{36}$$

for $t \geq 0$, with $1 < H < 2$.

B_H^2 is Gaussian, continuous, zero-mean, and self-similar.

Second-order increments of 2-fBm (2-fGn), which are stationary, are defined as:

$$\Delta_l^{(2)} B_H^2(t) = B_H^2(t+2l) - 2B_H^2(t+l) + B_H^2(t). \tag{37}$$

Their autocorrelation function is given by:

$$\begin{aligned} G_{H,2}^{(2)}(\tau) &= E\left[\Delta_l^{(2)} B_H(t+\tau)\Delta_l^{(2)} B_H(t)\right] \\ &= \frac{C_H}{2}\left[|\tau+2l|^{2H} - 4|\tau+l|^{2H} + 6|\tau|^{2H} - 4|\tau-l|^{2H} + |\tau-2l|^{2H}\right], \end{aligned} \tag{38}$$

and the corresponding spectrum $I(\omega)$ is given by:

$$I(\omega) = l^4\left(\frac{sin\frac{\omega l}{2}}{\frac{\omega l}{2}}\right)^4 |\omega|^{3-2H}. \tag{39}$$

4.3.2 Synthesis of 2-fGn Realizations

As 2-fBm is a nonstationary process, its synthesis is difficult. For this reason, we consider the synthesis of its second order increments which are stationary. The Cholesky decomposition method [23] can be used. The covariance matrix, R of 2-fGn is:

$$R[i, j] = (G_2\,[i-j])_{i,j=0,\ldots,N-1}\,. \tag{40}$$

It is a definite, positive, and Toeplitz matrix. R could be decomposed using the Cholesky decomposition method as:

$$R = LL^T. \tag{41}$$

where L is a lower triangular matrix, and T stands for the matrix transposition operator. Consider the transform $V = LY$, with Y being a white Gaussian noise vector. Then, $E[VV^T] = R$, proving that V has the same second-order statistics as 2-fGn.

For a Gaussian process, a synthesis method is exact if the second-order statistics of the generated data are equal to those of the process. This is clearly the case for the Cholesky method described above since second-order statistics of V are the same as those of the 2-fGn process.

For the reconstruction of the 2-fBm signal, B_H^2 depends not only on the 2-fGn process but on the first order increment, $\Delta^{(1)} B_H^2$ as well. In order to take advantage of the Toeplitz matrix, we prefer to replace in simulations the 2-fBm by its approximation, which has same second-order increments such that $\Delta^{(1)} B_H^2(0) = 0$. Moreover, the 2-fGn process is stationary and, thus, is easier to use in practice than its related 2-fBm nonstationary model. The approximated $B_H^2(m)$ process is recovered from $\Delta^{(l)} B_H^2(m)$ in the following way. Start with $\Delta^{(2)} B_H^2(m)$, which are the first increments of $B_H^2(m)$ deduced from (27):

$$\Delta^{(1)} B_H^2(m) = \Delta^{(2)} B_H^2(m-1) + \Delta_1^{(1)} B_H^2(m-1), \tag{42}$$

with the initial condition $\Delta^{(1)} B_H^2(0) = 0$. This leads to

$$B_H^2(m) = \Delta^{(1)} B_H^2(m-1) + B_H^2(m-1). \tag{43}$$

with $B_H^2(0) = 0$.

For $N = 1024$, realizations of 2-fGn, the increments of 2-fBm, and 2-fBm are presented for $H = 1.2$, $H = 1.5$, and $H = 1.8$ in Figs. 3, 4, and 5 respectively. The general appearance of 2-fGn (top drawings) looks like the classical fGn, whereas increments of 2-fBm (middle drawings) are very much like fBm. On the other hand, 2-fBm (bottom drawings) exhibits smoother behaviors than classical fBm. Namely, they are able to model nonstationary signals having very soft shapes.

4.3.3 Analysis of 2-fBm Realizations

Among estimators for the H parameter of fBm, the maximum likelihood estimator (MLE) is often considered to be the best one. It has been shown in [9] that it is asymptotically efficient for fBm (i.e. unbiased and reaches the Cramer-Rao lower bound (CRLB)) and that estimates are asymptotically Gaussian distributed, with

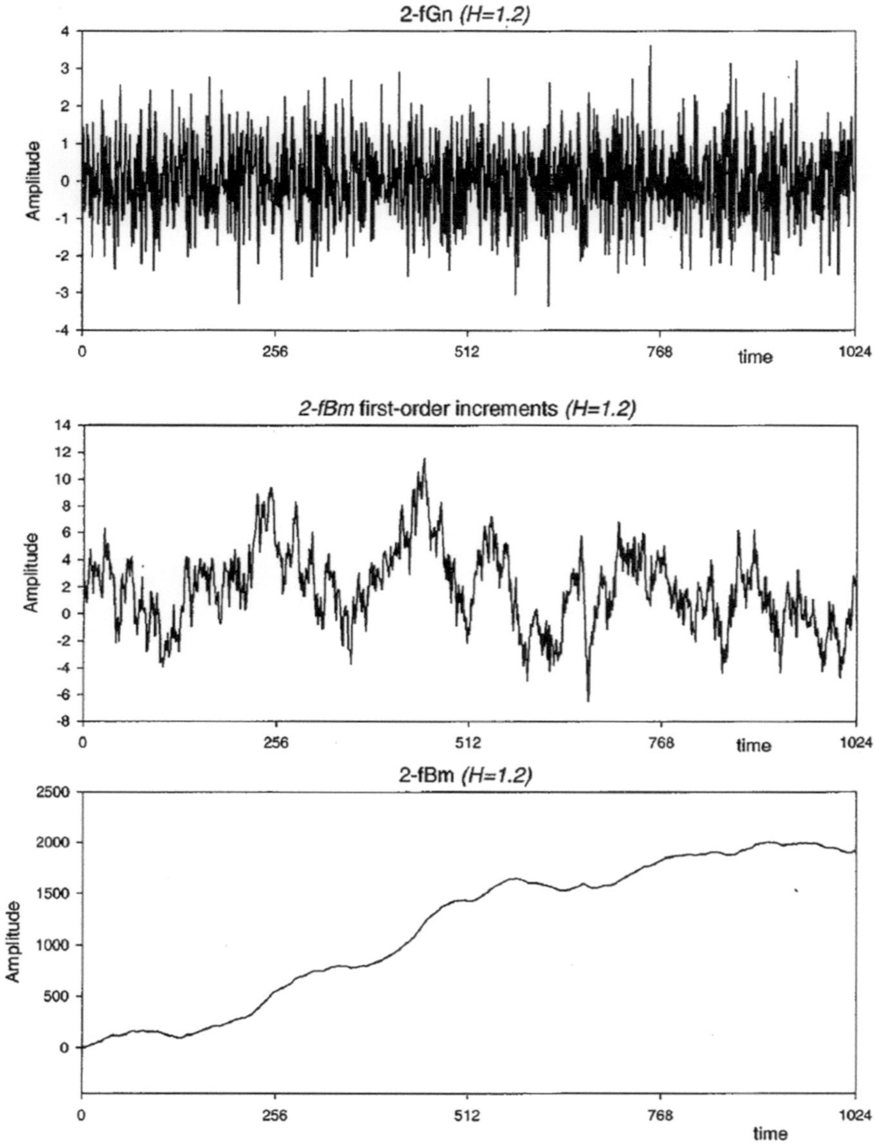

Fig. 3 A realization of 1024 samples for $H = 1.2$. (From top to bottom) 2-fGn, 2-fBm first-order increments, and 2-fBm

a rate of convergence $\sqrt{N}$. The implementation of MLE for 2-fGn is completely described in [32].

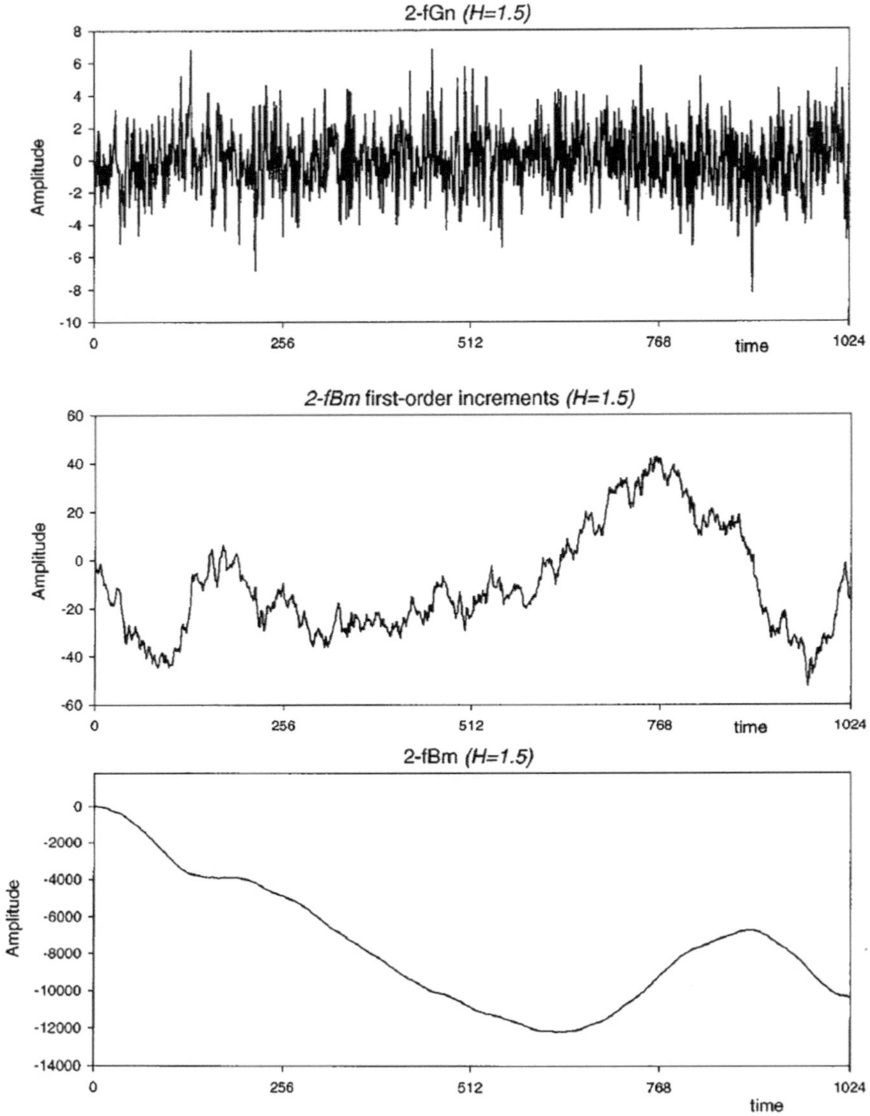

Fig. 4 A realization of 1024 samples for $H = 1.5$. (From top to bottom) 2-fGn, 2-fBm first-order increments, and 2-fBm

5 Application

Osteoporosis is considered as a public health issue [19]. Prevention of fracture results in finding which populations are at risk for fracture. Osteoporosis is defined as a disease characterized by low bone mass and microarchitectural alterations of bone tissue,

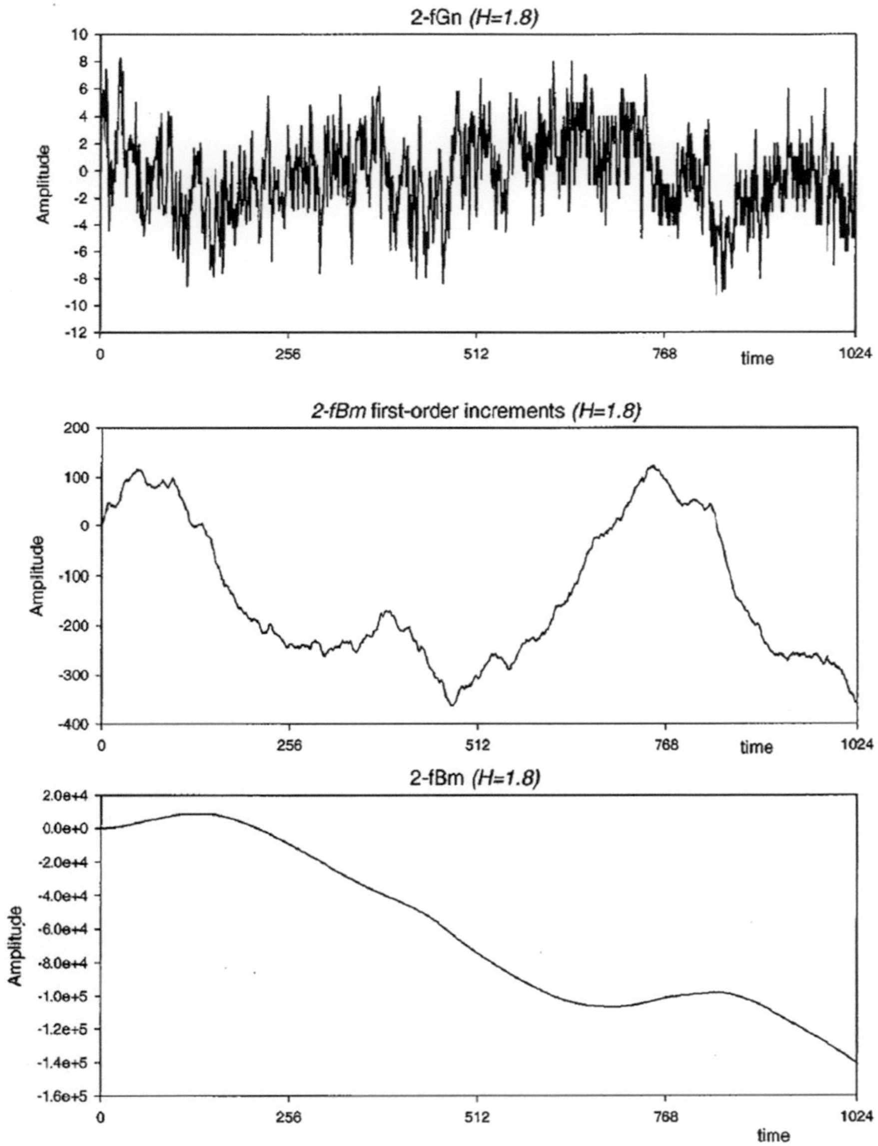

Fig. 5 A realization of 1024 samples for $H = 1.8$. (From top to bottom) 2-fGn, 2-fBm first-order increments, and 2-fBm

leading to enhance bone fragility and consequent increase in fracture risk [29]. The bone mass density (BMD) is evaluated in the clinical routine while the microarchitectural modifications are not. The development of a useful microarchitecture indicator providing an appropriate risk factor of osteoporotic fractures would lead to a better diagnosis of osteoporosis [10]. This indicator should be independent from BMD and thus yield complementary information versus BMD to osteoporotic bone changes.

Texture analysis applied to trabecular bone images offers the ability of exploiting the information present on conventional radiographs.

Calcaneus radiographs were performed after a standardized protocol. We used X-ray clinical apparatus with a tungsten tube and an aluminum filter of 1-mm thickness. The tube voltage was fixed to 36 kV and the exposure condition was 18 mA, with an exposure time of 0.08 s. The source-calcaneus distance was settled at 1 m, and the calcaneus was placed in contact with the sensor. An image of a region of interest (256 × 256 pixels) from a calcaneus radiography is extracted thanks to two anatomical marks Fig. 6. This region was scanned with 256 gray levels to obtain a trabecular bone texture image as presented in Fig. 7. Our technique was tested on a

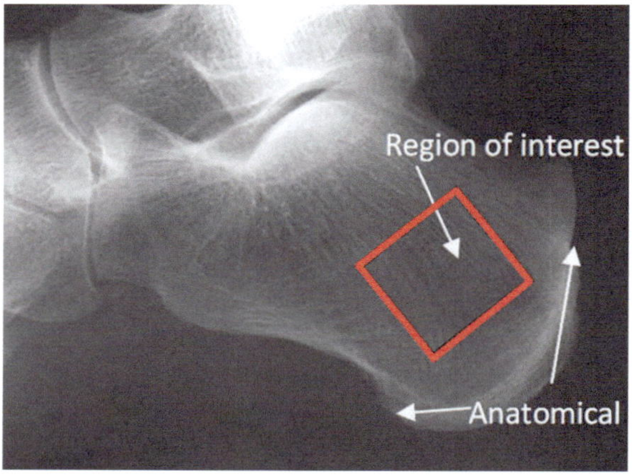

Fig. 6 Calcaneus radiograph and its region of interest (red square)

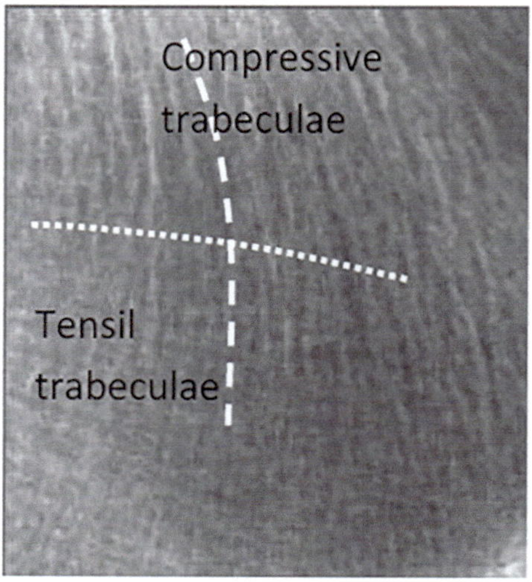

Fig. 7 Region of interest measuring 256 × 256 pixels (2.7 × 2.7 cm^2) used for processing

population composed of 77 radiographs provided by the medical stuff at the hospital of Orleans. Among these subjects, there were 38 patients with osteoporotic fractures (vertebral fracture) and 39 control cases. Because age has an influence on bone density and on trabecular bone texture analysis, the fracture cases were age-matched with the control cases.

The calcaneus (heel bone) as shown in Fig. 6 from a bone radiograph, presents a complex and textured image which is well suited to measure trabecular bone modifications. This bone is submitted to compression and tension forces produced by the walking and by the gravity. The evolution of the orientations of the trabeculae enables quantifying the degree of deterioration of the bone. For these reasons, first, we have analyzed these data with the classical fBm model. To take into account the trabecular bone modifications transposed on the radiographic images, we performed an oriented fractal analysis. For each direction, we analyzed several lines with the same orientation. The result in one direction was the average of the values for each line. This unidirectional evaluation was repeated in 36 directions. The final result for an image, is calculated as the mean value of the results of the 36 directions. The maximum likelihood estimator (MLE) [23, 27] was used to obtain the value of the H parameter of each line. The MLE was chosen because it is an asymptiotically unbiased estimator which has the lowest theoretical variance. Furthermore, this method was tested on synthetic data and shows excellent results for short data length [18]. The mathematical background of this method can be found in [23].

However, A precise analysis of the mean periodogram performed on the lines of trabecular bone images presents two distinct frequency regimes as shown on Fig. 8. These two regimes are separated by a cut-off frequency γ. The low frequencies $|f| < \gamma$, or, equivalently, the large scales correspond to the intertrabecular area. While the high frequencies $|f| > \gamma$, or, equivalently, the small scales correspond to the trabecular area. This finding confirm that the pfBm model is more suited for such data than the standard fBm model. According to Fig. 8, γ was fixed to value of 26. This can be explained by the repartition of the bone trabeculae which mean thickness is about 100–200 μm separated by a distance of about 400–500 μm. According to

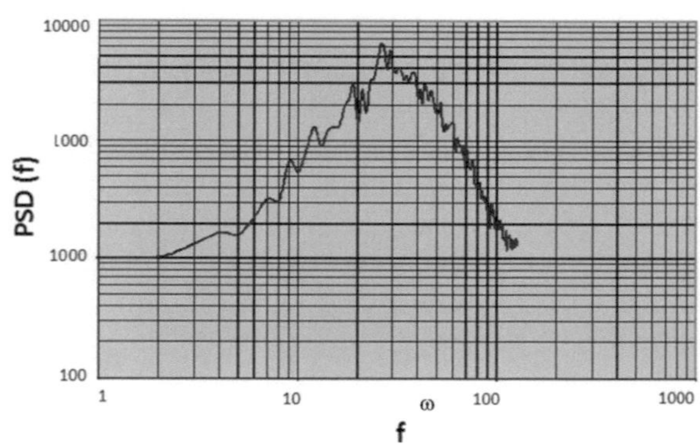

Fig. 8 A representative mean periodogram (power spectral density or PSD) of the lines of an X-ray image

the sensor's resolution used in this experiment, $\gamma = 26$ corresponds to the mean intertrabecular distance.

Then, the two extended models (pfBm and nfBm) were applied on these data to characterize the trabecular bone modifications, and check if it yields evaluating bone changes. The parameters H_o and H_i were measured by linear regression on the spectral density of the increments of the lines of the different images. Indeed, the non uniform changes due to osteoporosis induce variations of the degree of anisotropy.

A bilateral hypothesis test [38] was applied to evaluate the discrimination power of the H, H_o and H_i parameters. The nfBm model, with $n = 2$ was used when values over 1 for the H_o or H_i parameters were found. In this case, the 2fGn was used. Any obtained value is said to separate the 2 populations if the Student's $| t |$ value exceeds 2.69 for a high level of significance of 0.01. For the H parameter of the classical fBm, t is equal to 4.09. For H_o and H_i, we found, 0.909 and 5.39 respectively.

This demonstrates clearly that the H_i parameter in high frequencies discriminates better the two populations. In conclusion, over a reasonable range of spatial frequencies, a fractal-based model has been validated and can be used to describe trabecular bone data in a concise scale invariant fashion.

6 Conclusion

In this work, the main properties of two extensions of the fractional Brownian motion model are presented and studied.

First, the piece-wise fractional Brownian model (pfBm) with parameters $0 < H_o, H_i < 1$, and $\gamma > 0$ is presented. This process presents two distinct frequency or self-similar regimes: at low frequencies $|\omega| < \gamma$ or, equivalently, at large scale, the process can be considered to be self-similar with parameter H_o, whereas for high frequencies $|\omega| > \gamma$ or small scale, H_i governs the process. For $H = H_o = H_i$ or limit cases $\gamma \to 0$ and $\gamma \to \infty$, the pfBm process reduces to the classical fractional Brownian motion such that $0 < H < 1$. In addition, the pfBm is real, continuous, Gaussian, and nonstationary process but has continuous, Gaussian, and stationary increments.

Second, a generalization of the fractional Brownian motion called nth-order fractional Brownian motion (n-fBm) is investigated. The H parameter governing the n-fBm can vary in the range $]n-1, n[$, where n can be any strictly positive integer. Properties of this new nonstationary processes have been presented. Their nth-order increments are stationary and, thus, were called nth-order fractional Gaussian noises. The n-fBm offers a larger framework than the classical fBm to model $1/f^{\alpha}$ nonstationary signals with α within $]1, +\infty[$. These new processes allow the study of a wider range of behavior than the classical fBm/fGn. Synthesis methods were discussed for such processes.

An application of these processes was applied to the field of biomedical imaging. Trabecular bone radiographs from two populations composed of osteoporotic patients on one side and healthy subjects on the other side have been characterized using the standard fBm and its extensions, pfBm and $n$fbm. Results demonstrates that the pfBm and n-fBm models, more suited for such data, better discriminate the two populations than the classical fBm. This fractal evaluation of trabecular bone architecture may have applications in a variety of tasks of clinical interest.

References

1. Abry, P., Veitch, D.: Wavelet analysis of long-range-dependent traffic. IEEE Trans. Inform. Theory **44**, 1–15 (1998)
2. Ayache, A., Lévy-Véhel, J.: The generalized multifractional Brownian motion. Statist. Inference Stoch. Process. **3**(1–2), 7–18 (2000)
3. Bel, L., Oppenheim, G., Robbiano, L., Viano, M.C.: Distribution processes with stationary fractional increments. Proc. ESAIM 43–54 (1998)
4. Benassi, A., Deguy, S.: Multi-Scale Fractional Brownian Motion: Definition and Identification, Clermont-Ferrand. Blaise Pascal University Press, France (1999)
5. Beran, J.: Statistics for Log-Memory Processes, London. Chapman and Hall, U.K (1994)
6. Berger, J.M., Mandelbrot, B.B.: A new model for the clustering of errors on telephone circuits. IBM J. Res. Dev. **7**, 224–236 (1963)
7. Cintract, M.: Définition d'un modèle fractal par zone: Synthèse, analyze et application, Ph.D. dissertation, Univ. d'Orléans, Orleans, France (1999)
8. Cintract, M., Harba, R.: Définition d'un modèle fractal par zone: Synthèse, analyze et application. In: Proceedings of AGIS, pp. 289–294 (1997)
9. Dahlaus, R.: Efficient parameter estimation for self-similar processes. Ann. Stat. **17**(4), 1749–1766 (1989)
10. Dempster, D.W.: The contribution of trabecular architecture to cancellous bone quality. J. Bone Miner. Res. **15**, 20–23 (2000)
11. Deriche, M., Tewfik, A.H.: Signal modeling with filtered discrete fractional noise processes. IEEE Trans. Signal Process. **41**(9), 1239–2849 (1993)
12. Feder, J.: Fractals. Plenum, New York (1988)
13. Fournier, A., Fussell, D., Carpenter, L.: Computer rendering of stochastic models. Commun. ACM **25**(6), 371–384 (1982)
14. Harba, R., Cintract, M., Zabat, M., Van Hamme, H.: Fractal estimation in a given frequency range: Application to smectite images. EUSIPCO Signal Process. 2485–2488 (1998)
15. Harba, R., Jacquet, G., Jennane, R., Loussot, T., Benhamou, C.L., Lespessailles, E., Tourlière, D.: Determination of fractal scales on trabecular bone x-ray images. Fractals **2**, 451–456 (1994)
16. Hurst, H.E., Black, R.P., Simaka, Y.M.: Long Term Storage: An Experimental Study, London. Constable, U.K. (1965)
17. Jennane, R., Harba, R.: Fractional Brownian motion: A model for image texture. In: Proceedings of EUSIPCO Signal Processing, pp. 1389–1392 (1994)
18. Jennane, R., Harba, R., Jacquet, G.: Méthodes d'analyse du mouvement brownien fractionnaireÃ¢â,¬Â¯: théorie et résultats comparatifs. Traitement du Signal **18**, 419–436 (2001)
19. Johnell, O.: The socioeconomic burden of fractures: today and in the 21st century. Am. J. Med. **103**(2A), 25S-26S (1997)
20. Kaplan, L.M., Kuo, C.C.: Extending self-similarity for fractional Brownian motion. IEEE Trans. Signal Process. **42**, 3526–3530 (1994)
21. Kashyap, R.L., Lapsa, P.M.: Synthesis and estimation of random fields using long-correlation models. IEEE Trans. Pattern Anal. Mach. Intell. PAMI-6 800–809 (1984)
22. T. Loussot, R. Harba, G. Jacquet, C.L. Benhamou, E. Lespesailles, A. Julien, An oriented fractal analysis for the characterization of texture: Application to bone radiographs, EUSIPCO Signal Process. (1996), 371–374

23. T. Lundahl, W.J. Ohley, S.M. Kay, R. Siffert, Fractional Brownian Motion: A Maximum Likelihood Estimator and Its Application to Image Texture, IEEE-Trans. On Med. Imag., MI-5(3) (1986), 518–523
24. Mandelbrot, B.B.: Fractals: Form, Chance and Dimension. Freeman, New York (1977)
25. Mandelbrot, B.B.: The Fractal Geometry of Nature. Freeman, New York (1982)
26. Mandelbrot, B.B.: Stochastic models for the Earth's relief, the shape and the fractal dimension of the coastlines and the number-area rule for islands. Proc. Nat. Acad. Sci. **72**(10), 3825–3828 (1975)
27. Mandelbrot, B.B., Van Ness, J.W.: Fractional Brownian motion, fractional noises and applications. SIAM Rev. **10**, 422–438 (1968)
28. A. Manzanera, T. Bernard, F. Preteux, B. Longuet, Medial Faces from a Concise 3D Thinning Algorithm, IEEE International Conference on Computer Vision (ICCV'99) (1999), 337–343
29. NIH, Consensus development conference: diagnosis, prophylaxis, and treatment of osteoporosis., Am J Med, 94 (1993), 646–650
30. Peleg, S., Naor, J., Hartley, R., Avnir, D.: Multiple resolution texture analysis and classification. IEEE-Trans. Pattern Anal. Machine Intell. **4**, 518–523 (1984)
31. Pentland, A.P.: Fractal-based description of natural scenes. IEEE-Trans. Pattern Anal. Machine Intell. **6**, 661–674 (1984)
32. Perrin, E., Harba, R., Berzin-Joseph, C., Iribarren, I., Bonami, A.: nth-order fractional Brownian motion and fractional Gaussian noises. IEEE Trans. Signal Process. **49**(5), 1049–1059 (2001)
33. Perrin, E., Harba, R., Jennane, R., Iribarren, I.: Piecewise Fractional Brownian Motion. IEEE Trans. Signal Process. **53**(3), 1211–1215 (2005)
34. Perrin, E., Harba, R., Jennane, R., Iribarren, I.: Fast and exact synthesis for 1-d fractional Brownian motion and fractional gaussian noises. IEEE Signal Process. Lett. **9**(11), 382–384 (2002)
35. Pfeeifer, P., Avinir, D., Farin, A.: Ideally irregular surfaces of dimension greater than two in theory and practice. Surface Sci. **126**, 569–572 (1983)
36. Reed, I.S., Lee, P.C., Truong, T.K.: Spectral representation of fractional Brownian motion in dimensions and its properties. IEEE Trans. Inf. Theory **41**(5), 1439–1451 (1995)
37. Schaefer, D.W., Martin, J.E., Wiltzius, P., Cannell, D.S.: Fractal geometry of colloid aggregates. Phys. Rev. Lett. **52**(5), 2371–2374 (1984)
38. Spiegel, M.R.: Theory and problems of probability and statistics. Schaum, McGraw-Hill, New York (1975)
39. Yaglom, A.M.: Correlation Theory of Stationary and Related Random Functions. Springer-Verlag, New York (1987)

Recent Advances on Accuracy, Stability and Optimality of B-Bases

Juan Manuel Peña

Abstract Some old and new results relating accuracy, stability and optimal bases are revisited. We focus on the relationship of these fields in the framework of totally positive bases and matrices. In fact, for some subclasses of totally positive matrices, which are often collocation, wronskian or gramian matrices of totally positive bases, high relative accurate algebraic algorithms have been found. Moreover, for spaces with a totally positive basis, the concept of B-bases corresponds to the optimally stable bases, and these bases also satisfy many other optimality properties, which play an important role in Approximation Theory and Computer Aided Geometric Design.

1 Introduction

This paper surveys some recent advances on stability, optimal bases and high relative accuracy when working with some classes of matrices. Performing an algorithm with high relative accuracy (HRA) is a very desirable goal. Recent research in Numerical Linear Algebra has shown that certain classes of matrices allow us to perform many computations to HRA, independently of the size of the condition number. For instance, the computation of their singular values, eigenvalues or inverses. These classes of matrices include some subclasses of totally positive matrices, which are often collocation, wronskian or gramian matrices of totally positive bases. Let us recall that a matrix is called totally positive (TP) if all its minors are nonnegative, and a basis of a univariate space of functions is called totally positive (TP) if all its collocation matrices. TP matrices present applications to many fields, in particular to Approximation Theory and Computer Aided Geometric Design (C.A.G.D.) The basis with optimal shape preserving properties in C.A.G.D. is the normalized B-basis of the space. Some other recent optimal properties of B-bases are also presented.

J. M. Peña (✉)
Dep. Applied Mathematics/IUMA, Universidad de Zaragoza, Zaragoza, Spain
e-mail: jmpena@unizar.es

© The Author(s), under exclusive license to Springer Nature Switzerland AG 2025
M. J. Ibáñez-Pérez et al. (eds.), *Mathematical And Computational Modelling, Approximation and Simulation*, SEMA SIMAI Springer Series 41,
https://doi.org/10.1007/978-3-032-01114-5_3

The paper is organized as follows. Section 2 recalls some basic concepts relating the errors obtained when computing with floating point arithmetic. In a general framework, the forward error bound (difference between the computed and the exact solution) is obtained through the backward error and the condition number of our problem. However, in some problems it is possible to find a parametrization of the data and an algorithm leading to small forward error bounds in spite of a bad conditioning with its initial parametrization. In these cases, we can assure HRA. We also introduce the Bernstein basis of polynomials, whose optimal properties have been extended to other B-bases.

Section 3 presents the problem of optimal stability for the evaluation, where B-bases have played a crucial role. In fact, B-bases and, beyond Total Positivity, more general bases present optimal stability properties for the evaluation among all bases of nonnegative functions (see [51, 66, 68, 69]). The stability of the corresponding evaluation algorithms has been also deeply studied (see [17, 19, 20, 22–24, 27, 29, 52, 70, 71]).

Section 4 presents B-bases and the optimal shape preserving properties of normalized B-bases. It shows some applications of TP matrices to the field of C.A.G.D.. We recall that shape preserving representations of curves are associated with (normalized) TP bases, which are bases such that their collocation matrices are (stochastic) TP (see [67]). Other important applications of TP matrices to C.A.G.D. are related with the problem of finding bases with optimal shape preserving properties (which correspond to the concept of normalized B-bases) and with the problem of recognizing normalized TP bases (and so, shape preserving representations). Examples of normalized B-bases are the Bernstein basis and the B-spline basis. Normalized B-bases satisfy more optimal properties (see [9, 18]), and we recall some of them.

For the Bernstein basis, we also have optimal conditioning of its collocation matrices with respect to the collocation matrices of another normalized totally positive basis (see [21]). In Sect. 5 some extensions of this result to general B-bases are recalled.

2 Errors, High Relative Accuracy and Optimal Bases

Let us start with some basic notations and definitions. Given a real number x that can be calculated by an algorithm, we denote by $\hat{x}$ the corresponding computed number with floating point arithmetic. Then the absolute error performed to compute $\hat{x}$ is given by $E_{abs}(\hat{x}) = |x - \hat{x}|$ and the corresponding relative error when $x \neq 0$ is

$$E_{rel}(\hat{x}) = \frac{|x - \hat{x}|}{|x|}.$$

Analogously, if x is a real vector that can be calculated by an algorithm, let us denote by $\hat{x}$ the corresponding computed vector with floating point arithmetic. Then the relative error performed to compute $\hat{x}$ (with x a nonzero vector) is given by

$$E_{rel}(\hat{x}) = \frac{\|x - \hat{x}\|}{\|x\|}.$$

However, if x has nonzero components and we do not want to miss the computed error corresponding to the components of x with least absolute value, then we can consider a componentwise relative error given by $\max_i \frac{|x_i - \hat{x}_i|}{|x_i|}$.

When we apply an algorithm and perform the computations with floating point arithmetic, since we do not know the exact error performed with our computations, it is convenient to try to obtain upper bounds of this error, usually known as *forward* error bounds. However, it is usually difficult to obtain directly such bounds, so that an alternative approach that has been very successful in the field of numerical linear algebra and other fields tries to obtain the forward error bounds through the backward errors. If we consider that our computed solution is the exact solution of a perturbed problem, the *backward* error measures the distance between the perturbed problem and the initial problem. For example, given $y = f(x)$, a continuous real function, and $\hat{y}$ a numerical approximation to f at a point x, consider the set of values $x + \triangle x$ for which $\hat{y}$ is the exact value:

$$\hat{y} = f(x + \triangle x),$$

and consider the least $| \triangle x|$, which is called *backward* error. If for all x, the value $| \triangle x|$ is small (in the context of our problem), then we say that our method is *backward* stable. Backward stability plays an important role for designing a good algorithm.

The backward error depends on the numerical method that we have used and it is well known that the growth factor of an algorithm is an indicator of its stability (cf. [45]). Recall that the *growth factor* of a numerical algorithm is usually defined as the quotient between the maximal absolute value of all the elements that occur during the performance of the algorithm and the maximal absolute value of all initial data. The optimal growth factor for a numerical method is 1 and so optimal methods under this viewpoint are methods with growth factor 1.

Backward and forward errors are related by the *conditioning* of the problem, which measures the effect of data perturbations on the solution of the problem. Roughly speaking, when for a given problem we have defined the corresponding forward error, backward error and the condition number, one tries to prove the relation:

$$\text{forward error} \leq \text{condition number} \times \text{backward error},$$

which allows us to get a forward error bound through the backward error (cf. [47]). Although the computed solution has a small backward error, it can be amplified by the condition number leading to a large forward error. Hence, in contrast to the backward error, which depends of the used method, the conditioning can become an intrinsic cause to obtain a nice forward error bound. In conclusion, in this framework, in order to get a small forward error we need a small conditioning and using a method with small backward error.

Let us illustrate the previous approach with the problem of solving TP linear systems, that is, linear systems whose coefficient matrices are TP. It is known that solving TP linear systems through Gauss elimination is backward stable (cf. [3]). However, TP matrices are often very ill conditioned and so the forward errors are large. However, in many cases it is possible to get small forward errors with these ill conditioned TP linear systems though an alternative approach that we now introduce.

In some problems it is possible to find a parametrization of the data and an algorithm leading to small forward error bounds in spite of a bad conditioning with its initial parametrization. The desired goal is to guarantee *high relative accuracy* (HRA). There exists a sufficient condition to assure the HRA of an algorithm that we now recall. Given an algorithm using only additions of numbers of the same sign, multiplications and divisions, and assuming that each initial real datum is known to HRA, then it is well–known that the output of that algorithm can be computed to HRA (cf. [35, p. 52]). Moreover, in (well–implemented) floating point arithmetic HRA is also preserved even when we perform true subtractions when the operands are original (and so, exact) data (cf. p. 53 of [35]). So, the sufficient condition to assure the HRA of an algorithm is satisfied if it only uses additions of numbers of the same sign, multiplications, divisions and subtractions (additions of numbers of different sign) of the initial data. This condition is called "no inaccurate cancellation" (NIC). An example of a simple problem for which an HRA algorithm cannot be found is provided by the evaluation of three real numbers $x + y + z$ (see [34]). For some subclasses of TP matrices, HRA algorithms have been found, as we shall see later. However, this is not always possible for any structured class of matrices. For instance, accurate linear algebra for the problem of calculating determinants or minors is impossible on the class of Toeplitz matrices (see corollaries 3.43 and 3.45 of [34]). Let us recall that a Toeplitz matrix B has the following simple structure:

$$B = \begin{pmatrix} a_0 & a_1 & \cdots & a_{n-2} & a_{n-1} \\ a_{-1} & a_0 & \ddots & & a_{n-2} \\ \vdots & \ddots & \ddots & \ddots & \vdots \\ a_{-n+2} & & \ddots & \ddots & a_1 \\ a_{-n+1} & a_{-n+2} & \cdots & a_{-1} & a_0 \end{pmatrix}.$$

In the problem of evaluating a real function of a finite dimensional vector space of functions, the conditioning depends on the basis of the space that we use. The problem of finding bases with minimal condition number has been analyzed for bases of nonnegative functions, which play an important role in many problems of approximation theory or computer aided geometric design (cf. [38]). In this framework, the optimal bases for the evaluation of a function are bases b of nonnegative functions that cannot be expressed, up to permutation and positive scaling, in the form vK, where v is another basis of nonnegative functions and K is a nonnegative matrix (see [9, 68, 69]). This topic will be considered in the next section. Examples of these optimal bases are the Bernstein basis [39] for the space of polynomials on a compact interval, the B-spline basis [66] or the bases of spaces of real multivariate functions given in [51]. Let us recall that the *Bernstein basis* $(b_0^n(t), \dots, b_n^n(t))$ of the space P_n of polynomials of degree at most n on $[0, 1]$ is defined by

$$b_i^n(t) = \binom{n}{i} t^i (1-t)^{n-i}, \quad t \in [0, 1], \ i = 0, 1, \dots, n, \tag{1}$$

and it is the most used polynomial basis in computer aided geometric design. Other optimal properties of Bernstein basis and their generalizations will be considered in later sections.

3 Optimal Bases for the Evaluation

In the previous section, we have mentioned the bases of nonnegative functions that are optimal, with respect to the corresponding condition number, for the problem of evaluating a real function of a finite dimensional vector space of functions. Let us now recall this condition number. Given a basis $u = (u_0, \dots, u_n)$ of a real vector space $\mathcal{U}$ of real functions defined on a subset S of $\mathbf{R}^m$ $(m \geq 1)$ and a function $f \in \mathcal{U}$, we can write $f(x) = \sum_{i=0}^n c_i u_i(x)$ for all $x \in S$, where $c_i \in \mathbf{R}$ for all $i = 0, \dots, n$. The stability of the basis u with respect to the evaluation at a point is measured by the function $C_u : \mathcal{U} \times I \to \mathbf{R}_+$ given by

$$C_u(f, x) := \sum_{i=0}^n |c_i u_i(x)|. \tag{2}$$

Given two bases of nonnegative functions u, v, let A the matrix of change of basis such that $v = uA$. The following result compares the conditioning of the bases by means of the nonnegativity of A. The result can be proved with the adaptation to the condition number (2) of the proof of Lemma 3.1 of [51], which proves a similar result for a relative condition number derived from the previous one.

Lemma 3.1 Let $\mathcal{U}$ be a finite dimensional vector space of functions defined on a subset S of $\mathbf{R}^m$. Let u, v be two bases of nonnegative functions of $\mathcal{U}$. Then

$$C_u(f, x) \leq C_v(f, x), \quad \forall f \in \mathcal{U}, \quad \forall x \in S \tag{3}$$

if and only if the matrix A such that $v = uA$ is nonnegative.

□

In the case of the space P_n of polynomials of degree at most n on an interval $[0, 1]$, Theorem 3 of [39] leads to the optimality result of the Bernstein basis given by (1).

Theorem 3.1 Let $b = (b_0^n, \ldots, b_n^n)$ be the Bernstein basis. Then there does not exist (up to reordering and positive scaling) another basis $u = (u_0, \ldots, u_n)$ of nonnegative functions in P_n such that $C_u(p, t) \leq C_b(p, t)$ for all $t \in [0, 1]$ and $p \in P_n$. □

Similar results to the previous one are satisfied, for instance, by the B-spline basis [66] or, more generally, by some B-bases (see next section, [68, 69]) and by the bases of spaces of real multivariate functions given in [51].

4 Bases with Optimal Shape Preserving Properties

Let us start by recalling some basic facts on methods of computer aided geometric design (C.A.G.D.) using control polygons. Given a sequence $U_0, \ldots, U_n$ of points in $\mathbf{R}^k$, we define a curve $\gamma(t) = \sum_{i=0}^n U_i u_i(t)$, $t \in I$. We shall denote by $U_0 \cdots U_n$ the polygonal arc with vertices $U_0, \ldots, U_n$. This is usually called the *control polygon* of γ and the points U_i, $i = 0, \ldots, n$, are called control points. In C.A.G.D. it is usually required that the functions u_i, $i = 0, \ldots, n$, are nonnegative and $\sum_{i=0}^n u_i(t) = 1$ for all $t \in I$ (that is, the system $U = (u_0, \ldots, u_n)$ is *normalized*). A normalized system of nonnegative functions is usually called a *blending* system. Now we shall recall some properties which are convenient for design purposes, following mainly the notation of [38]. An important property for curve design is the *convex hull property*: for any control polygon, the curve lies always in the convex hull of the control polygon. Let us remark that the convex hull property holds if and only if U is a blending system. These geometric properties correspond to some properties of the collocation matrices of the system of functions U. Given a system of functions $U = (u_0, \ldots, u_n)$ defined on $I \subseteq \mathbf{R}$, the *collocation matrix* of U at $t_0 < \cdots < t_m$ in I is given by

$$M \begin{pmatrix} u_0, \ldots, u_n \\ t_0, \ldots, t_m \end{pmatrix} := (u_j(t_i))_{i=0,\ldots,m;\, j=0,\ldots,n}.$$

Clearly, U is blending if and only if all its collocation matrices are stochastic (that is, nonnegative and such that the sum of each row is one).

In order to have a more precise guide of the curve, and to put together several pieces of curves, it is desirable for the designer to have a very precise control of what happens at the ends of the curve. This leads to the *endpoint interpolation property*: the first control point always coincides with the start point of the curve and the last control point always coincides with the final point of the curve.

In interactive design we also want that the shape of a parametrically defined polynomial curve mimics the shape of its control polygon; thus we can predict or manipulate the shape of the curve by suitably choosing or changing the control polygon. It is well-known (cf. [7]) that when the basis is normalized totally positive the curve imitates the shape of its control polygon, due to the variation diminishing properties of the TP matrices. A system of functions is *totally positive* (TP) if all its collocation matrices are TP. We say that a system is *normalized totally positive* (NTP) if it is TP and normalized. If U is NTP then the curve γ inherits many shape properties of the control polygon. For instance, any line intersects the curve no more often than it intersects the control polygon. In particular, a planar curve or polygon is convex if and only if it crosses any line of the plane no more than two times. So, if the control polygon is planar and convex then the curve generated is also planar and convex. For NTP bases it also holds that the length, number of inflections and angular variation of the curve are bounded above by those of the control polygon (see [7] for a more precise statement of these properties). From Bemerkung II.4 of [73], we obtain that a blending system of functions satisfying the variation diminishing and the endpoint interpolation properties is necessarily TP. Therefore, if a system of functions is such that the curves generated by any control polygon satisfy the convex hull, variation diminishing and endpoint interpolation properties simultaneously, then it is an NTP system.

Now we shall deal with the problem of comparing the shape preserving properties of different NTP systems. Our goal is to find a basis with optimal shape preserving properties: we want a normalized totally positive basis such that the control polygon of a curve with respect to it is closer in shape to the curve than the control polygon with respect to any other normalized totally positive basis of the space. Due again to the variation diminishing property of the TP matrices (see Sect. 2 of [7] and Sect. 2 of [9]), the precise formulation of this problem is finding an NTP basis $B = (b_0, \ldots, b_n)$ such that any other (normalized) TP basis U is of the form $U = (b_0, \ldots, b_n)K$, where K is a (stochastic) TP matrix. The concept of (normalized) *B-basis* (introduced in [8]) gives always an affirmative answer to the previous problem by Theorem 4.2 of [8]: a space with a normalized totally positive basis has always a unique normalized B-basis B, which is the basis with optimal shape preserving properties, as we shall explain later. For the space of polynomials of degree less than or equal to n on $[a, b]$, the Bernstein basis is the optimal (that is, the normalized B-basis, as was proved in [7]). This was conjectured by Goodman and Said in [46] and an affirmative answer was given by Carnicer and Peña in [7]. For the corresponding space of polynomial splines, the B-spline basis is the normalized B-basis and so, optimal (see Theorem 4.6 of [8]). The monomial basis of polynomials is a B-basis of the space of polynomials of degree at most n on the nonnegative real numbers $[0, \infty]$. B-bases have been also found for many other spaces of functions (see [67] and many papers by Carnicer, Mainar and Peña (cf. [4–6]).

The following characterization of a B-basis is a consequence of Corollary 3.10 of [8] and Proposition 3.11 of [8].

Theorem 4.1 Let $(u_0, \ldots, u_n)$ be a TP basis of a space U of univariate functions on I. Then $(u_0, \ldots, u_n)$ is a *B-basis* if and only if for any other TP basis $(v_0, \ldots, v_n)$ of U the matrix K of change of basis such that $(v_0, \ldots, v_n) = (u_0, \ldots, u_n)K$ is TP. □

In general, a space with a totally positive basis has always B-bases (see Remark 3.8 of [8]). As for the problem of uniqueness of B-bases, we have the following result, which corresponds to Corollary 3.9 (iii) of [8]:

Proposition 4.1 Let $(c_0, \ldots, c_n)$ be a B-basis of a space of functions $\mathcal{U}$. A basis of $\mathcal{U}$ is a B-basis if and only if it is of the form $(d_0c_0, \ldots, d_nc_n)$ with $d_i > 0$ for all $i = 0, \ldots, n$.

About the uniqueness of normalized B-bases, we have an affirmative answer, which corresponds to Theorem 4.2 (i) of [8].

Proposition 4.2 If a space of functions has a normalized totally positive basis, then it has a (unique) normalized B-basis.

Now let us justify the optimal shape preserving properties of the normalized B-basis. As shown in [9], a basis is a normalized B-basis if and only if it satisfies the least variation diminishing, the endpoint interpolation and the convex hull properties simultaneously.

Let us show now that the unique normalized B-basis of a given space has optimal properties in this geometric framework. Since the normalized B-basis is least variation diminishing, we may hope that the control polygon of a curve with respect to the normalized B-basis is the closest in shape to the curve than the control polygons with respect to any other reasonable basis for curve design. Let us collect now some shape properties established in [7, 46], which applied to the normalized B-basis, confirm that it has optimal shape preserving properties.

Let γ be a curve generated by the control polygon $P_0 \cdots P_n$ with respect to a normalized totally positive basis. Let $B_0 \cdots B_n$ be the control polygon with respect to the normalized B-basis. Then the following properties hold:

(i) If $P_0 \cdots P_n$ is convex, then so are $B_0 \ldots B_n$ and the curve γ, and $B_0 \cdots B_n$ lies between $P_0 \cdots P_n$ and γ.
(ii) Length $\gamma \leq$ length $B_0 \cdots B_n \leq$ length $P_0 \cdots P_n$.
(iii) If $P_0 \cdots P_n$ turns through an angle $< \pi$, then $I(\gamma) \leq I(B_0 \cdots B_n) \leq I(P_0 \cdots P_n)$, where $I(\beta)$ denotes the number of inflexions of a curve β.
(iv) $\theta(\gamma) \leq \theta(B_0 \cdots B_n) \leq \theta(P_0 \cdots P_n)$, where $\theta(\beta)$ denotes the angular variation of a curve β.

In [9] there was a survey of other optimal properties which are satisfied by B-bases. We are now recalling new optimality properties found for B-bases and in the last section we shall also deal with optimal properties concerning their collocation matrices.

Let us now consider other interesting property in C.A.G.D. called the *progressive iteration approximation* property (see [18]), which is satisfied by NTP bases. Given a sequence of points $(P_i)_{i=0}^n$ such that the ith point is assigned to a parameter value t_i for $i = 0, 1, \ldots, n$ and a basis $(u_0, \ldots, u_n)$, we construct a starting curve $\gamma^0(t) = \sum_{i=0}^n P_i^0 u_i(t)$ with $P_i^0 = P_i$ for all $i \in \{0, 1, \ldots, n\}$. Then, computing the adjusting vector $\Delta_i^0 = P_i - \gamma^0(t_i)$ we can take $P_i^1 = P_i^0 + \Delta_i^0$, for $i = 0, 1, \ldots, n$, and construct a new curve as $\gamma^1(t) = \sum_{i=0}^n P_i^1 u_i(t)$. Iterating this process we can get a sequence of curves $\{\gamma^k\}_{k=0}^\infty$. The progressive iteration approximation property holds when this curve sequence converges to the polynomial curve interpolating the given initial sequence of points. This property holds for NTP bases and we proved in Theorem 4 of [18] the optimal convergence speed of the normalized B-basis.

Theorem 4.2 The normalized B-basis of a space $\mathcal{U}$ with an NTP basis provides a progressive iterative approximation with the fastest convergence rate among all NTP bases of $\mathcal{U}$. □

Other new optimal properties satisfied by B-bases are related to their collocation matrices and will be presented in the last section of the paper.

5 Totally Positive Matrices and HRA

Let us recall that TP matrices are real, nonnegative matrices whose minors are all nonnegative. They are also called totally nonnegative matrices and they present many applications to several fields, including C.A.G.D.. If all minors of a matrix are positive, then the matrix is called *strictly totally positive matrix* (STP matrix). TP and STP matrices have a long history and many applications (see [2, 37, 40, 41, 48, 72]) and have been studied mainly by researchers of those applications. The parametrization of TP matrices leading to HRA algorithms is provided by their bidiagonal factorizations, which are in turn closely related to an elimination procedure known as Neville elimination (NE). NE is a procedure to make zeros in a column of a matrix by adding to each row an appropriate multiple of the precedent one and it has been very useful for the study of TP matrices (see [42–44]). In particular, NE can be used to get the factorization of a nonsingular TP matrix A as a product of bidiagonal matrices, which in turn provides the adequate parametrization to obtain algorithms with HRA for many algebraic computations, such as the calculations of all eigenvalues, all singular values, the inverse or the solution of some linear systems (the linear systems $Ax = b$, where b has alternating consecutive signs). In fact, if the bidiagonal factorization can be computed to HRA, then these computations can be performed with HRA with the algorithms of [49].

Given a nonsingular matrix $A = (a_{ij})_{1\le i,j\le n}$, the *Neville elimination* (NE) procedure consists of $n-1$ steps and leads to the following sequence of matrices:

$$A =: A^{(1)} \to \widetilde{A}^{(1)} \to A^{(2)} \to \widetilde{A}^{(2)} \to \cdots \to A^{(n)} = \widetilde{A}^{(n)} = U, \tag{4}$$

where U is an upper triangular matrix. The matrix $\widetilde{A}^{(k)} = (\widetilde{a}_{ij}^{(k)})_{1\le i,j\le n}$ is obtained from the matrix $A^{(k)} = (a_{ij}^{(k)})_{1\le i,j\le n}$ by a row permutation that moves to the bottom the rows with a zero entry in column k below the main diagonal. For nonsingular TP matrices, it is always possible to perform NE without row exchanges (see [42]). If a row permutation is not necessary at the kth step, we have that $\widetilde{A}^{(k)} = A^{(k)}$. The entries of $A^{(k+1)} = (a_{ij}^{(k+1)})_{1\le i,j\le n}$ can be obtained from $\widetilde{A}^{(k)} = (\widetilde{a}_{ij}^{(k)})_{1\le i,j\le n}$ using the formula:

$$a_{ij}^{(k+1)} = \begin{cases} \widetilde{a}_{ij}^{(k)} - \frac{\widetilde{a}_{ik}^{(k)}}{\widetilde{a}_{i-1,k}^{(k)}} \widetilde{a}_{i-1,j}^{(k)}, & \text{if } k \leq j < i \leq n \text{ and } \widetilde{a}_{i-1,k}^{(k)} \neq 0, \\ \widetilde{a}_{ij}^{(k)}, & \text{otherwise}, \end{cases} \tag{5}$$

for $k = 1, \ldots, n-1$. The (i, j) *pivot* of the NE of A is given by

$$p_{ij} = \widetilde{a}_{ij}^{(j)}, \quad 1 \leq j \leq i \leq n.$$

If $i = j$ we say that p_{ii} is a *diagonal pivot*. The (i, j) *multiplier* of the NE of A, with $1 \leq j \leq i \leq n$, is defined as

$$m_{ij} = \begin{cases} \frac{\widetilde{a}_{ij}^{(j)}}{\widetilde{a}_{i-1,j}^{(j)}} = \frac{p_{ij}}{p_{i-1,j}}, & \text{if } \widetilde{a}_{i-1,j}^{(j)} \neq 0, \\ 0, & \text{if } \widetilde{a}_{i-1,j}^{(j)} = 0. \end{cases}$$

The multipliers satisfy that

$$m_{ij} = 0 \Rightarrow m_{hj} = 0 \quad \forall h > i.$$

Pivots and multipliers of the NE of A and A^T characterize nonsingular TP and STP matrices, as the following result shows. It follows from Theorems 4.1 and 4.2 of [44] and p. 116 of [44].

Theorem 5.1 A matrix A is nonsingular TP (STP, respectively) if and only if the NE of A and A^T can be performed without row exchanges, all the multipliers of the NE of A and A^T are nonnegative (positive, respectively) and all the diagonal pivots of the NE of A are positive. □

Nonsingular TP matrices can be expressed as a product of nonnegative bidiagonal matrices using again the pivots and multipliers of the NE of A and A^T. The following theorem (see Theorem 4.2 and p. 120 of [44]) presents this factorization, which is called the *bidiagonal decomposition*.

Theorem 5.2 (cf. Theorem 4.2 of [44]) Let $A = (a_{ij})_{1 \leq i,j \leq n}$ be a nonsingular TP matrix. Then A admits the following representation:

$$A = F_{n-1} F_{n-2} \cdots F_1 D G_1 \cdots G_{n-2} G_{n-1}, \tag{6}$$

where D is the diagonal matrix $diag(p_{11}, \ldots, p_{nn})$ with positive diagonal entries and F_i, G_i are the nonnegative bidiagonal matrices given by

$$F_i = \begin{pmatrix} 1 & & & & & & \\ 0 & 1 & & & & & \\ & \ddots & \ddots & & & & \\ & & 0 & 1 & & & \\ & & & m_{i+1,1} & 1 & & \\ & & & & \ddots & \ddots & \\ & & & & & m_{n,n-i} & 1 \end{pmatrix}, \tag{7}$$

$$G_i = \begin{pmatrix} 1 & 0 & & & & & \\ & 1 & \ddots & & & & \\ & & \ddots & 0 & & & \\ & & & 1 & \widetilde{m}_{i+1,1} & & \\ & & & & 1 & \ddots & \\ & & & & & \ddots & \widetilde{m}_{n,n-i} \\ & & & & & & 1 \end{pmatrix}, \tag{8}$$

for all $i \in \{1, \dots, n-1\}$. If, in addition, the entries m_{ij} and $\widetilde{m}_{ij}$ satisfy

$$\begin{aligned} m_{ij} = 0 &\Rightarrow m_{hj} = 0 \quad \forall h > i, \\ \widetilde{m}_{ij} = 0 &\Rightarrow \widetilde{m}_{hj} = 0 \quad \forall h > i, \end{aligned} \tag{9}$$

then the decomposition is unique. □

In the bidiagonal decomposition given by (6), (7) and (8), the entries m_{ij} and p_{ii} are the multipliers and diagonal pivots, respectively, corresponding to the NE of A (see Theorem 4.2 of [44] and the comment below it) and the entries $\widetilde{m}_{ij}$ are the multipliers of the NE of A^T (see p. 116 of [44]). In [49] the following matrix notation $\mathcal{BD}(A)$ was introduced to represent the bidiagonal decomposition of a nonsingular TP matrix

$$(\mathcal{BD}(A))_{ij} = \begin{cases} m_{ij}, & \text{if} \quad i > j, \\ \widetilde{m}_{ji}, & \text{if} \quad i < j, \\ p_{ii}, & \text{if} \quad i = j. \end{cases} \tag{10}$$

The bidiagonal factorization has been used to compute accurately with these matrices (see [49]). In fact, if we have the $\mathcal{BD}(A)$ of a nonsingular TP matrix with HRA, then we can perform many computations of A with HRA, such as computing its inverse or computing its eigenvalues or its singular values (cf. [49, 50]).

There are some subclasses of nonsingular TP matrices for which the bidiagonal factorization can be obtained to HRA, and so, the computations mentioned previously can be also performed with HRA. We now mention some of these subclasses of matrices. For instance, for Vandermonde positive matrices [36], for Cauchy-Vandermonde positive matrices [64], arising in rational interpolation and that include the famous example of ill-conditioned matrices given by Hilbert matrices.

A remarkable example is provided by Pascal matrices $P = (p_{ij})_{1\le i,j\le n}$ with $p_{ij} := \binom{i+j-2}{j-1}$, which have a long history and arise in important applications in filter design and image and signal processing, probability, combinatorics, numerical analysis and electrical engineering, among other fields. The $\mathcal{BD}(P)$ can be obviously computed with HRA because it is very simple (cf. [1], where the conditioning of these matrices was also analyzed):

$$\mathcal{BD}(P) = \begin{pmatrix} 1 & \cdots & 1 \\ \vdots & & \vdots \\ 1 & \cdots & 1 \end{pmatrix}.$$

Bidiagonal factorizations and HRA computations were also assured for other classes of matrices arising in Combinatorics, such as the class of Jacobi-Stirling matrices (see [26]) or for several generalizations of Pascal matrices (see [13]).

Let us now focus on HRA computations with *collocation* matrices. We have already mentioned Vandermonde matrices, which are the collocation matrices of the monomial basis of polynomials (a B-basis of the space of polynomials of degree at most n on the nonnegative real numbers). Our next example is provided by the Bernstein basis on $[0, 1]$ (see (1)) and, more generally, by the Bernstein basis on $[a, b]$. Their corresponding collocation matrices are called *Bernstein-Vandermonde matrices*. The bidiagonal factorization and computations with HRA of Bernstein-Vandermonde matrices were obtained in [62]. The Said-Ball basis is another interesting basis used in C.A.G.D. and their collocation matrices are called *Said-Ball-Vandermonde matrices*. Their bidiagonal decompositions and corresponding computations with HRA were obtained in [63].

Bidiagonal factorizations and HRA computations have also been obtained for *singular* collocation matrices of TP bases, as can be seen in [10].

Now we consider the corresponding rational bases and matrices, whose bidiagonal factorizations and computations with HRA were considered in [25]. Given a basis $u = (u_0^n, \dots, u_n^n)$ of nonnegative functions on $[a, b]$ and a sequence of strictly positive weights $(w_i)_{i=0}^n$, we can construct a rational basis $r = (r_0^n, \dots, r_n^n)$ defined by

$$r_i^n(t) = \frac{w_i\, u_i^n(t)}{W(t)}, \quad t \in [a, b], \quad i \in \{0, 1, \dots, n\}, \tag{11}$$

where $W(t) = \sum_{j=0}^n w_j\, u_j^n(t)$. If the initial basis u is the Bernstein basis, then the corresponding rational basis r is called the *rational Bernstein basis* and is the normalized B-basis of its corresponding space. In C.A.G.D., the usual representation of a polynomial curve is the so called Bernstein-Bézier form, that is, these curves are expressed in terms of the Bernstein basis (1). The Bernstein basis is a rational Bernstein basis with all weights equal to 1: $w_i = 1$ for all $i = 0, \dots, n$. The corresponding square collocation matrices of the rational Bernstein basis at a sequence of parameters $0 < t_0 < t_1 < \ldots < t_n < 1$, given by $(r_j^n(t_i))_{0 \le i,j \le n}$, where functions r_i^n are given by (11) with $u_i^n = b_i^n$ for $i = 0, 1, \ldots, n$, will be called *rational Bernstein-Vandermonde* (RBV) matrices. In [25], the HRA calculations with RBV matrices and rational bases derived from the Said-Ball basis through their bidiagonal decompositions in terms of the diagonal pivots and multipliers of their Neville elimination and the multipliers of the Neville elimination of their transposes were obtained. They were also obtained for more general classes of rational bases (see [54]).

Bidiagonal decompositions and computations with HRA have been also obtained for other bases used in C.A.G.D. and closely related with the Bernstein basis. For instance, for the q-Bernstein basis [28], for the Lupaş basis [30] or the Bernstein-like bases (which are the normalized B-bases) of the spaces mixing algebraic, trigonometric and hyperbolic polynomials [4–6]. Other general bases used in C.A.G.D. for whose collocation matrices HRA computations were assured through their bidiagonal decompositions can be seen in [53, 65] (in this last case for h-Bernstein-Vandermonde matrices). HRA computations for TP matrices of q-integers were obtained in [14].

Another source of TP collocation matrices with HRA computations is provided by the collocation matrices of orthogonal polynomials. A first example that can be mentioned is given by *Laguerre* polynomials, which form a classical family of orthogonal polynomials and present many applications. The larger family of generalized Laguerre polynomials presents important applications in quantum mechanics and they are orthogonal polynomials on $[0, \infty)$ with respect to the weight function $x^\alpha e^{-x}$. Theorem 2 of [11], assures that, given the parameters $(0 >) t_0 > t_1 > \ldots > t_n$, many algebraic computations with the collocation matrices of generalized Laguerre polynomials can be performed with HRA, as well as the strict total positivity and the bidiagonal factorization of these matrices.

Another example of TP collocation matrices with HRA computations is provided by the collocation matrices of Bessel polynomials arise in many fields such as partial differential equations, number theory, algebra and statistics. They form an orthogonal sequence of polynomials and are also closely related to the reverse Bessel polynomials, with many applications in Electrical Engineering, in network analysis of electrical circuits. Bessel polynomials also occur naturally in the theory of traveling spherical waves and are very important for some problems of static potentials, signal processing and electronics. The zeros of Bessel polynomials and generalized Bessel polynomials also play a crucial role in applications in Electrical Engineering and are related with the length of the parameter domain where cycloidal spaces admit shape preserving representations ([6]). Given a sequence of parameters $0 < t_0 < t_1 < \cdots < t_{n-1}$, we call the collocation matrix of the Bessel polynomials $(B_0, \ldots, B_{n-1})$ at this sequence of parameters,

$$M = M\begin{pmatrix} B_0, \ldots, B_{n-1} \\ t_0, \ldots, t_{n-1} \end{pmatrix} = (B_{j-1}(t_{i-1}))_{1 \le i,j \le n},$$

a *Bessel matrix*. Theorem 4 of [12] and shows that the Bessel matrices are STP and that some usual algebraic problems with these matrices can be solved to HRA.

Jacobi polynomials form another classical family of orthogonal polynomials. The bidiagonal decomposition and HRA computations for their collocation matrices were obtained in [55]. In this paper, the bidiagonal decomposition and HRA computations for their *wronskian matrices* were also constructed. For wronskian matrices of the monomial basis of the space of polynomials and of the basis of exponential polynomials, the bidiagonal decomposition and HRA computations were also obtained (see [56]). About HRA computations with wronskian matrices of Bessel and Laguerre polynomials and with wronskian matrices of Bernstein and related bases, see [57, 58]. *Gramian matrices* of bases, with their bidiagonal decomposition and associated HRA calculations have analyzed in [59–61].

Bidiagonal decompositions and HRA algebraic computations were also obtained for Schoenmakers-Coffey matrices [32], arising in Finance, and more generally for the important class of *Green* matrices [33]. Finally, let us come back to the class of Toeplitz matrices mentioned in Sect. 2 as an example of a class of structured matrices for which accurate linear algebra for the problem of calculating determinants or minors is impossible. However, if we restrict ourselves to its subclass of *tridiagonal Toeplitz* matrices, HRA computations can be assured, as shown in [16].

6 Optimal Properties of Collocation Matrices of B-Bases

In Sects. 3 and 4 we have already commented some optimal properties of normalized B-bases. In Sect. 5 we have seen that crucial algebraic operations can be sometimes performed with HRA when using TP collocation matrices. However, even in these cases, we can assure HRA for solving a TP linear system $Ax = b$ only when b has alternating consecutive signs. But there are also some optimal conditioning properties of the collocation matrices of normalized B-bases. Let us start by recalling the case of the Bernstein basis of polynomials studied in [21].

Given a nonsingular matrix $A = (a_{ij})_{1\le i,j\le n}$, the classical condition number is

$$\kappa_\infty(A) := \|A\|_\infty \, \|A^{-1}\|_\infty.$$

Moreover, denoting by $|A|$ the matrix whose (i, j)-entry is $|a_{ij}|$, the Skeel condition number is

$$\mathrm{Cond}(A) := \|\, |A^{-1}| \, |A| \,\|_\infty.$$

The following result corresponds to Theorem 2.1 of [21]. It shows that the collocation matrices of the Bernstein basis are the best conditioned among all the corresponding collocation matrices of NTP bases of the space of polynomials of degree at most n on $[0, 1]$, P_n, and a similar result using the Skeel condition number of the transposes of the collocation matrices.

Theorem 6.1 Let $(b_0^n, \ldots, b_n^n)$ be the Bernstein basis of P_n, let $(v_0, \ldots, v_n)$ be another NTP basis of P_n. For a sequence $0 \le t_0 < \cdots < t_n \le 1$, let $V := M\begin{pmatrix} v_0,\ldots,v_n \\ t_0,\ldots,t_n \end{pmatrix}$ and $B := M\begin{pmatrix} b_0^n,\ldots,b_n^n \\ t_0,\ldots,t_n \end{pmatrix}$. Then:

$$\kappa_\infty(B) \le \kappa_\infty(V), \quad \mathrm{Cond}(B^T) \le \mathrm{Cond}(V^T).$$

In [31], the previous result was extended to the collocation matrices of any normalized B-basis and, in [15], to any collocation matrix of the tensor product of normalized B-bases. In [31] it was also proved that the minimal eigenvalue and the minimal singular value of a collocation matrix of a normalized B-basis are always greater than or equal to the minimal eigenvalue and the minimal singular value of the corresponding collocation matrix of any NTP basis of its space. In [31] it was shown a similar result for the tensor product of the collocation matrices of normalized B-bases with respect to the tensor product of the corresponding collocation matrices of other NTP bases. In a future work, we shall explore optimal properties of the Wronskian matrices of TP bases.

Acknowledgements This work has been partially supported by MCIU/AEI through the Spanish research grants PID2022-138569NB-I00 (MCI/AEI) and RED2022-134176-T (MCI/AEI), and by Gobierno de Aragón (E41_23R).

References

1. Alonso, P., Delgado, J., Gallego, R., Peña, J.M.: Conditioning and accurate computations with Pascal matrices. J. Comput. Appl. Math. **252**, 21–26 (2013)
2. Ando, T.: Totally positive matrices. Linear Algebra Appl. **90**, 165–219 (1987)
3. de Boor, C., Pinkus, A.: Backward error analysis for totally positive linear systems. Numer. Math. **27**, 485–490 (1977)
4. Carnicer, J.M., Mainar, E., Peña, J.M.: Critical length for design purposes and Extended Chebyshev spaces. Constr. Approx. **20**, 51–71 (2004)
5. Carnicer, J.M., Mainar, E., Peña, J.M.: On the critical lengths of cycloidal spaces. Constr. Approx. **39**, 573–583 (2014)
6. Carnicer, J.M., Mainar, E., Peña, J.M.: Critical lengths of cycloidal spaces are zeros of Bessel functions. Calcolo **54**, 1521–1531 (2017)
7. Carnicer, J.M., Peña, J.M.: Shape preserving representations and optimality of the Bernstein basis. Adv. Comput. Math. **1**, 173–196 (1993)
8. Carnicer, J.M., Peña, J.M.: Totally positive bases for shape preserving curve design and optimality of B-splines. Comput. Aided Geom. Des. **11**, 633–654 (1994)
9. Carnicer, J.M., Peña, J.M.: Total positivity and optimal bases. In: Gasca, M., Micchelli, C.A. (eds.) Total Positivity and Its Applications, vol. 359, pp. 133–155. Math. Appl. Kluwer Acad. Publ. (1996)
10. Delgado, J., Koev, P., Marco, A., Martínez, J.J., Peña, J.M., Persson, P.-O., Spasov, S.: Bidiagonal decompositions of vandermonde-type matrices of arbitrary rank. J. Comput. Appl. Math. **426**, Paper 115064 (2023)
11. Delgado, J., Orera, H., Peña, J.M.: Accurate computations with Laguerre matrices. Numer. Linear Algebra Appl. **26**, e2217 (2019)

12. Delgado, J., Orera, H., Peña, J.M.: Accurate algorithms for Bessel matrices. J. Sci. Comput. **80**, 1264–1278 (2019)
13. Delgado, J., Orera H., Peña, J.M.: Accurate bidiagonal decomposition and computations with generalized pascal matrices. J. Comput. Appl. Math. **391**, Paper 113443 (2021)
14. Delgado, J., Orera, H., Peña, J.M.: High relative accuracy with matrices of q-integers. Numer. Linear Algebra Appl. **28**, e2383 (2021)
15. Delgado, J., Orera H., Peña, J.M.: Optimal properties of tensor product of B-bases. Appl. Math. Lett. **121**, Paper 107473 (2021)
16. Delgado, J., Orera, H., Peña, J.M.: Characterizations and accurate computations for tridiagonal Toeplitz matrices. Linear Multilinear Algebra **70**, 4508–4527 (2022)
17. Delgado, J., Peña, J.M.: A corner cutting algorithm for evaluating rational Bézier surfaces and the optimal stability of the basis. SIAM J. Sci. Comput. **29**, 1668–1682 (2007)
18. Delgado, J., Peña, J.M.: Progressive iterative approximation and bases with the fastest convergence rates. Comput. Aided Geom. Des. **24**, 10–18 (2007)
19. Delgado, J., Peña, J.M.: Error analysis of efficient evaluation algorithms for tensor product surfaces. J. Comput. Appl. Math. **219**, 156–169 (2008)
20. Delgado, J., Peña, J.M.: Running Relative Error for the Evaluation of Polynomials. SIAM J. Sci. Comput. **31**, 3905–3921 (2009)
21. Delgado, J., Peña, J.M.: Optimal conditioning of Bernstein collocation matrices. SIAM J. Matrix Anal. App. **31**, 990–996 (2009)
22. Delgado, J., Peña, J.M.: Running error for the evaluation of rational Bézier surfaces. J. Comput. Appl. Math. **233**, 1685–1696 (2010)
23. Delgado, J., Peña, J.M.: Running error for the evaluation of rational Bézier surfaces through a robust algorithm. J. Comput. Appl. Math. **235**, 1781–1789 (2011)
24. Delgado, J., Peña, J.M.: On the evaluation of rational triangular Bézier surfaces and the optimal stability of the basis. Adv. Comput. Math. **38**, 701–721 (2013)
25. Delgado, J., Peña, J.M.: Accurate computations with collocation matrices of rational bases. Appl. Math. Comput. **219**, 4354–4364 (2013)
26. Delgado, J., Peña, J.M.: Fast and accurate algorithms for Jacobi-Stirling matrices. Appl. Math. Comput. **236**, 253–259 (2014)
27. Delgado, J., Peña, J.M.: Accurate evaluation of Bézier curves and surfaces and the Bernstein-Fourier algorithm. Appl. Math. Comput. **271**, 113–122 (2015)
28. Delgado, J., Peña, J.M.: Accurate computations with collocation matrices of q-Bernstein polynomials. SIAM J. Matrix Anal. Appl. **36**, 880–893 (2015)
29. Delgado, J., Peña, J.M.: Algorithm 960: POLYNOMIAL: An object-oriented Matlab library of fast and efficient algorithms for polynomials. ACM Trans Math Softw. **42**, Art. 23 (2016)
30. Delgado, J., Peña, J.M.: Accurate computations with Lupaş matrices. Appl. Math. Comput. **303**, 171–177 (2017)
31. Delgado, J., Peña, J.M.: Extremal and optimal properties of B-bases collocation matrices. Numer. Math. **146**, 105–118 (2020)
32. Delgado, J., Peña, G., Peña, J.M.: Accurate and fast computations with positive extended Schoenmakers-Coffey matrices. Numer. Linear Alg. Appl. **23**, 1023–1031 (2016)
33. Delgado, J., Peña, G., Peña, J.M.: Accurate and fast computations with Green matrices. Appl. Math. Lett. **145**, Paper 108778 (2023)
34. Demmel, J., Dumitriu, I., Holtz, O., Koev, P.: Accurate and efficient expression evaluation and linear algebra. Acta Numer. **17**, 87–145 (2008)
35. Demmel, J., Gu, M., Eisenstat, S., Slapnicar, I., Veselic, K., Drmac, Z.: Computing the singular value decomposition with high relative accuracy. Linear Algebra Appl. **299**, 21–80 (1999)
36. Demmel, J., Koev, P.: The accurate and efficient solution of a totally positive generalized Vandermonde linear system. SIAM J. Matrix Anal. Appl. **27**, 142–152 (2005)
37. Fallat, S.M., Johnson, C.R.: Totally Nonnegative Matrices. Princeton University Press, Princeton and Oxford (2011)
38. Farin, G.: Curves and Surfaces for Computer Aided Geometric Design. Academic, Boston (1988)

39. Farouki, R.T., Goodman, T.N.T.: On the optimal stability of Bernstein basis. Math. Comp. **65**, 1553–1566 (1996)
40. Gantmacher, F.P., Krein, M.G.: Oscillation Matrices and Kernels and Small Vibrations of Mechanical Systems, revised AMS Chelsea, Providence, RI (2002)
41. Gasca, M., Micchelli, C.A. (eds.): Total positivity and its applications. Mathematics and its Applications, vol. 359. Kluwer Acad. Publ. Dordrecht (1996)
42. Gasca, M., Peña, J.M.: Total positivity and Neville Elimination. Linear Algebra Appl. **165**, 25–44 (1992)
43. Gasca, M., Peña, J.M.: A matricial description of Neville elimination with applications to total positivity. Linear Algebra Appl. **202**, 33–45 (1994)
44. Gasca, M., Peña, J.M.: On factorizations of totally positive matrices. In: Gasca, M., Micchelli, C.A. (eds.) Total Positivity and Its Applications, vol. 359, pp. 109–130. Math. Appl. Kluwer Acad. Publ. (1996)
45. Golub, G.H., Van Loan, C.F.: Matrix Computations, 3rd edn. The John Hopkins University Press (1996)
46. Goodman, T.N.T., Said, H.B.: Shape preserving properties of the generalized Ball basis. Comput. Aided Geometric Des. **8**, 115–121 (1991)
47. Higham, N.J.: Accuracy and Stability of Numerical Algorithms, 2nd edn. SIAM, Philadelphia, PA (2002)
48. Karlin, S.: Total Positivity, vol. I. Stanford University Press, Stanford (1968)
49. Koev, P.: Accurate computations with totally nonnegative matrices. SIAM J. Matrix Anal. Appl. **29**, 731–751 (2007)
50. Koev, P.: http://math.mit.edu/~plamen/software/TNTool.html
51. Lyche, T., Peña, J.M.: Optimally stable multivariate bases. Adv. Comput. Math. **20**, 149–159 (2004)
52. Mainar, E., Peña, J.M.: Error analysis of corner cutting algorithms. Numer. Algor. **22**, 41–52 (1999)
53. Mainar, E., Peña, J.M.: Accurate computations with collocation matrices of a general class of bases. Numer. Linear Algebra Appl. **25**, e2184 (2018)
54. Mainar, E., Peña, J.M., Rubio, B.: Accurate bidiagonal decomposition of collocation matrices of weighted φ-transformed systems. Numer. Linear Algebra Appl. **27**, e2295 (2020)
55. Mainar, E., Peña, J.M., Rubio, B.: Accurate computations with collocation and Wronskian matrices of Jacobi polynomials. J. Sci. Comput. **87**, Paper 77 (2021)
56. Mainar, E., Peña, J.M., Rubio, B.: Accurate computations with Wronskian matrices. Calcolo **58**, Paper 1 (2021)
57. Mainar, E., Peña, J.M., Rubio, B.: Accurate and efficient computations with Wronskian matrices of Bernstein and related bases. Numer. Linear Algebra Appl. **29**, e2423 (2021)
58. Mainar, E., Peña, J.M., Rubio, B.: Accurate computations with Wronskian matrices of Bessel and Laguerre polynomials. Linear Algebra Appl. **647**, 31–46 (2022)
59. Mainar, E., Peña, J.M., Rubio, B.: Total positivity and accurate computations with Gram matrices of Bernstein bases. Numer. Algor. **91**, 841–859 (2022)
60. Mainar, E., Peña, J.M., Rubio, B.: Accurate computations with matrices related to bases $\{t^i e^{\lambda t}\}$. Adv. Comput. Math. **48**, 38 (2022)
61. Mainar, E., Peña, J. M., Rubio, B.: Accurate computations with Gram and Wronskian matrices of geometric and Poisson bases. Rev. Real Acad. Cienc. Exactas Fis. Nat. Ser. A-Mat. **116**, 126 (2022)
62. Marco, A., Martínez, J.J.: A fast and accurate algorithm for solving Bernstein-Vandermonde linear systems. Linear Algebra Appl. **422**, 616–628 (2007)
63. Marco, A., Martínez, J.J.: Accurate computations with Said-Ball-Vandermonde matrices. Linear Algebra Appl. **432**, 2894–2908 (2010)
64. Marco, A., Martínez, J.J., Peña, J.M.: Accurate bidiagonal decomposition of totally positive Cauchy-Vandermonde matrices and applications. Linear Algebra Appl. **517**, 63–84 (2017)
65. Marco, A., Martínez, J.J., Viaña, R.: Accurate bidiagonal decomposition of totally positive h-Bernstein-Vandermonde matrices and applications. Linear Algebra Appl. **579**, 320–335 (2019)

66. Peña, J.M.: B-splines and optimal stability. Math. Comp. **66**, 1555–1560 (1997)
67. Peña, J.M. (ed.): Shape Preserving Representations in Computer Aided Geometric Design. Nova Science Publishers, Commack (NY) (1999)
68. Peña, J.M.: On the optimal stability of bases of univariate functions. Numer. Math. **91**, 305–318 (2002)
69. Peña, J.M.: A note on the optimal stability of bases of univariate functions. Numer. Math. **103**, 151–154 (2006)
70. Peña, J.M., Sauer, T.: On the multivariate Horner scheme. SIAM J. Numer. Anal. **37**, 1186–1197 (2000)
71. Peña, J.M., Sauer, T.: On the multivariate Horner scheme II: Running error analysis. Computing **65**, 313–322 (2000)
72. Pinkus, A.: Totally Positive Matrices. Cambridge Tracts in Mathematics, Num. 181. Cambridge University Press. Cambridge (2010)
73. Schmeltz, G.: Variationsreduzierende Kurvendarstellungen und Krümmungskriterien für Bézierflächen. Thesis. Fachbereich Mathematik, Technische Hochschule Darmstadt (1992)

Intrinsic Supersmoothness of Piecewise Functions

Tatyana Sorokina

Abstract Many spaces of multivariate piecewise functions (or splines) with prescribed global smoothness exhibit additional, unprescribed, or intrinsic supersmoothness at certain faces where the pieces join together. Intrinsic supersmoothness is a consequence of the geometry of the underlying partition. In this survey, we provide insights into this phenomenon, review and generalize known results, and demonstrate how intrinsic supersmoothness affects other properties of spline spaces.

1 Introduction and Preliminaries

Piecewise functions of prescribed global smoothness $r \geq 0$ over partitions Δ of domains Ω in $\mathbb{R}^n$ are used ubiquitously in applied mathematics: computer-aided geometric design and manufacturing, numerical solutions of partial differential equations (PDEs), and machine learning among many others.

Most often, such functions are piecewise polynomials, or *splines*. For a fixed partition Δ, and a fixed integer $d \geq 0$, these functions form vector spaces $S_d^r(\Delta)$ of splines of degree d and smoothness r:

$$S_d^r(\Delta) = \{s \in C^r(\Omega) : s\big|_T \in P_d^n, \quad \text{for all} \quad T \in \Delta\}, \tag{1}$$

where P_d^n is the $\binom{d+n}{n}$-dimensional space of polynomials of degree $\leq d$ in n variables. In many applications, partitions Δ are *simplicial*.

A simplicial partition in $\mathbb{R}^n$ is a collection of n-simplices with the following property: for any two n-simplices T_1 and T_2, we require

$$T_1 \cap T_2 \neq \emptyset \quad \Rightarrow \quad T_1 \cap T_2 = F, \quad \text{where } F \text{ is an } m\text{-dimensional face of both } T_1 \text{ and } T_2 \text{ for some } 0 \leq m \leq n.$$

T. Sorokina (✉)
Towson University, Towson, MD, USA
e-mail: tsorokina@towson.edu

© The Author(s), under exclusive license to Springer Nature Switzerland AG 2025
M. J. Ibáñez-Pérez et al. (eds.), *Mathematical And Computational Modelling, Approximation and Simulation*, SEMA SIMAI Springer Series 41,
https://doi.org/10.1007/978-3-032-01114-5_4

The most basic building block of a simplicial partition is a cell of a vertex: if v is a vertex in a simplicial partition, then we call the collection of all simplices sharing v a *cell* around v. If v is an interior point of this collection, then we say that this cell around v is an *interior cell.*

The term "intrinsic supesmoothness" was coined in [13], where it was proved that many spaces of multivariate polynomial splines possess additional, unprescribed, or intrinsic supersmoothness $\rho > r$ at certain faces of the underlying partition where polynomial pieces join together. The only partitions considered in [13] are simplicial partitions. Since [13], nonpolynomial piecewise functions, and non-simplicial partitions have been studied in the context of intrinsic supersmoothness, and we review many of these results in this survey.

It is very important to note that many spline spaces used in practical applications have additional (super) smoothness at certain faces of the underlying partition embedded in their definitions, see [9] for numerous examples. This is typically done to reduce the number of parameters in the model. However, there is a fundamental difference between additional smoothness explicitly defined and known in advance versus unprescribed intrinsic supersmoothness, typically unknown and requiring discovery.

Why should we study intrinsic supersmoothness? Besides its intrinsic intriguing nature, it has significant implications on many properties of splines. The most fundamental question about vector spaces $S_d^r(\Delta)$ defined in (1) is their dimensions. Despite decades of extensive research, this question for general partitions is fully answered only for univariate splines. There are relatively good lower and upper bounds for bivariate splines on triangulations, and many exact dimension formulae for specific partitions and specific spline spaces, see [9] and the references therein. As the computation of these dimensions falls outside the purview of this survey, we do not present the exact formulae here. It is worth noting, however, that the primary components in these formulae include the polynomial degree d, specific characteristics of the underlying partition Δ, and varying smoothness values across faces of Δ. Therefore, discovering intrinsic supersmoothness usually improves both lower and upper bounds on the dimension of $S_d^r(\Delta)$, or simplifies the partition Δ and allows determination of the exact dimension. Intrinsic supersmoothness not only plays an important role to derive conclusions about dimension of spline spaces, but also helps in the construction of normalized B-spline bases, see [17].

Additionally, intrinsic supersmoothness directly affects interpolating properties: it usually restricts the choices of interpolating sets. For example, if we know that intrinsic supersmoothness of three pieces of a C^1-continuous surface meeting at a single point v is C^2-continuous at this point, then we cannot interpolate all second order derivatives for each piece separately. It has to be the same set of values for all three pieces.

Intrinsic supersmoothness plays a role in the construction of finite elements in numerical PDEs. It can predict extra (and usually undesirable) smoothness of certain elements. Finally, knowing when intrinsic supersmoothness occurs, helps choosing suitable partitions to avoid it. This is equally important in computer-aided geometric design and finite element constructions.

The paper is organized as follows. In Sect. 2, we use simple examples to demonstrate how intrinsic supersmoothness arises as a key difference between univariate and multivariate polynomial splines. Section 3 contains two results that shed some light on the analytical fundamentals of supersmoothness of piecewise functions in more than one variable. In Sect. 4 we examine most known types of intrinsic supersmoothness, and improve some of the previously established results. In Sect. 5 we demonstrate how intrinsic supersmoothness can be used to determine dimensions of polynomial spline spaces. Throughout this survey, we aim to avoid duplicating proofs from the original papers unless the result can be improved; however, we do comment on the ideas and key components of the proofs.

An essential tool that can be used to check and conjecture intrinsic supersmoothness of bivariate and trivariate polynomial splines is the software designed by P. Alfeld for computing dimensions and finding minimal determining sets. It is available on http://www.math.utah.edu/~pa.

2 Motivational Examples

In this section, we study in detail several spaces of univariate and bivariate polynomial splines on simple simplicial partitions Δ_n^k, where k stands for the spatial dimension and n stands for the number of pieces in the partition. These examples illustrate the impact of the geometry of underlying partitions on the properties of piecewise functions defined over them. Polynomial splines serve as an instructive class of piecewise functions for such investigations, since they are linear spaces, and, therefore, by examining their dimensions, we can infer deeper properties of non-polynomial piecewise functions.

2.1 A Case of Univariate Smooth Quadratic Splines

We begin with three univariate spline spaces $S_2^1(\Delta_n^1([a,b]))$, $n \in \{1,2,3\}$, defined on three partitions of an interval $[a,b] \subset \mathbb{R}$, depicted in Fig. 1, where the super-index one in $\Delta_n^1([a,b])$ shows the spatial dimension, while the sub-index n stands for a number of sub-intervals in the partition of $[a,b]$:

$$\Delta_1^1([a,b]) := [a,b], \quad \Delta_2^1([a,b]) := [a,v_1] \cup [v_1,b],$$
$$\Delta_3^1([a,b]) := [a,v_1] \cup [v_1,v_2] \cup [v_2,b].$$

For $\Delta_2^1([a,b])$, we assume that v_1 is any interior point in (a,b), while for $\Delta_3^1([a,b])$, we assume that $v_1 < v_2$ are any two interior points in (a,b).

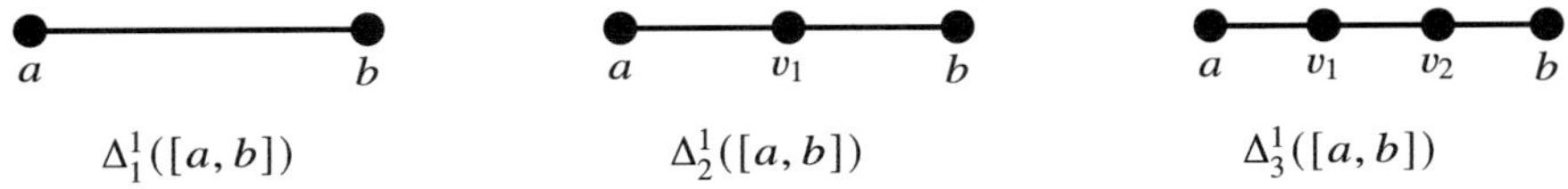

Fig. 1 Partitions $\Delta^1_n([a,b])$ of the interval $[a,b]$ for $n = 1, 2, 3$

The dimension of any space of quadratic splines over $\Delta^1_1([a,b])$ is just the dimension of the univariate quadratic polynomials. Thus, dim $S^1_2(\Delta^1_1([a,b])) = \dim \mathbb{P}^1_2 = 3$. To count the dimension of $S^1_2(\Delta^1_2([a,b]))$ we note that two univariate polynomials, p_1 on $[v_0, v_1]$ and p_2 on $[v_1, v_2]$, have six degrees of freedom (or coefficients). To build a C^1-smooth spline, we need to *glue* p_1 and p_2 at the break point v_1: that is, we need to match the values and the first derivatives of p_1 and p_2 at v_1. Thus, we subtract two degrees of freedom from six, and dim $S^1_2(\Delta^1_2([a,b])) = 4$. Using a similar count, we can verify that dim $S^1_2(\Delta^1_3([a,b])) = 5$. In fact, if we are to split the interval even more, then the following can be proved

$$\dim S^1_2(\Delta^1_n([a,b])) = 3n - 2(n-1) = n + 2. \tag{2}$$

Formula (2) essentially shows that univariate spline approximation is good news: by adding break points we enlarge the space $S^1_2(\Delta^1_n([a,b]))$ and, thus, create a necessary condition for building a sequence of smooth quadratic splines approximating a smooth function defined on the interval. This is the crucial benefit of spline approximation as opposed to the polynomial one: the polynomial degree remains fixed and low (quadratic in this case) as the dimension of the approximating space grows. Note that dimension formula (2) can be easily generalized to $S^r_d(\Delta^1_n([a,b]))$, the space of univariate splines of degree $\leq d$ and smoothness r, (a nice exercise to try!).

2.2 A Case of Bivariate Smooth Quadratic Splines

We now try to recreate the case of the previous subsection for a 2D domain and splines in two variables. We define three bivariate spline spaces on three partitions $\Delta^2_n(T)$, $n \in \{1, 2, 3\}$, of a triangle $T := [v_0, v_1, v_2] \subset \mathbb{R}^2$, see Fig. 2, where the super-index two in $\Delta^2_n(T)$ shows the spatial dimension, while the sub-index n stands for a number of subtriangles in the partition of T:

$$\Delta^2_1(T) := [v_0, v_1, v_2], \quad \Delta^2_2(T) := [v_0, v, v_2] \cup [v_2, v, v_1],$$
$$\Delta^2_3(T) := [v_0, v, v_1] \cup [v_1, v, v_2] \cup [v_2, v, v_0].$$

For $\Delta^2_2(T)$, we assume that v is any interior point in (v_0, v_1), while for $\Delta^2_3(T)$, we assume that v is any point strictly interior to T. The dimension of any space of quadratic splines over $\Delta^2_1(T)$ is just the dimension of quadratic polynomials

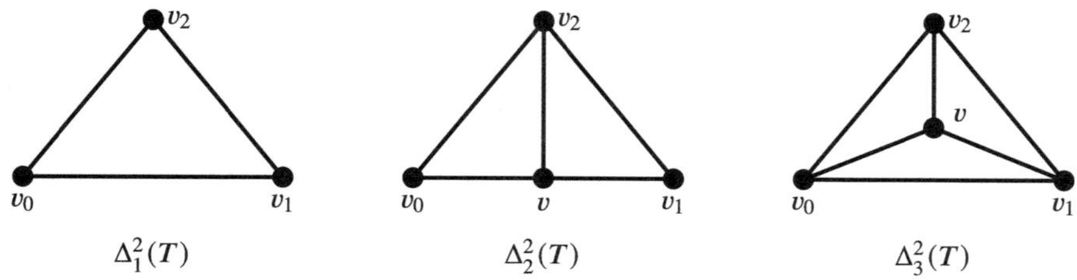

Fig. 2 Partitions $\Delta_n^2(T)$ of the triangle $T = [v_0, v_1, v_2]$, for $n = 1, 2, 3$

in two variables. Thus, $\dim S_2^1(\Delta_1^2(T)) = \dim \mathbb{P}_2^2 = 6$. To count the dimension of $S_2^1(\Delta_2^2(T))$ we note that two bivariate polynomials, p_1 on $[v_0, v, v_2]$ and p_2 on $[v_2, v, v_1]$, have twelve degrees of freedom (or coefficients). To build a C^1-smooth spline we need to match the values and the first derivatives of p_1 and p_2 on the edge $e := [v, v_2]$. We observe that $p_1|_e$ and $p_2|_e$ are quadratic univariate polynomials. Thus, $p_1|_e = p_2|_e$ requires imposing three conditions (linear equations) on the twelve coefficients. Once p_1 and p_2 are glued continuously across e, the first order directional derivatives of p_1 and p_2 along e also coincide, since $p_1|_e = p_2|_e$ is the same quadratic univariate polynomial. Thus, we only need to make sure that directional derivatives of p_1 and p_2 along another direction not collinear to e match on e. This requires two more conditions (linear equations) since first order derivatives of p_1 and p_2 restricted to e are linear polynomials. Thus, $\dim S_2^1(\Delta_2^2(T)) = 12 - 3 - 2 = 7$, which seems like a step in the right direction: the dimension goes up from six to seven as the number of subtriangles increases from one to two.

We now add one more triangle as in Fig. 2(right), and obtain an interior cell around v. The counting argument becomes less obvious since not all conditions seem linearly independent. The expectation that the dimension of $S_2^1(\Delta_3^2(T))$ is greater than seven is unfortunately incorrect. In fact, as we learn in the next subsection, the situation is even worse than that: $\dim S_2^1(\Delta_3^2(T)) = 6 < 7$. This fact implies that

$$S_2^1(\Delta_3^2(T)) = S_2^1(\Delta_1^2(T)) = \mathbb{P}_2^2, \tag{3}$$

or, equivalently, every spline in $S_2^1(\Delta_3^2(T))$ is just a polynomial, see Lemma 2.1.

2.3 *First Example of Intrinsic Supersmoothness*

An alternative way of interpreting (3) is to state that every spline $s \in S_2^1(\Delta_3^2(T))$ has supersmoothness two at the interior vertex v. This is clearly the case of intrinsic, not prescribed, and even undesirable supersmoothness. Thus, we should expect that this unprescribed supersmoothness should be also present for splines higher polynomial degrees on the same partition: for any $s \in S_d^1(\Delta_3^2(T))$, for any $d \geq 2$. Indeed, let

$\{p_i\}_{i=1}^3$ be three polynomial pieces of $s \in S_d^1(\Delta_3^2(T))$ defined on the three subtriangles of $\Delta_3^2(T)$, see Fig. 2(right). Let $\{t_i\}_{i=1}^3$ be the three quadratic Taylor polynomials of $\{p_i\}_{i=1}^3$ centered at v. Then the quadratic spline formed by $\{t_i\}_{i=1}^3$ is C^1 smooth, and, thus, by (3) must be a single polynomial. Therefore, s is C^2 smooth at v. Using the language of formulae, we could say that $S_d^1(\Delta_3^2(T))$ coincides with one of its subspaces:

$$S_d^1(\Delta_3^2(T)) = S_d^1(\Delta_3^2(T)) \cap C^2(v).$$

At this point, we are ready to observe that all arguments above had nothing to do with the boundary of T. In fact, we can remove this boundary, and consider $\Delta_3^2(T)$ as three noncollinear rays emanating from v. To signify this, we change the notation from $\Delta_3^2(T)$ to $\Delta_3^2(v)$. What if we have more than three rays emanating from v? Then, indeed we have the following result, see [13].

Lemma 2.1 *Let $\Delta_n^2(v)$ be a partition of $\mathbb{R}^2$ formed by n noncollinear edges emanating from a single vertex v. Then*

$$S_d^r(\Delta_n^2(v)) = \mathbb{P}_d^2, \quad \textit{where } d \le r + \left\lfloor \frac{r+1}{n-1} \right\rfloor.$$

The original proof of Lemma 2.1 in [13] is not very illuminating. We review its outline here for completeness. Clearly, $\mathbb{P}_d^2 \subseteq S_d^r(\Delta_n^2(v))$ for any $r \le d$. In Chap. 9 of [9] we find a formula for the dimension of splines on interior cells—essentially partitions of $\mathbb{R}^2$ formed by rays emanating from one point—and compute the dimension of $S_\rho^r(\Delta_n^2(v))$ for $\rho = r + \left\lfloor \frac{r+1}{n-1} \right\rfloor$. This dimension happens to be equal to the dimension of bivariate polynomials of degree $\le \rho$, which completes the proof.

Historically, the first observation of intrinsic supersmoothness was made in [5] for the space $S_d^r(\Delta_3^2(v))$, that was noted to be C^{r+1} at the interior split point v. The proof of this fact in [5] relies on direct matching of the derivatives using the special geometry of the partition. This proof does not immediately generalize to splines of on $\Delta_n^2(v)$.

Note that Lemma 2.1 does not mention the word supersmoothness (yet). Let us look at a special case of $n = r = 3$. Then we have $S_5^3(\Delta_3^2(v)) = \mathbb{P}_5^2$, and it is tempting to say that splines, say in $S_9^3(\Delta_3^2(v))$, have supersmoothness five at the interior split point v, see Fig. 2(right). The reader should feel a slight unease with this statement since a C^3 smooth function of two variables can be C^5 at a certain point only if it is at least C^4 smooth in a neighborhood of that point. Therefore, a spline in $S_9^3(\Delta_3^2(v))$ cannot be truly C^5 smooth at v, however, it can certainly have all derivatives of order up to five of the three pieces of polynomials p_1, p_2 and p_3 have matching values at v. We are now ready to state the first theorem that brought intrinsic supersmoothness to light in [13].

Theorem 2.1 *Let $\Delta_n^2(v)$ be partition of $\mathbb{R}^2$ formed by n noncollinear rays emanating from a single vertex v. Then*

$$S_d^r(\Delta_n^2(v)) = S_d^r(\Delta_n^2(v)) \cap C^\rho(v), \quad \textit{where} \ \rho = r + \left\lfloor \frac{r+1}{n-1} \right\rfloor,$$

where $C^\rho(v)$ is understood as matching of the derivatives up to order ρ at the point v if $r+2 \le \rho < d$. We then say that every $s \in S_d^r(\Delta_n^2(v))$ has intrinsic supersmoothness ρ at v. □

We note that both Lemma 2.1 and Theorem 2.1 can be stated in terms of interior cells with noncollinear edges, see [13]. We chose not to do it here to emphasize that the boundary of the cell is irrelevant to this type of supersmoothness.

3 Basic Intrinsic Supersmoothness

In this section we consider a simple set up of an open disk $\Omega \subset \mathbb{R}^2$ that is split into two sets Ω_1 and Ω_2 by either a line segment ℓ or a curve γ, see Fig. 3. Additionally, two (at least continuous) on Ω, not necessarily polynomial functions, f_1 and f_2, are "glued continuously" to form a piecewise function $F(x, y) \in C^0(\Omega)$ as follows:

$$F(x, y) := \begin{cases} f_1(x, y) \text{ if } (x, y) \in \Omega_1, \\ f_2(x, y) \text{ if } (x, y) \in \Omega_2. \end{cases} \tag{4}$$

The idea is that if intrinsic supersmoothness is an analytical property, then it has to show in a form of jump from continuity of $F(x, y)$ to differentiability of $F(x, y)$ at certain points of ℓ and γ. Both results below are somewhat of a mathematical folklore that seems to be known to some researchers, but does not belong to any defined subfield of mathematics. Theorem 3.1 is obvious in the context of computer-aided geometric design, while consequences of Theorem 3.1 seem to be well understood in the finite element (numerical PDEs) settings.

Fig. 3 Partitions of an open disk Ω into two regions Ω_1 and Ω_2 used in Lemma 3.1 and Theorem 3.1

3.1 Intrinsic Supersmoothness Along a Line

Lemma 3.1 *Let ℓ be a segment of a straight line that divides the open disk Ω into two subsets Ω_1 and Ω_2 as in Fig. 3(left). For $k \geq 0$, let $f_1 \in C^k(\Omega)$ and $f_2 \in C^k(\Omega)$ be continuously glued along ℓ, that is, let $F \in C^0(\Omega)$, where F is a piecewise function defined in (4). Then the directional derivatives $D_\ell^m F$, in the direction ℓ, for $m \leq k$, exist on ℓ, or, equivalently:*

$$D_\ell^m f_1(v) = D_\ell^m f_2(v), \quad \forall v \in \ell, \quad \forall m \leq k,$$

where D_ℓ is a directional derivative in the direction of ℓ. □

The proof immediately follows from the observation that restrictions of f_1 and f_2 on ℓ are identical C^k-continuous functions in one variable. We also encourage the reader to compare Lemma 3.1 with Theorem 4.5 in Sect. 4.2 of this survey.

3.2 Intrinsic Supersmoothness at a Singular Point

We wave some differential geometry describing the curve γ in this subsection, and refer the reader to [11] for full details. The main idea is that the curve γ has no self intersections and it splits the disk Ω into two regions as in Fig. 3(right).

Theorem 3.1 *Let $\gamma \subset \mathbb{R}^2$ be the trace of a Jordan arc that divides the open disk Ω into two subsets Ω_1 and Ω_2 as in Fig. 3(right). Let γ be not smooth at $P \in \gamma$. Let f_1, f_2 be C^1 functions on Ω continuously glued along γ, that is, let $F \in C^0(\Omega)$, where F is a piecewise function defined in (4). Then the piecewise function F is differentiable at P, that is, $\nabla f_1(P) = \nabla f_2(P)$.* □

To gain an insight into Theorem 3.1, we consider a very special case: $\Omega := \mathbb{R}^2$, the point P is the origin, the curve γ consists of the positive rays of x and y axes, Ω_1 is the first quadrant, Ω_2 is the union of the second, third and fourth quadrants, f_1 and f_2 are linear polynomials. The reader should immediately deduce that $f_1 \equiv f_2$. Then Theorem 3.1 is true for polynomials of any degree by passing to linear Taylor polynomials. The full proof of Theorem 3.1 is not obvious, and uses one of the most delicate results of analysis: the implicit function theorem. We refer the reader to [11] for the detailed proof.

4 Known Types of Intrinsic Supersmoothness

4.1 Interior Vertex of a Cell

This type of supersmoothness has been extensively studied by several authors, see [6, 11–14]. We begin with a stronger version of Theorem 2.1 that can be found in [6]:

Theorem 4.1 *Suppose the n edges of the interior cell $\Delta_n^2(v)$ around a vertex v have $m \le n$ different slopes. Then the following intrinsic supersmoothness is observed:*

$$S_d^r(\Delta_n^2(v)) = S_d^r(\Delta_n^2(v)) \cap C^\rho(v), \quad where \quad \rho = \begin{cases} r + \left\lfloor \frac{r+1}{n-1} \right\rfloor, & m = n, \\ r, & m < n, \end{cases}$$

If $\rho \ge r + 2$, then $C^\rho(v)$ is understood as matching of the derivatives. □

Next we shift our exploration into the so-called *semi-algebraic* splines: the edges of $\Delta_n^2(v)$ are no longer segments of straight lines; they are irreducible algebraic arcs, see Fig. 4. Such splines were first introduced in [3], and further studied in [4, 12]. Note that the same algebraic curve can define two edges. This is the direct analog of collinear straight edges in Theorem 4.1, and it explains the double indexing on the edges in Theorem 4.2 below.

Theorem 4.2 *Let $\{\tau_{i_j}\}_{j=0}^r$, $1 \le r \le n$, be the edges of $\Delta_n^2(v)$ defined by the polynomials $\{g_{i_j}\}_{j=0}^r$ with non-zero gradients at v. If $\{\nabla g_{i_j}(0,0)\}_{j=0}^r$ are pairwise linearly independent, then every $s \in S_d^{r-1}(\Delta_n^2(v))$ has supersmoothness r at v. If $r = 0$, or if all edges in $\Delta_n^2(v)$ are defined by polynomials with zero gradients, then every $s \in S_d^0(\Delta_n^2(v))$ has supersmoothness one at v.* □

The proof of Theorem 4.2 in [12] relies on some facts from algebraic geometry and real analysis. Intrinsic supersmoothness in Theorem 4.2 is true C^r continuity at v.

We now look at splines in more than two variables. The result that we present here is proved in [6]. We first describe the construction of a family of interior cells $\Delta_A^n(T)$

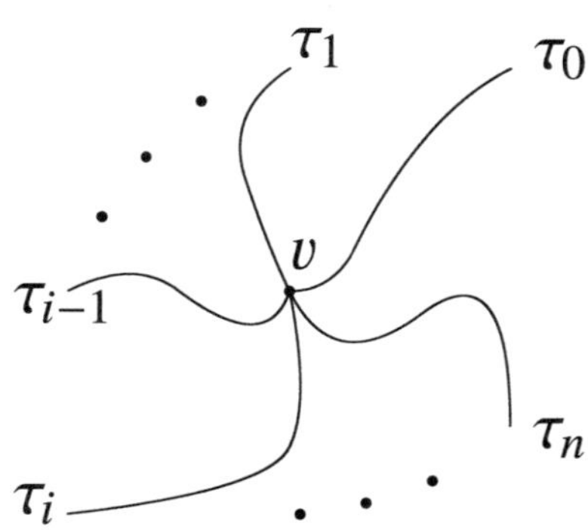

Fig. 4 An interior cell defined by irreducible algebraic arcs

in $\mathbb{R}^n$ that is usually called the *Alfeld split* of a n-simplex T: choose an arbitrary strictly interior to T point v, then connect v to each of the $n+1$ vertices of T. This splits T into $n+1$ subsimplices, whose bases are the facets of T. Note that $\Delta_A^2(T)$ is our familiar split in Fig. 2(right). The trivariate version $\Delta_A^3(T)$ was studied in [1], where supersmoothness $C^3(v)$ in the sense of matching of the derivatives was proved by making use of the special geometry of the split (Fig. 5).

Theorem 4.3 *Let $\Delta_A^n(T)$ the Alfeld split of an n-simplex T with the interior split point v. Then the following intrinsic supersmoothness is observed:*

$$S_d^r(\Delta_A^n(T)) = S_d^r(\Delta_A^n(T)) \cap C^\rho(v), \quad \textit{where}$$

$$\rho = \begin{cases} (r+1)(n+1)/2 - 1, & \textit{if } r \textit{ is odd;} \\ r(n+1)/2, & \textit{if } r \textit{ is even,} \end{cases}$$

and intrinsic supersmoothness is understood as matching of the derivatives for $\rho \geq r+2$. □

Another split of a simplex in $\mathbb{R}^n$ considered in [6] is the so-called *facet split* $\Delta_F^n(T)$. It is obtained from $\Delta_A^n(T)$ by splitting further as follows: for each facet of T choose an arbitrary strictly interior to F point v_F, then construct an Alfeld split $\Delta_A^{n-1}(F)$ using v_F. Finally, cone off v to obtain $n(n+1)$ subsimplices. We call the facet split *aligned* if for each facet F in T the following three points are collinear: v_F, v, and the vertex of T opposite F. The dimension of splines on the aligned facet split was computed in [10].

Theorem 4.4 *Let $\Delta_F^n(T)$ the aligned facet split of an n-simplex T with the interior point v. Then the following intrinsic supersmoothness is observed:*

$$S_d^r(\Delta_F^n(T)) = S_d^r(\Delta_F^n(T)) \cap C^\rho(v), \quad \textit{where}$$

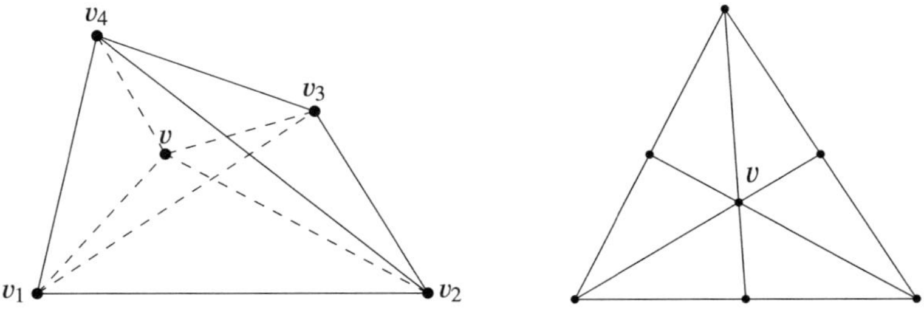

Fig. 5 Alfed split $\Delta_A^3(T)$ of a tetrahedron $T := [v_1, v_2, v_3, v_4]$ with the interior point v (left), and an aligned facet split $\Delta_F^2(T)$ of a triangle (right)

$$\rho = \begin{cases} (r+1)n/2 - 1, & \text{if } r \text{ is odd;} \\ rn/2, & \text{if } r \text{ is even.} \end{cases}$$

and intrinsic supersmoothness is understood as matching of the derivatives for $\rho \geq r+2$. *For* $r = 1$, *the result holds even if the facet split is not aligned.* □

The proofs of Theorems 4.3 and 4.4 as well as Theorems 2.1 and 4.1 rely on a very similar approach: find the degree ρ for which the dimension of the spline space of interest is equal to that of a polynomial, see, for example, a more detailed explanation after Lemma 2.1. Unfortunately, this approach only works if the dimensions of the splines spaces are known. For splines in more than two variables, the dimension of spaces of splines on an arbitrary interior cell is not known. Therefore, this approach cannot lead to generalizing Theorem 4.1 to splines in n variables. We refer the reader to [6] for some more results on rather special types of interior cells in $\mathbb{R}^n$.

4.2 Along and Across an Edge

A version of Lemma 3.1 holds for semi-algebraic splines in two variables.

Theorem 4.5 *Let* Δ *be a partition of an open disk into two regions by an arc defined by the polynomial equation* $g(x, y) = 0$. *Then the directional derivative of any spline* $s \in S_d^r(\Delta)$ *of order* $r+1$ *in the direction tangent to the arc exists at every point* (x, y) *on the arc where* $\nabla g(x, y) \neq \mathbf{0}$. □

We note that both Lemma 3.1 and Theorem 4.5 describe intrinsic supesmoothness **along** an edge, not **across** an edge. Thus, every spline in every spline space has this type of supermoothness. However, it is the "across" an edge type of intrinsic supersmoothness that plays a crucial role in defining properties of these spaces. We now turn our attention to studying this type of supersmoothness. For a given interior cell $\Delta^2(v)$ around a vertex $v \in \mathbb{R}^2$, we say that an edge e emanating from v has a collinear counterpart if there is a different from e edge emanating from v that has the same slope as v. For example, in Fig. 8(left), the edges $[q_1, v_1]$ and $[t_1, v_1]$ have collinear counterparts in $\Delta_5^2(v_1)$, while the edge $[v_2, v_1]$ does not.

Theorem 4.6 *Let* $\Delta^2(v)$ *be an interior cell, and let smoothness* $r \geq 1$. *Suppose the number of different slopes* $m \leq r+2$. *Let* $\widetilde{\Delta}^2(v)$ *be the interior cell obtained from* $\Delta^2(v)$ *by removing the edges with no collinear counterparts. Then*

$$S_\rho^r(\Delta^2(v)) = S_\rho^r(\widetilde{\Delta}^2(v)), \quad \textit{where} \quad \rho = r + \left\lfloor \frac{r+1}{m-1} \right\rfloor.$$

The proof of this result in [14] is as follows. Clearly, $S_\rho^r(\widetilde{\Delta}^2(v)) \subseteq S_\rho^r(\Delta^2(v))$. In Chap. 9 of [9] we find the dimensions of both vector spaces, and conclude that they are equal. Thus, $S_\rho^r(\Delta^2(v)) = S_\rho^r(\widetilde{\Delta}^2(v))$. We could interpret Theorem 4.6 as

C^ρ intrinsic supersmoothness result across the edges with no collinear counterparts. In this case, we should be able to generalize it to higher polynomial degrees. To this end we need to refine our definition of supersmoothness at v as follows.

Definition 4.1 Let $\Delta_n^2(v)$ be an interior cell around v, and let e be an interior edge in $\Delta_n^2(v)$. Let T_1 and T_2 be the two triangles in $\Delta_n^2(v)$ sharing e. We say that a spline $s \in S_d^r(\Delta_n^2(v))$ has supersmoothness ρ at v relative e, and denote it by $C^\rho(v_e)$, if $s|_{T_1 \cup T_2} \in C^\rho(v)$. □

Note that as usual, if $\rho \geq r+2$, then C^ρ is understood as matching of the derivatives. Using Definition 4.1 and passing to the Taylor polynomials on T_1 and T_2, we obtain the next result, see [14] for a detailed proof.

Theorem 1 *Let $\Delta^2(v)$ be an interior cell, and let smoothness $r \geq 1$. Suppose the number of different slopes $m \leq r+2$. Then for any interior edge e without a collinear counterpart, we have*

$$S_d^r(\Delta^2(v)) = S_d^r(\Delta^2(v)) \cap C^\rho(v_e), \quad \text{where} \quad \rho = r + \left\lfloor \frac{r+1}{m-1} \right\rfloor.$$

We now look at splines in more than two variables. For simplicity, we first discuss $\Delta_A^3(v)$, the Alfeld split of a tetrahedron T in $\mathbb{R}^3$. Let v_1 be a vertex of T, and $e := [v_1, v]$ be one of the interior edges of $\Delta_A^3(v)$. Consider a cross-section of $\Delta_A^3(v)$ by a plane, say, perpendicular to e and passing through any point u strictly interior to e. This cross-section is clearly $\Delta_A^2(u)$. Therefore, the instrinsic supersmoothness μ at the vertex u in $\Delta_A^2(u)$ becomes the intrinsic supersmoothness μ across the edge e in $\Delta_A^3(v)$. In particular, we also have C^μ at v_1. This explains why the macro-element $S_5^1(\Delta_A^3(v))$ constructed in [1], and studied in Sect. 18.3 of [9], has to have supersmoothness two at every vertex of T. The same cross-section argument works for splines on any $\Delta_A^n(v)$, and we obtain the following companion to Theorem 4.3.

Theorem 4.7 *Let $\Delta_A^n(T)$ the Alfeld split of an n-simplex T with the interior split point v. Then any spline $s \in S_d^r(\Delta_A^n(T))$ has intrinsic supersmoothness μ across any interior edge of $\Delta_A^n(T)$, where*

$$\mu = \begin{cases} n(r+1)/2 - 1, & \text{if } r \text{ is odd;} \\ nr/2, & \text{if } r \text{ is even.} \end{cases}$$

We note that while it is certainly possible to generalize Lemma 2.1, Theorems 4.5 and 4.7 for k-dimensional interior faces of corresponding partitions in $\mathbb{R}^n$, such generalizations fall outside the scope of this survey.

4.3 Spline Vector Fields

We now consider a bivariate vector spline field, that is a pair (u, v) of bivariate splines, where both u and v are polynomial splines defined on the same partition and are of the same degree and smoothness. Special subspaces of such spline spaces are used in numerical solutions of PDEs. For example, divergence-free subspaces are good candidates for finite element families for solving Stokes type equations: let $(u, v) \in [S_d^r(\Delta)]^2$ satisfies an additional condition: $\frac{\partial u}{\partial x}|_T + \frac{\partial u}{\partial y}|_T = 0$ on each $T \in \Delta$, then we call the vector spline field (u, v) divergence-free.

Theorem 4.8 *Let $\Delta_n^2(v)$ be a cell with n noncollinear edges. Let (u, v) be any divergence-free vector spline of degree $\le d$ and smoothness $r \le d$. Then both u and v have intrinsic supersmoothness $r + \left\lfloor \frac{r+2}{n-1} \right\rfloor$ at the interior vertex of the cell v.* □

The proof of Theorem 4.8 in [2] uses the fact that a divergence-free vector field (u, v) has a potential function $s \in S_{d+1}^{r+1}(\Delta_n^2(v))$. Thus, we can apply Theorems 2.1 or 4.1 to s, and the result follows.

4.4 Mixed Derivatives

As the reader deduced from the previous sections, symmetry of the underlying partition seems to be an enemy of intrinsic supersmoothness. Here we show that even in the least favorable situation, when each interior edge has a collinear counterpart, there is still some residual intrinsic supersmoothness.

Theorem 4.9 *Let $\Delta_{2\ell}^2(v)$ be an interior cell with ℓ pairs of collinear edges meeting at v with distinct slopes $u_1, \ldots, u_\ell$, and no other edges. Then for any $s \in S_d^{\ell-1}(\Delta_{2\ell}^2(v))$ any ℓ-th order mixed derivative defined below exists:*

$$\frac{\partial^\ell s}{\partial u_{i_1} \cdots \partial u_{i_\ell}}(v), \quad \textit{where} \quad u_{i_j} \neq u_{i_k} \quad \textit{for all} \quad j \neq k.$$

Note that Theorem 4.9 holds for $\ell = 1$, that is, for continuous splines on $\Delta_{2\ell}^2(v)$, which is equivalent to a special case of Lemma 3.1. The proof of Theorem 4.9 in [14] is done by induction on ℓ, and it can be easily generalized to the case of splines in more than two variables.

4.5 Along a Hidden Line

In this subsection we present an example of a less studied type of intrinsic supersmoothness: it is associated with a very specific space and partition. We hope that the reader might be more successful in finding a general result for this type of supersmoothness.

Theorem 4.10 *Let $\Delta_4^2(v)$ be a cell formed by four edges with distinct slopes meeting at the point v. Then there exists a unique straight line ℓ passing through v with the property that for any spline $\in S_2^1(\Delta_4^2(v)$, the restriction of s on ℓ is a univariate quadratic polynomial.* □

Note that in general a restriction of $s \in S_2^1(\Delta_4^2(v)$ on any straight line passing through v is a C^1 quadratic univariate spline, not a polynomial. Thus, we can treat the hidden line ℓ as a line of C^2 intrinsic supersmoothness. In addition, note that ℓ in Theorem 4.10 does not depend on s, it depends on the geometry of the cell $\Delta_4^2(v)$ only. The proof of Theorem 4.10 in [14] is a straightforward computation of the slope of ℓ.

We provide an example to illustrate Theorem 4.10 in Fig. 6. The cell $\Delta_4^2(v)$ is formed by four edges emanating from the origin v with the following equations:

$$y = 0, \quad x = 0, \quad y + x = 0, \quad y - x = 0,$$

and the polynomial pieces of $s \in S_2^1(\Delta_4^2(v))$ are denoted by $p_i(x, y), i \in \{1, \ldots, 4\}$, see Fig. 6(left). Assuming that $p_1(x, y)$ is given, using the C^1 smoothness conditions we obtain the following formulae for s:

$$\begin{aligned} p_2(x, y) &= p_1(x, y) - 2cy^2, \\ p_3(x, y) &= p_1(x, y) - 2c(x^2 + y^2), \\ p_4(x, y) &= p_1(x, y) - c(x - y)^2, \end{aligned} \tag{5}$$

thus, making $S_2^1(\Delta_4^2(v))$ a seven-dimensional vector space: six coefficients for $p_1(x, y)$ and the coefficient c in (5). It is easy to check by direct substitution that the line ℓ is given by the equation $2y - x = 0$ and depicted as a dashed line in Fig. 6(right) is the only line satisfying the condition that $s|_\ell$ is a univariate quadratic polynomial.

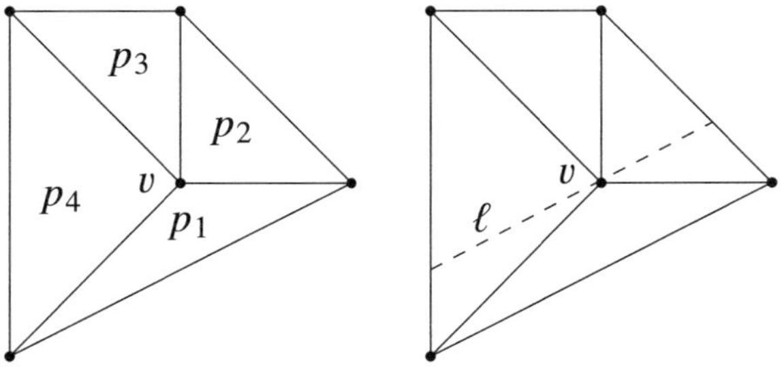

Fig. 6 The partition $\Delta_4^2(v)$ (left), and the hidden dashed line ℓ from Theorem 4.10(right)

4.6 Harmonic Splines

Harmonic polynomials p in two real variables satisfy the Laplace equation $\Delta p = 0$. It is reasonable to expect that piecewise harmonic polynomials (or *harmonic splines*) could be a useful tool in constructing finite elements for numerical solutions of Laplace and Poison equations. However, our next result shows that only continuous harmonic splines can be used for that purpose.

Theorem 4.11 *Let ℓ be a segment of a straight line that divides the open disk Ω into two subsets Ω_1 and Ω_2, as in Fig. 3. Let p_1 and p_2 be two harmonic polynomials C^k-continuously glued along ℓ, that is, let $F \in C^k(\Omega)$, $k \geq 1$, where F is a piecewise function defined in (4). Then $p_1(x, y) \equiv p_2(x, y)$.* □

Theorem 4.11 can be proved in many ways: using real analysis, by passing to complex variables, or using Bernstein-Bézier analysis, as in [2]. It can also be generalized to the case of more that two variables. Theorem 4.11 shows that all smooth harmonic splines are just polynomials and, thus, only C^0 harmonic splines form reasonably good approximating spaces, see [16] and the references therein for constructions of continuous harmonic finite elements.

4.7 Individual Smoothness Conditions

In Bernstein-Bézier analysis of bivariate splines, see [9] for a comprehensive treatment, smoothness conditions across an edge e are written as a system of linear equations involving local coefficients. For example C^1-smoothness conditions of a spline of degree d across an edge e are represented as a system of $d - 1$ linear equations with a total of $3d + 1$ unknowns. Each equation, that we shall call *an individual smoothness condition*, has only four coefficients. Note that, in general, these individual smoothness conditions cannot be interpreted as smoothness at certain points on e. It turns out that many spline spaces also have increased intrinsic supersmoothness of some individual smoothness conditions. Full description of such results goes beyond the scope of this survey, and we refer interested readers to [14] for a complete study.

5 How to Compute Dimensions of Spline Spaces Using Intrinsic Supersmoothness

We conclude the survey with two applications of intrinsic supersmoothness. The goal of this section is to demonstrate the benefit of knowing intrinsic supersmoothness to compute dimensions of spline spaces on triangulations in $\mathbb{R}^2$. For similar results on $\mathbb{R}^n$, we refer the reader to [6, 13].

5.1 Powell-Sabin Refinement of an Arbitratry Triangulation

It is known, see [9] for details, that construction of macro-elements and finite elements on an arbitrary triangulation Δ entails additional refinements of Δ. The reason is that ensuring smoothness conditions for macro-elements on unrefined traingulations requires splines of high polynomial degree. For example the minimal degree for a C^1 macro-element on an arbitrary unrefined Δ is five, see Sect. 6.1 in [9]. We start our discussion by defining some well-known refinements.

Definition 5.1 The Clough-Tocher (CT) refinement Δ_{CT} of an arbitrary triangulation Δ of a polygonal domain in $\mathbb{R}^2$ is obtained as follows: for each triangle $T \in \Delta$, choose an arbitrary split point v_T in the interior of T. Connect v_T to the vertices of T. □

The Clough-Tocher split of one triangle is, of course, just $\Delta_3^2(v)$ depicted in Fig. 2(right), and also the Alfeld split of T. From Lemma 2.1 we know that

$$S_d^r(\Delta_{CT}) = S_d^r(\Delta), \quad \text{for all} \quad d \leq \rho, \quad \rho = r + \left\lfloor \frac{r+1}{2} \right\rfloor. \tag{6}$$

and, thus, only C^r splines of degree higher than ρ on the Clough-Tocher refinement are worth considering, see Sect. 8.2 in [9] for the corresponding Clough-Tocher macro-elements. Note that for $r = 1$, we have $\rho = 2$, and, indeed, the corresponding CT macro-element has polynomials degree three (much better that five on unrefined triangulations!) We next refine Δ_{CT} hoping to further lower the polynomial degree d.

Definition 5.2 The unconstrained Powell-Sabin (PS) refinement Δ_{PS} of an arbitrary triangulation Δ of a polygonal domain in $\mathbb{R}^2$ is obtained as follows:

1. For each triangle $T \in \Delta$, choose an arbitrary split point v_T in the interior of T. Connect v_T to the vertices of T. This, of course, results in the Clough-Tocher Δ_{CT} refinement from Definition 5.1.
2. For each interior edge $e \in \Delta$ shared by triangles T_1 and T_2, choose a split point v_e in the interior of e such that the three points $v_1 := v_{T_1}$, $v_2 := v_{T_2}$, and v_e are not collinear. Connect v_e to v_1 and v_2. □

If for each interior edge the three points v_1, v_2 and v_e are collinear, then the corresponding refinement is known as the Powell-Sabin 6 (PS6) split. We shall call it a *constrained Powell-Sabin refinement*. Note that a constrained Powell-Sabin refinement can contain unaligned facet splits, as in Theorem 4.4; and an unconstrained PS might have all individual facet splits aligned.

Theorem 5.1 *Let Δ_{PS} be an unconstrained Powell-Sabin refinement of an arbitrary triangulation Δ as in Definition 5.2. Then for all $r \geq 1$*

$$S_d^r(\Delta_{PS}) = S_d^r(\Delta), \quad \textit{for all } d \leq \rho, \quad \rho = r + \left\lfloor \frac{r+1}{2} \right\rfloor.$$

Proof The proof proceeds in two steps.
Step 1. Consider the interior cell $\Delta_4^2(v_e)$ for some interior edge $e \in \Delta$ shared by two triangles T_1 and T_2 of the original triangulation Δ, see Fig. 7(left). This cell has four edges: two of them are collinear and form e, while the other two, $[e_v, v_1]$ and $[e_v, v_2]$, are noncollinear. Thus, the total number of slopes at v_e is three. Applying Theorem 4.6 to each $\Delta_4^2(v_e)$, we obtain that for $d \leq \rho$, the noncollinear edges $[e_v, v_1]$ and $[e_v, v_2]$ are extraneous and can be removed from Δ_{PS}, see Fig. 7(middle). Thus, we have shown that

$$S_d^r(\Delta_{PS}) = S_d^r(\Delta_{CT}), \quad d \leq \rho.$$

Step 2. Next we apply (6) and obtain that for $d \leq \rho$, the three interior edges in each $\Delta_3^2(v_T)$ are extraneous and can be removed from Δ_{CT}, see Fig. 7(right). The proof is complete. □

Theorem 5.1 shows that to build macro-elements of lower degree on Powell-Sabin partitions, one needs to work with constrained refinements, see Sect. 8.4 in [9] for relevant constructions. For example, for $r = 1$, on a constrained Powell-Sabin refinement, it is possible to construct a quadratic C^1 macro-element (as opposed to a cubic one on the Clough-Tocher refinement). For analogs of Theorem 5.1 in the trivariate case, see [13].

Remark 5.1 It has been observed in [7] that in the case of smooth cubic splines $S_3^1(\Delta_{PS})$, imposing C^2 supersmoothness at the vertices of Δ yields intrinsic C^2 supersmoothness across the edges of Δ despite the presence of another vertex of Δ_{PS} in between the original vertices of Δ, see the end of Sect. 6.2 in [7], where other supersmoothness relations are noted as well. In [8], the authors observe that some normalized B-splines constructed for $S_3^1(\Delta_{PS})$ also have intrinsically more supersmoothness than prescribed by the definition of the space. □

5.2 *Tohăneanu Partition*

A special case of the so-called *Tohăneanu partition* was first described in [18], and it represents an important example of triangulation with one strictly interior edge.

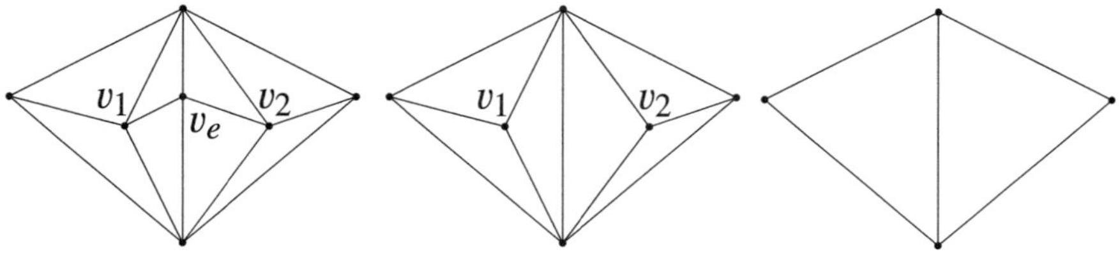

Fig. 7 The partition Δ_{PS} of two triangles (left), the partition Δ_{CT} of two triangles obtained after *Step 1* in Theorem 5.1(middle), the original partition Δ obtained after *Step 2* in Theorem 5.1(right)

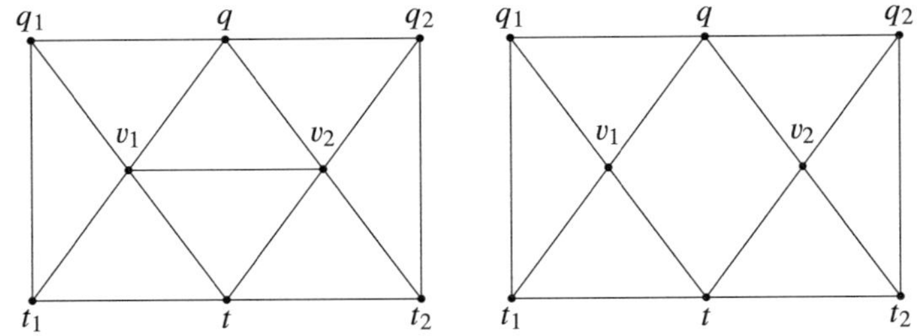

Fig. 8 The Tohăneanu partition Δ_T (left), and the cross-cut partition Δ (right)

Definition 5.3 Suppose four points q_1, q_2, t_2, t_1 in $\mathbb{R}^2$ form a convex quadrilateral, and let two points q and t be chosen in the interiors of two opposite edges $[q_1, q_2]$ and $[t_1, t_2]$, respectively, as in Fig. 8. Connect t and q with the vertices on the opposite sides of the quadrilateral. Denote

$$v_1 := [t_1, q] \cap [q_1, t], \quad v_2 := [t_2, q] \cap [q_2, t],$$

and connect v_1 and v_2. This creates a *Tohăneanu partition* Δ_T of the quadrilateral consisting of eight triangles, see Fig. 8(left). □

In [18], it was shown that for $d \geq 2r + 1$ the dimension of $S_d^r(\Delta_T)$ is equal to the so-called Schumaker's lower bound, see Theorem 9.9 in [9], while for $d \leq 2r$, it was proved that the dimension of $S_d^r(\Delta_T)$ is strictly greater than that bound. However, the exact dimension of $S_d^r(\Delta_T)$ for $d \leq 2r$ was not determined in [18]. As it was shown in [15], the difficulty was caused be the intrinsic supersmoothness across the interior edge $[v_1, v_2]$.

Theorem 5.2 *Let Δ_T be the Tohăneanu partition described in Definition 5.3, and depicted in Fig. 8(left). Let Δ be the partition obtained from the Tohăneanu partition by removing the interior edge $[v_1, v_2]$ as in Fig. 8(right). Then*

$$S_d^r(\Delta_T) = S_d^r(\Delta), \quad d \leq 2r.$$

Note that Δ, shown in Fig. 8(right), is no longer a triangulation. It is the so-called *cross-cut* partition whose dimension is well-known. The main idea of the proof is to consecutively apply Theorem 1 to the cells around v_1 and v_2, eventually building up to Theorem 4.6, see [15] for the proof and more details. Theorem 5.2 can be generalized to other partitions with one strictly interior edge.

References

1. Alfeld, P.: A trivariate Clough-Tocher scheme for tetrahedral data. Comput. Aided Geom. Des. **1**, 169–181 (1984)
2. Alfeld, P., Sorokina, T.: Linear differential operators on bivariate spline spaces and spline vector fields. BIT Numer. Math. **56**, 15–32 (2016)
3. DiPasquale, M., Sottile, F., Sun, L.: Semialgebraic splines. Comput. Aided Geom. Des. **55**, 29–47 (2017)
4. DiPasquale, M., Sottile, F.: Bivariate Semialgebraic splines. J. Approx. Theory **254**, 105392 (2020)
5. Farin, G.: Bézier polynomials over triangles. In: Report TR/91, Department of Mathematics, Brunel University, Uxbridge, UK (1980)
6. Floater, M.S., Hu, K.: A characterization of supersmoothness of multivariate splines. Adv. Comput. Math. **46**, 70 (2020)
7. Grošelj, J., Speleers, H.: Construction and analysis of cubic Powell-Sabin B-splines. Comput. Aided Geom. Des. **57**, 1–22 (2017)
8. Grošelj, J., Speleers, H.: Extraction and application of super-smooth cubic B-splines over triangulations. Comput. Aided Geom. Des. **103**, 102194 (2023)
9. Lai, M.-J., Schumaker, L.L.: Spline Functions on Triangulations. Cambridge University Press, Cambridge (2007)
10. Schenck, H., Sorokina, T.: Subdivision and spline spaces. Constr. Approx. **47**, 237–247 (2018)
11. Shekhtman, B., Sorokina, T.: Intrinsic supersmoothness. J. Concr. Appl. Math. **13**(3), 232–241 (2015)
12. Shekhtman, B., Sorokina, T.: A note on intrinsic supersmoothness of bivariate semialgebraic splines. Comput. Aided Geom. Des. **98**, 102137 (2022)
13. Sorokina, T.: Intrinsic supersmoothness of multivariate splines. Numer. Math. **116**, 421–434 (2010)
14. Sorokina, T.: Redundancy of smoothness conditions and supersmoothness of bivariate splines. IMA J. Numer. Anal. **34**(4), 1701–1714 (2014)
15. Sorokina, T.: Bivariate splines on Tohăneanu partitions. J. Approx. Theory **232**, 6–11 (2018)
16. Sorokina, T., Zhang, S.: An interpolated Galerkin finite element method for the Poisson equation. J. Sci. Comput. **92**, 47 (2022)
17. Speleers, H.: Construction of normalized B-Splines for a family of smooth spline spaces over Powell-Sabin triangulations. Constr. Approx. **37**(1), 41–72 (2013)
18. Tohăneanu, S.: Smooth planar r-splines of degree $2r$. J. Approx. Theory **132**, 72–76 (2005)

Contributed Papers

Analysis of Historical Average Temperatures for Weather Derivatives Valuation. The Case of Parma (Italy)

Alessandra Aimi, Chiara Guardasoni, Luis Ortiz-Gracia, and Francesco Zanni

Abstract This paper is a preliminary work for the analysis of strategies to valuate weather derivatives using a complete Italian archive of average temperature data. We set a model to forecast average temperatures over relatively small time intervals (up to one year) and describe the related parameters estimation. The analysis ends up with the valuation of a Call on HDD index.

1 Introduction

The impact of climate change on various aspects of our society is now macroscopically evident [1, 15, 20, 23] and, by consequence, the assessment of climate risk. In this context, the analysis of so-called climate-related financial risks, or the implications of climate change on the financial system, becomes crucial for the well-being of individuals and society as a whole. An international network of over 90 central banks, financial regulators and institutions, the Network for Greening the Financial System (NFGS), has identified as physical risks those arising from the impact on individuals of extreme weather events (hurricanes, droughts, fires, floods) or long-term climate change (increase in the overall temperature of the Earth's atmosphere) leading to important consequences. To mitigate the potential risks for individuals and the banking system associated with adverse weather conditions, insurances and derivatives

A. Aimi · C. Guardasoni (✉) · F. Zanni
Department of Mathematical, Physical and Computer Sciences, University of Parma, Parma, Italy
e-mail: chiara.guardasoni@unipr.it

L. Ortiz-Gracia
Department d'Econometria, Estadística i Economia Aplicada, Universitat de Barcelona (UB), Barcelona, Spain

A. Aimi · C. Guardasoni
Member of the INdAM-GNCS Research Group, Torino, Italy

L. Ortiz-Gracia
RISKcenter, Institut de Recerca en Economia Aplicada (IREA), Universitat de Barcelona (UB), Barcelona, Spain

© The Author(s), under exclusive license to Springer Nature Switzerland AG 2025
M. J. Ibáñez-Pérez et al. (eds.), *Mathematical And Computational Modelling, Approximation and Simulation*, SEMA SIMAI Springer Series 41,
https://doi.org/10.1007/978-3-032-01114-5_5

can be taken into consideration as climate-related financial instruments [16]. Insurances that compensate damages are more adequate to hedge against catastrophic events but derivatives seem to be more suitable to hedge against uncertainties in normal weather, such as hotter than usual seasons that may damage farms in summer or ski plants in winter [2, 13, 17]. We will focus on derivatives, that are commonly used as a means of transferring risk from a party that is exposed to the risk, to another which is willing to bear such risk through the global capital market in exchange for an advance payment of a premium.

Note that the weather is not the same throughout the world and therefore we need to suit specific contracts for specific areas and needs (Geographical Basis Risk [11]). Motivated by a project financed by Bank of Italy to investigate the valuation of these contracts focusing on the case of Italian state, this paper is the first in modeling daily average temperatures of Parma city (Italy) for pricing of financial weather derivatives. Other efforts in this sense were made in [7, 17] but using different data and different methods.

First of all, we need a model for the average daily temperature T_t, defined as the average of maximum and minimum daily temperatures measured in Celsius degrees. Our choice fell on Brownian motion driven dynamics that make possible analytical pricing. The reference model is the Ornstein-Uhlenbeck process detailed in [2, 6, 21]

$$T_t = s(t) + \tilde{T}_t, \tag{1}$$

$$d\tilde{T}_t = -k\tilde{T}_t dt + \sigma(t) dB_t, \tag{2}$$

where $s(t)$ is a function that aims at incorporating the trend and the seasonality of the process T_t, $\tilde{T}_t$ represents the stochastic fluctuations around $s(t)$, k is the stochastic parameter responsible for the "speed" of mean reversion towards $s(t)$, $\sigma(t)$ is the variance assumed to be periodic, and B_t is a Brownian motion.

In Sect. 2 we investigate the calibration of the function $s(t)$ that is considered as a deterministic function representing the mean seasonal variation i.e. the annual cycle or seasonality of the temperature approximated by an expansion in terms of sines and cosines and it may take into account also the eventuality of global warming, greenhouse effect and urbanization, by adding a linear trend. Therefore we approximated it by the usual form of Fourier series

$$s(t) \approx s_n(t) = a_0 + b_0 t + \sum_{i=1}^{n} a_i \sin\left(\frac{2\pi}{365} it\right) + \sum_{i=1}^{n} b_i \cos\left(\frac{2\pi}{365} it\right), \tag{3}$$

observing that other analogous trigonometric expansions can be taken into consideration as described in [18]. Note that we will see along the analysis of our data that the linear trend may turn out to be a too strong simplification when a long time series of temperature data is considered.

Referring to [5, 9], we expect a clear seasonality in temperature variations, after removing the mean reversion effects towards the seasonal mean $s(t)$, and, hence, even the volatility is modeled as suggested in [6] by a truncated Fourier series

$$\sigma^2(t) \approx \sigma_m^2(t) = c_0 + \sum_{i=1}^{m} c_i \sin\left(\frac{2\pi}{365}it\right) + \sum_{i=1}^{m} d_i \cos\left(\frac{2\pi}{365}it\right). \qquad (4)$$

The main focus is the analysis of Parma temperatures that are recorded in a very complete repository that goes back long time with respect to usually available repositories and, to the best of our knowledge, has not yet been analysed.

Section 3 is then devoted to HDD-call valuation, setting suitable HDD-indexes.

1.1 The Dataset

Assuming to calibrate the model (1)–(4), we have to estimate the following set of parameters:

- n, $\{a_i, b_i\}_{i=0,\dots,n}$ to define the seasonality function $s(t)$ in (3),
- m, c_0, $\{c_i, d_i\}_{i=1,\dots,m}$ to define the periodic variance function $\sigma(t)$ in (4),
- k, the speed of mean reversion in (2).

The data at disposal to perform this calibration came from a digital repository edited by P. Fantini for CIDEA (Interdepartmental Center for Energy and Environment [10]) at University of Parma that collects the records of daily observations acquired from the meteorological control units of the city of Parma (Italy) since 1st January 1878 up to 31st December 2022. For homogeneisation, the dataset has been cleaned of the temperatures coming from 29th of February in leap years.

2 Calibration of Temperature Model

2.1 The Linear Trend

Let us start with the analysis of the linear trend in the seasonality function by linear regression. In Fig. 1 on the left, the red line shows that, during the 145 years of recorded temperatures, we have observed an increase in temperature of 2.5 °C and assuming this trend to forecast the future temperatures we expect an increment of 1 °C over about every 56.9 years.

But, if we consider an approximation by piecewise constant trend over subintervals of 20 years, the resulting base temperatures are almost constant until 80 s and then suddenly rise (Fig. 1 on the right) suggesting different estimates of future scenarios [18].

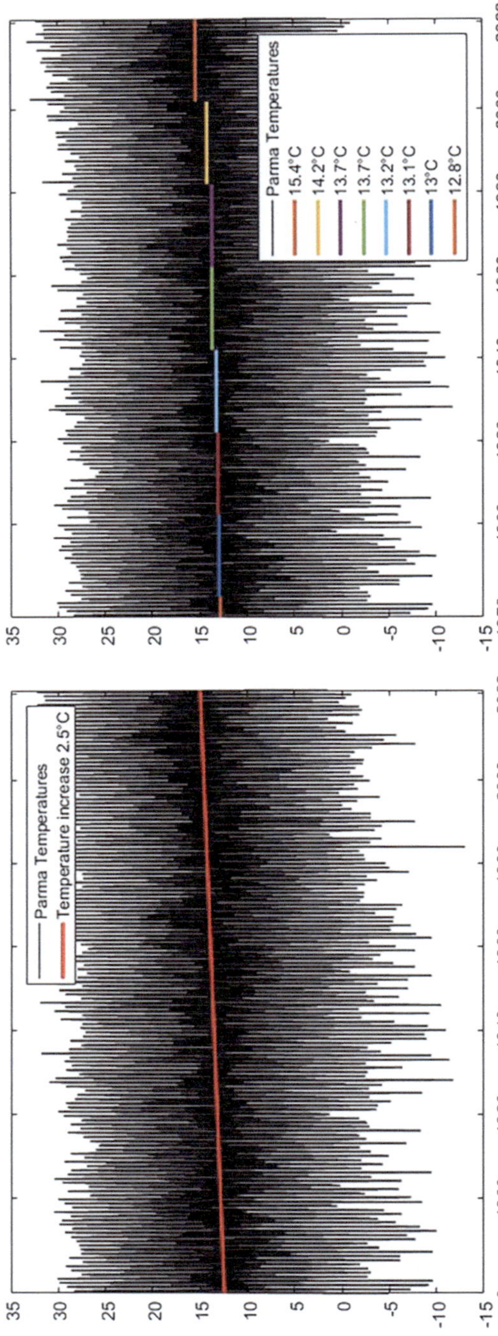

Fig. 1 On the left: linear trend evaluated on the whole dataset. On the right: constant trend evaluated over intervals of 20 years

In Fig. 2 we plot the constant base temperatures, assuming different intervals of years: the valuation over a small number of years produces some expected oscillations that could be due to the superposition of periodical climate phenomena like El Niño, but the behavior has noticeably changed in the last 40 years with a clear monotone trend. This is confirmed also by the last picture in Fig. 2 that represents the average of recorded temperatures AT(k), for $k = 0, \ldots, 125$ during the years $[2003 - k, 2022 - k]$.

We think, looking at Fig. 3 on the left, that a short term conservative prediction can be made, considering a linear regression based on the last 20 to 30 years. In this regard, in Fig. 3 on the right, we can see variation of coefficient b_0 in (3) backward moving the 25-year time window of regression: $\mathrm{b}_0(\mathrm{k})$, for $k = 0, \ldots, 125$ is the slope of the linear trend observed, by linear regression, during the years $[1998 - k, 2022 - k]$. Also for this parameter the high jump of the last years is evident.

In the end we will restrict the dataset to the last 25 years and fix the linear trend parameters $a_0 \approx 1.46 \cdot 10^{-4}$ and $b_0 \approx 8.23$ forecasting a temperature increment of about 1 °C in the next 18.8 years starting from a base temperature of 16 °C at the end of 2022, a quite different scenario from that inferred considering the whole series since 1878.

2.2 *Periodical Behavior of Seasonality Function*

Using non linear regression in least square sense, we estimated in (3) the parameters $\{a_i, b_i\}_{i=1,\ldots,n}$ checking the suitable n truncating the Fourier series. Table 1 and Fig. 4 suggest that a good compromise between accuracy and costs is truncation to $n = 2$ so further steps will be done truncating series (3) to $n = 2$.

The data and seasonality function approximations $s_1(t)$ and $s_2(t)$ obtained by calibration are collected in Fig. 4 emphasizing that, at the scale of the pictures, higher orders of Fourier expansion do not lead to perceptible differences.

2.3 *Calibration of Volatility Function*

It is now possible to address the estimation of volatility $\sigma(t)$. In Fig. 5, we represent the standard deviation $\hat{\sigma}_\ell$ in each day of the year $\ell = 1, \ldots, 365$, computed with the detrended and deseasonalized data of the last 25 years:

$$\hat{\sigma}_\ell^2 := \frac{1}{24} \sum_{j=1}^{25} \tilde{T}^2_{(j-1)365+\ell}, \qquad \ell = 1, \ldots, 365.$$

In Fig. 5 on the right, the volatility appears to be a little higher during spring season so we would like to insert a seasonal behavior also in modeling the continuous volatility

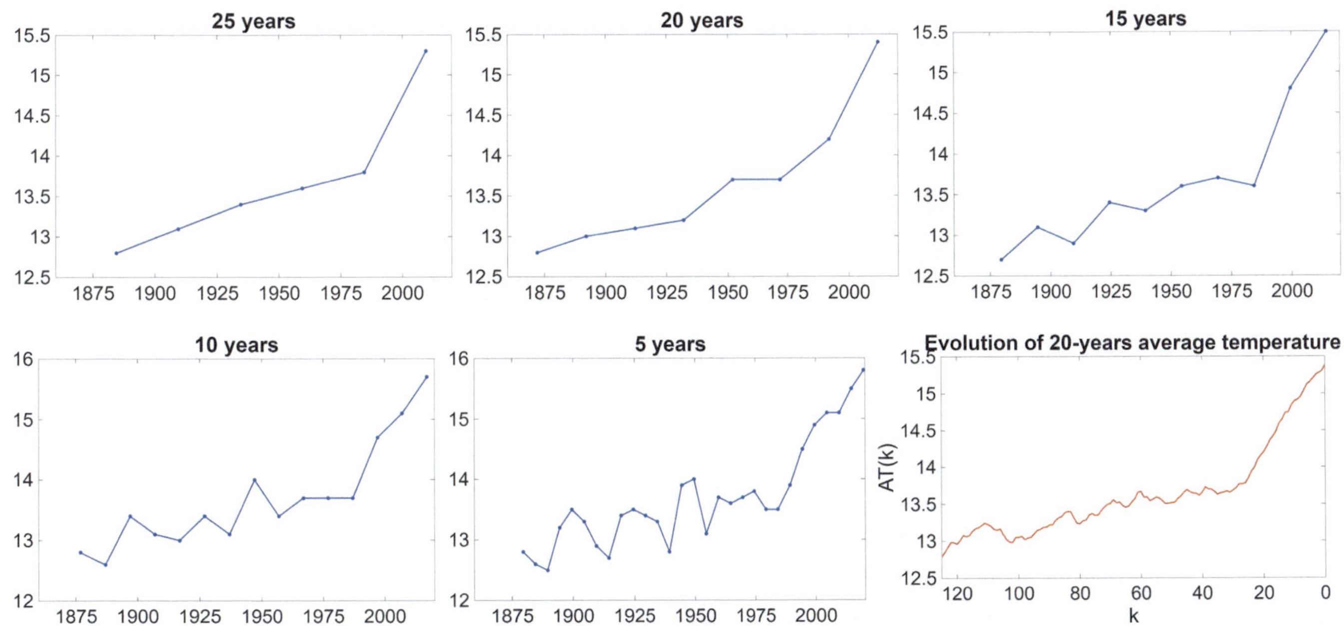

Fig. 2 Base constant trend evaluated over intervals of 25, 20, 15, 10, 5 years, depicted in blue. In orange, the evolution of average temperatures computed over a moving 20-years window

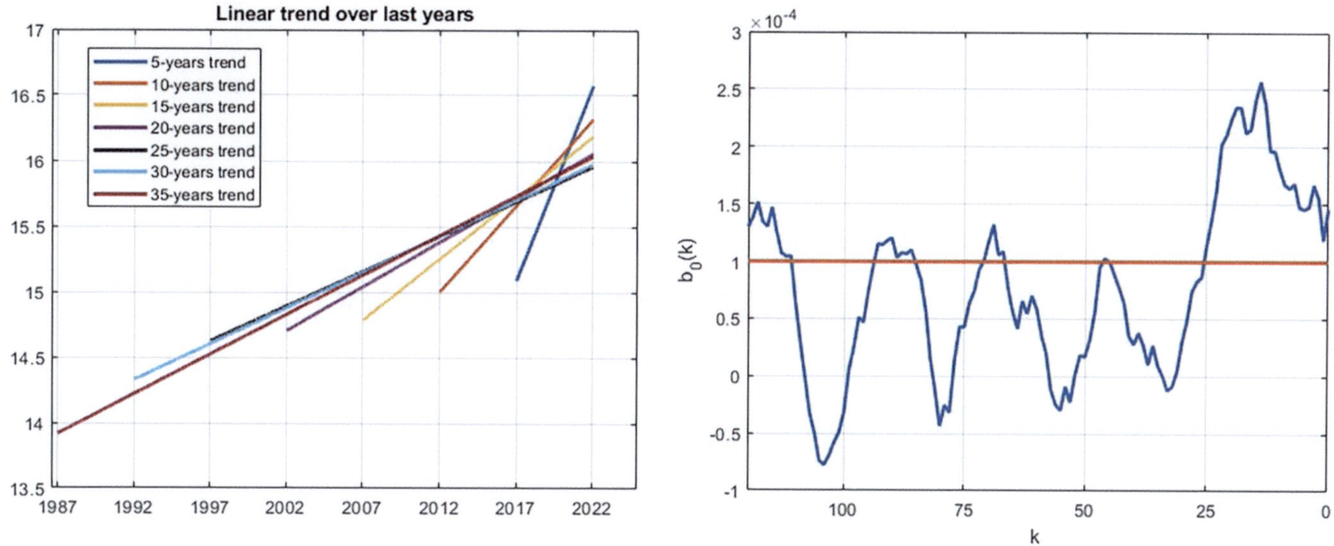

Fig. 3 On the left: linear trend evaluated over different time interval length. On the right: estimation of parameter b_0 in (3) over a moving window of 25 years

Table 1 Estimation of seasonality parameters and valuation of standard deviation from real data

	n=1	n=2	n=3	n=4
a_1	–3.02e+00	–3.02e+00	–3.02e+00	–3.02e+00
b_1	–1.08e+01	–1.08e+01	–1.08e+01	–1.08e+01
a_2		8.11e–01	8.11e–01	8.11e–01
b_2		–4.06e–01	–4.06e–01	–4.06e–01
a_3			–1.51e–01	–1.51e–01
b_3			–3.68e–01	–3.68e–01
a_4				3.18e–02
b_4				–8.09e–02
Standard deviation	2.88e+00	2.81e+00	2.79e+00	2.79e+00

in (2) fitting the model (4) suggested in [6]. We want however to report here that we do not observe a general increase in volatility in recent years (the high maximum in the right picture is mainly due to an outlier) but, in the past, higher volatility was concentrated in the winter season, as exemplified in Fig. 5 on the left taking into account the years interval [1958, 1982], while in recent years it seems to be more widespread.

To be sure to have eliminated any sort of periodicity from the residuals we check the sample autocorrelation functions normalized: of the standard deviation AC($\hat{\sigma}_\ell^2$) at first and, then, of AC($\hat{\sigma}_\ell^2 - \sigma_m^2(t_\ell)$)=:AC$_m$ for $m = 1, \ldots, 5$ defining t_ℓ as the equivalence classes of remainder after division of t by 365. It results that already AC$_1$ differs from AC$_2$ at most $8.7 \cdot 10^{-4}$ (look at Fig. 6) and the same holds for AC$_3$, AC$_4$ and AC$_5$. The values in AC$_1$ are about normally distributed around 0 as expected (Fig. 7), with some positive skewness, therefore, from now on, we will apply the approximation σ_1^2 to model $\sigma^2(t)$.

2.4 Estimation of Speed of Mean Reversion

Differently from what is done in the papers already cited as reference about temperature modelling, here the estimation of the speed of mean reversion k in the Ornstein-Uhlenbeck stochastic process detrended and deseasonalized (2) is done aiming at maximizing the likelihood function, as illustrated in the recent paper [3] but with application to PM_{10} particulates daily concentration in Turin metropolitan area (Italy).

The solution to stochastic differential equation (2) is

$$\tilde{T}_t = \tilde{T}_\tau e^{-k(t-\tau)} + \int_\tau^t e^{-k(t-\xi)} \sigma(\xi) dB_\xi, \tag{5}$$

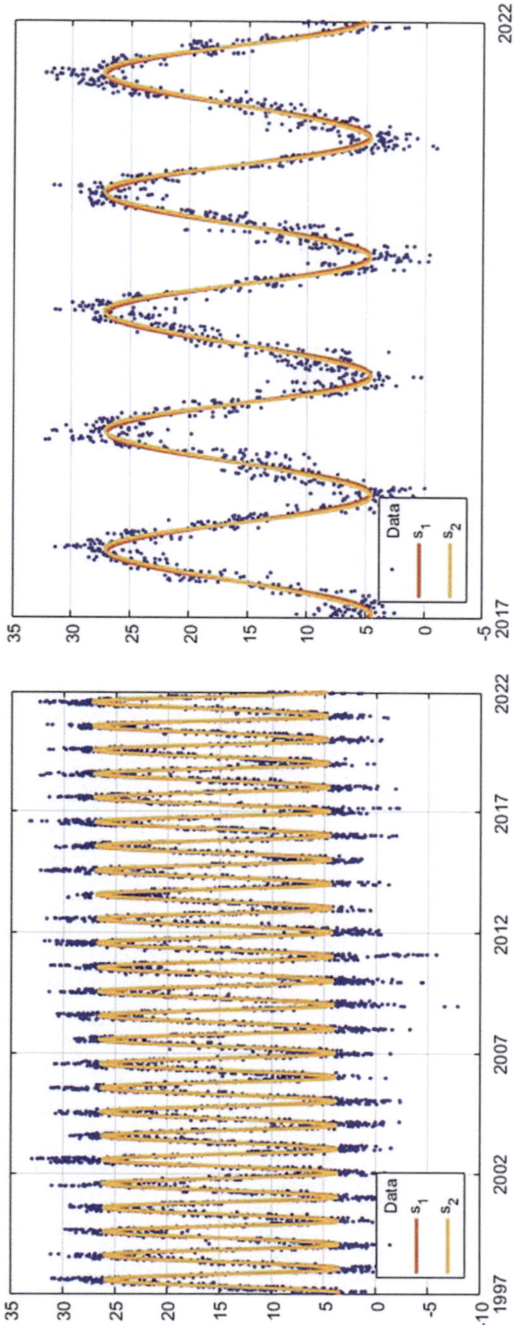

Fig. 4 On the left: superposition of data and calibrated seasonality functions. On the right: zoom over last years

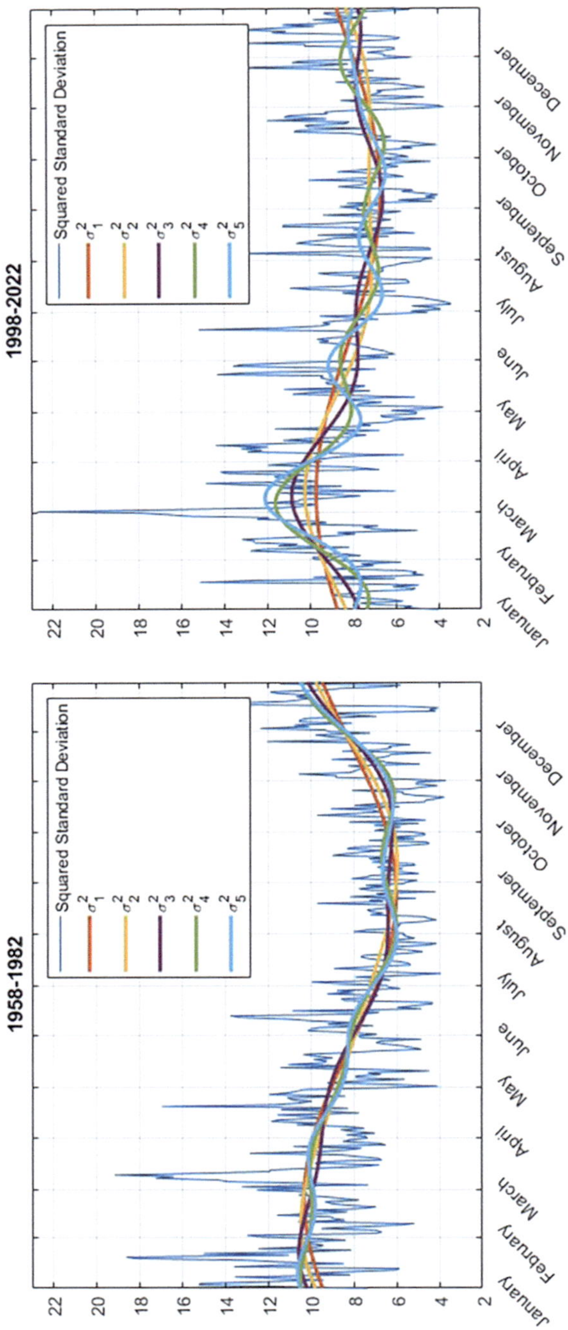

Fig. 5 On the left: volatility during years 1958–1982. On the right: volatility during years 1998–2022

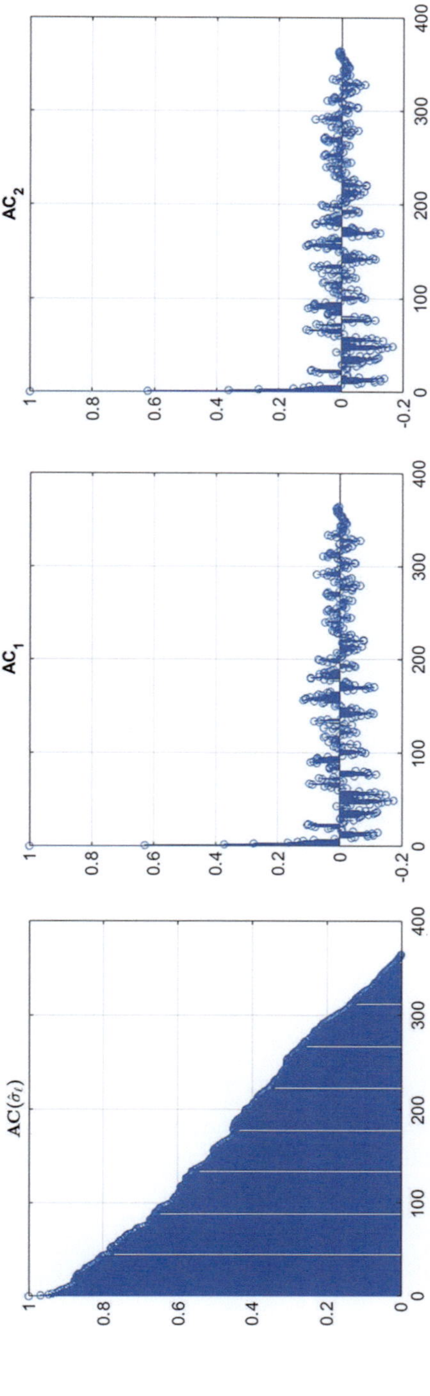

Fig. 6 From the left: discrete normalized autocorrelation of standard deviation $\hat{\sigma}_\ell^2$ sample, after the subtraction of σ_1^2, after subtraction of σ_2^2

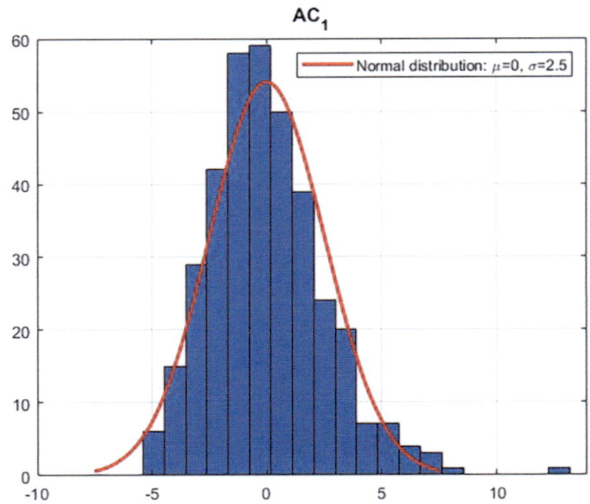

Fig. 7 Superposition of histogram of AC$_1$ values and fitting with a normal probability density function

then the transition probability density function (pdf) that defines the probability of having a temperature $\tilde{t}$ at time t starting from a temperature $\tilde{\tau}$ at time τ is denoted by $f(\tilde{t}, t; \tilde{\tau}, \tau)$ and satisfies the Kolmogorov partial differential equation

$$\frac{\partial f(\tilde{t}, t; \tilde{\tau}, \tau)}{\partial \tau} - k\tilde{\tau}\frac{\partial f(\tilde{t}, t; \tilde{\tau}, \tau)}{\partial \tilde{\tau}} + \frac{\sigma^2(\tau)}{2}\frac{\partial^2 f(\tilde{t}, t; \tilde{\tau}, \tau)}{\partial \tilde{\tau}^2} = 0, \tag{6}$$

and the Fokker-Planck equation

$$\frac{\partial f(\tilde{t}, t; \tilde{\tau}, \tau)}{\partial t} - k\frac{\partial}{\partial \tilde{t}}\left(\tilde{t} f(\tilde{t}, t; \tilde{\tau}, \tau)\right) - \frac{\sigma^2(t)}{2}\frac{\partial^2 f(\tilde{t}, t; \tilde{\tau}, \tau)}{\partial \tilde{t}^2} = 0, \tag{7}$$

from which it is possible to find an explicit expression

$$f(\tilde{t}, t; \tilde{\tau}, \tau) = \frac{1}{\sqrt{2\pi \int_\tau^t \sigma^2(\xi) e^{-2k(t-\xi)} d\xi}} \exp\left(\frac{-\left(\tilde{t} - \tilde{\tau} e^{-k(t-\tau)}\right)^2}{2\int_\tau^t \sigma^2(\xi) e^{-2k(t-\xi)} d\xi}\right), \tag{8}$$

so $\tilde{T}_t$ is normal distributed with mean $\tilde{\tau} e^{-k(t-\tau)}$ and variance $\int_\tau^t \sigma^2(\xi) e^{-2k(t-\xi)} d\xi$.

The conditional likelihood $\mathcal{L}(\tilde{T}_t, k)$ of the process $\tilde{T}_t$ defined in (5), dependent on the parameter k and evaluated by the discrete sampling $\tilde{T}_i$ of the process measured every day t_i during the last 25 years, i.e. for $i = (j-1)365 + \ell,\ j = 1, \ldots 25,\ \ell = 1, \ldots, 365$, is

$$\mathcal{L}(\tilde{T}_t, k) = \prod_{i=2}^{N=9125} f(\tilde{T}_i, t_i; \tilde{T}_{i-1}, t_{i-1}). \tag{9}$$

We are looking for the value k that maximizes $\mathcal{L}$ and, as usual, we pass to the logarithms

$$\arg\max_k \log\left(\mathcal{L}(\tilde{T}_t, k)\right) = \arg\max_k \sum_{i=2}^{N} \log\left(f(\tilde{T}_i, t_i; \tilde{T}_{i-1}, t_{i-1})\right) \tag{10}$$

$$= \arg\max_k \left\{ -\frac{N}{2}\log(2\pi) - \frac{1}{2}\sum_{i=2}^{N} \log\left(\int_{t_{i-1}}^{t_i} \sigma^2(\xi) e^{-2k(t_i-\xi)} d\xi\right) - \sum_{i=2}^{N} \frac{\left(\tilde{T}_i - \tilde{T}_{i-1}e^{-k}\right)^2}{2\int_{t_{i-1}}^{t_i} \sigma^2(\xi) e^{-2k(t_i-\xi)} d\xi} \right\},$$

then we derive $\log\left(\mathcal{L}(\tilde{T}_t, k)\right)$ with respect to the parameter k

$$\frac{d}{dk}\log\left(\mathcal{L}(\tilde{T}_t, k)\right) = \sum_{i=2}^{N} \frac{\int_{t_{i-1}}^{t_i} \bar{\sigma}^2(\xi)(t-\xi)d\xi}{\int_{t_{i-1}}^{t_i} \bar{\sigma}^2(\xi)d\xi} \tag{11}$$

$$-\sum_{i=2}^{N} \frac{\tilde{T}_{i-1}e^{-k}\left(\tilde{T}_i - \tilde{T}_{i-1}e^{-k}\right)\int_{t_{i-1}}^{t_i} \bar{\sigma}^2(\xi)d\xi + \left(\tilde{T}_i - \tilde{T}_{i-1}e^{-k}\right)^2 \int_{t_{i-1}}^{t_i} \bar{\sigma}^2(\xi)(t-\xi)d\xi}{\left(\int_{t_{i-1}}^{t_i} \bar{\sigma}^2(\xi)d\xi\right)^2},$$

having set $\bar{\sigma}^2(\xi) = \sigma^2(\xi)e^{-2k(t_i-\xi)}$ and we apply the Newton method to find k such that

$$\frac{d}{dk}\log\left(\mathcal{L}(\tilde{T}_t, k)\right) = 0. \tag{12}$$

Looking at Fig. 8 on the left, we observe a maximum for the likelihood that we easily find by applying Newton method for roots searching to its derivative, initializing the algorithm with $k = 0$. The resulting value is $k \approx 0.4876$.

2.5 *Simulation of Temperatures*

In this section we show an example of temperatures simulation computed by an Euler discretization for (1) and (2) advancing with a one-day time step. Observe, from Figs. 9 and 10, that, despite the modeling of seasonality function, shown in

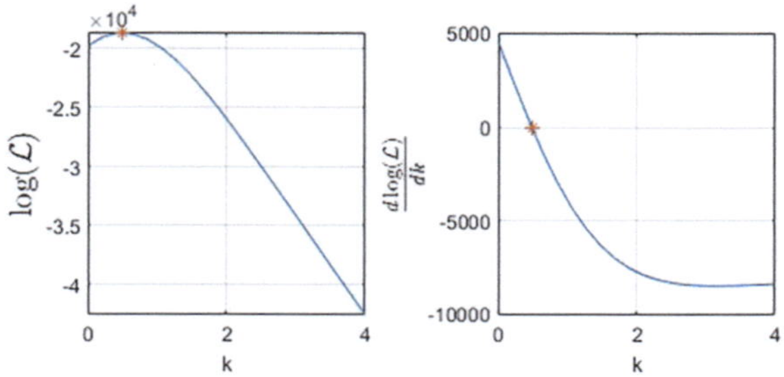

Fig. 8 On the left: graph of likelihood logarithm as defined in (10) with highlight of maximum value. On the right: graph of its derivative as defined in (11) with highlight of the root

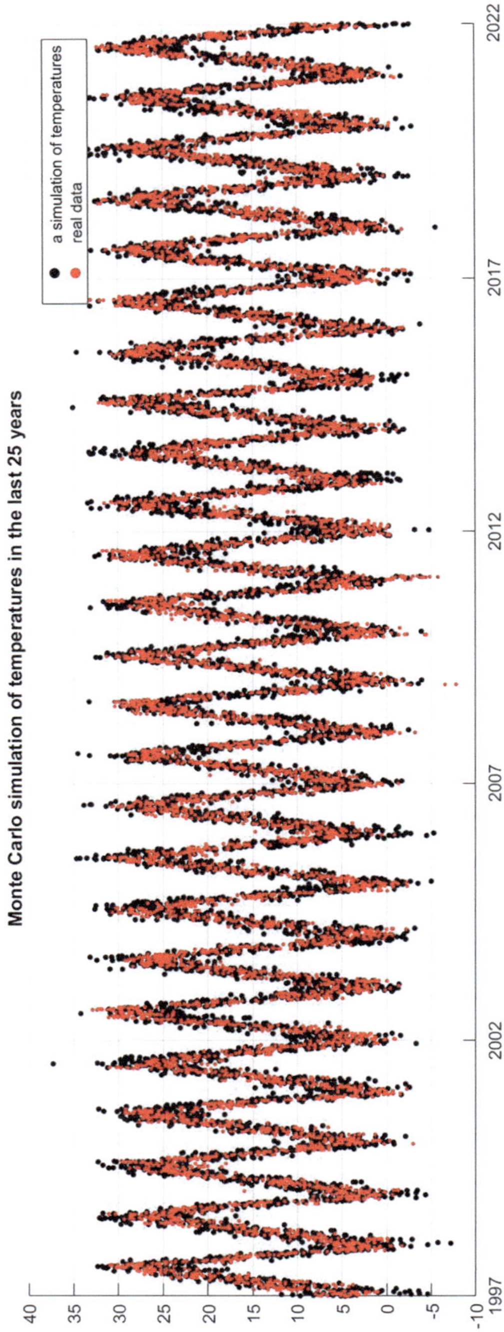

Fig. 9 In black, a simulation of temperatures over the 25 years of analysis computed by the calibrated model. In red the real data

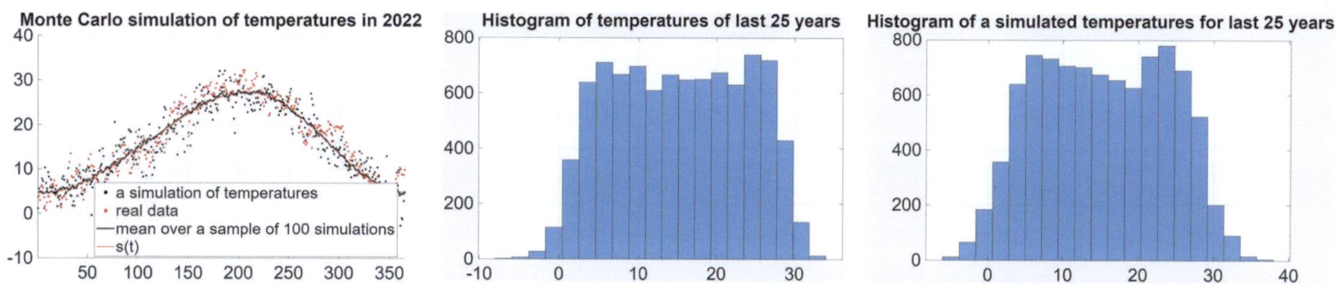

Fig. 10 On the left, focus on year 2022: in black, a stochastic simulation of temperatures and the mean related to 10^2 simulations and, in red, the real data and the seasonality function. In the center: histogram of the temperature data of the last 25 years. On the right: histogram of a simulation of temperatures for the last 25 years

Fig. 4, seems to exclude extreme temperatures, these extremes are then recovered by the addition of randomness.

3 Valuation of Weather Derivatives

3.1 Set Weather Index

The CME Group Inc. (Chicago Mercantile Exchange & Chicago Board of Trade) received approval from the Commodity Futures Trading Commission (CFTC) to list Heating and Cooling Degree Day futures in August 1999.

The standard HDD (Heating Degree Days) and CDD (Cooling Degree Days) are the most common indexes, useful also for agriculture purposes (https://agri4cast.jrc.ec.europa.eu/DataPortal/Index.aspx?o=d), on which weather derivatives are based as they give the measure of uncommonly cold winters or warm summers in an interval of time defined by $[t_{start}, t_{end}]$, in terms of deviation from a benchmark 18 °C (or 65° Fahrenheit) [24]:

$$\text{HDD} := \sum_{i=t_{start}}^{t_{end}} \max\left(18\,^\circ\text{C} - \bar{T}_i, 0\right), \tag{13}$$

$$\text{CDD} := \sum_{i=t_{start}}^{t_{end}} \max\left(\bar{T}_i - 18\,^\circ\text{C}, 0\right). \tag{14}$$

According to the European Environment Agency different thresholds may be considered such as that stated by Eurostat through the Joint Research Center Agri4Cast [19]

$$\text{HDD}_{EU} := \sum_{i=t_{start}}^{t_{end}} \max\left(18\,^\circ\text{C} - \bar{T}_i, 0\right) H[15\,^\circ\text{C} - \bar{T}_i], \tag{15}$$

$$\text{CDD}_{EU} := \sum_{i=t_{start}}^{t_{end}} \max\left(\bar{T}_i - 21\,^\circ\text{C}, 0\right) H[\bar{T}_i - 24\,^\circ\text{C}], \tag{16}$$

implying[1], for example, that, if the daily mean air temperature is 26 °C, for that day the value of the CDD index is 5 °C (26 °C–21 °C) but if the daily mean air temperature is 22 °C, for that day the CDD index is 0 °C.

In Fig. 11 there is the comparison of simulated and real HDD indexes extracted from our data over two different periods. The model is calibrated with our data over

[1] $\bar{T}_i$, in the following, will be either the simulated temperature or the measured average temperature at day i.
$H[\cdot]$ is the Heaviside function.

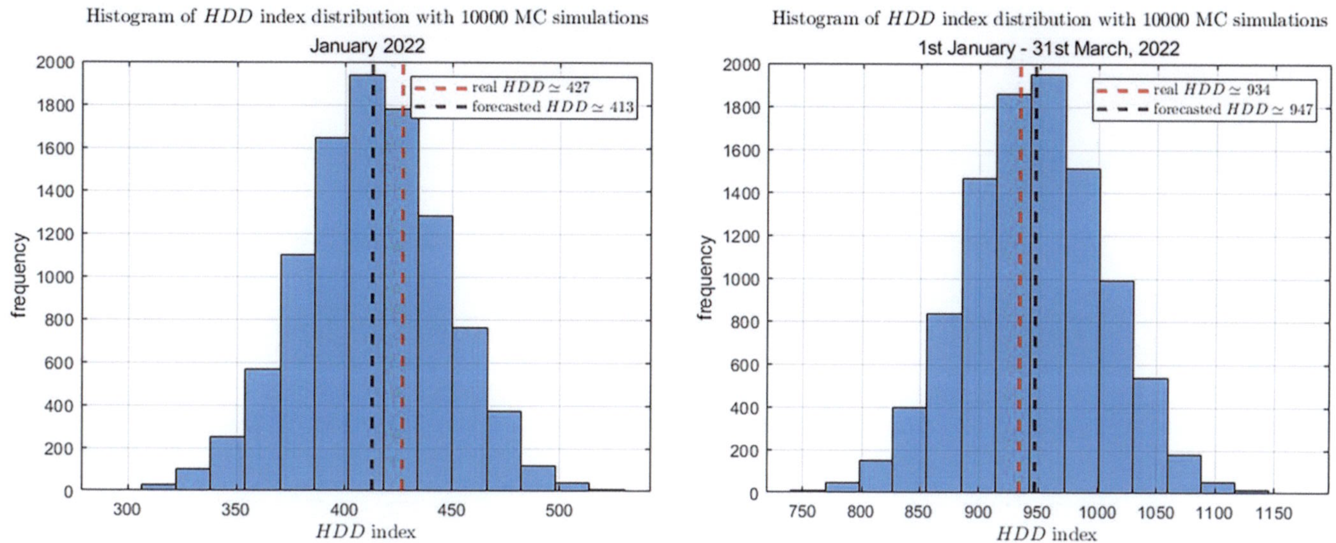

Fig. 11 On the left: histogram of HDD for the month of January 2022 simulated with 10^4 Monte Carlo simulations. On the right: the same histogram for the simulation of HDD from 1st January 2022 to 31st of March 2022

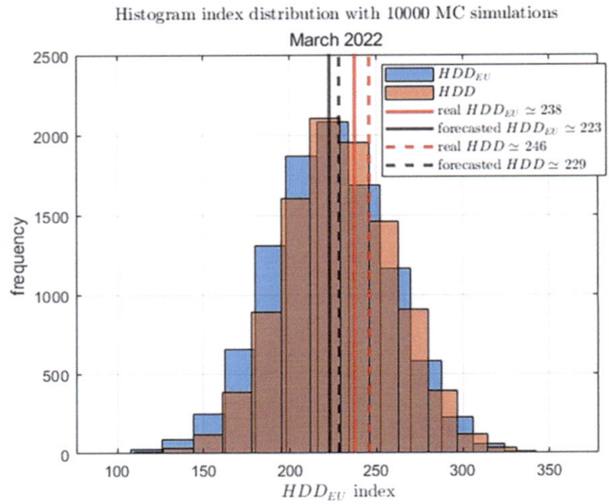

Fig. 12 Comparison between the indexes (13) and (15) real and simulated in March 2022

the time interval [1st January 1998, 31st December 2021]. On the left of Fig. 11, HDD-indexes are computed along January 2022, considering Monte Carlo simulations with 10^4 paths that produce the forecast value identified by the black dashed line; the true value observed in January 2022 is represented by the red dashed line. Analogously, on the right, the same indexes are evaluated over the 3-month period 1st of January–31st of March, 2022. Note that the relative errors between the forecasted indexes and the real ones is in the range of a few percentage points: 3.3% in January 2022 and 1.4% in the months January–March 2022. Considering the month of January, the average temperature in 2022 was below 15° every day, therefore (13) and (15) provide the same result, the situation is slightly different if we consider the month of March, look at Fig. 12.

3.2 Call on HDD Index

The payout for a Call contract on HDD index, related to the month of January, is defined as

$$Call(\text{HDD}) := \lambda \max(\text{HDD} - \text{HDD}_{strike}, 0), \tag{17}$$

where both the tick size λ, the amount of money that translates temperatures into cash flows, and HDD_{strike} are set in the contract. Often this kind of options state also a maximum payout, to reduce the risks that extreme weather conditions would cause, but we will here ignore this possibility as an option with a maximum payout can be constructed from two options without maximum payouts. The value of the contract can be obtained by computing the expectation of expression (17).

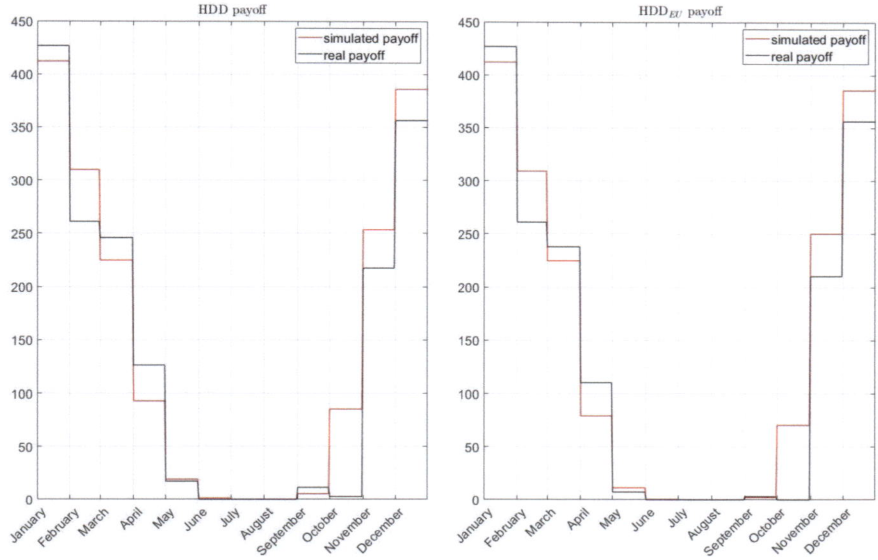

Fig. 13 Expected payoffs for call contracts lasting one month. On the left: using formula (13). On the right: using formula (15)

We set $\lambda = 1$. As written in [4], *there is no consensus on how to choose* HDD_{strike}*. However, it is a market practice to use quantiles, and particularly historical quantiles of the index, to define this strike.* However the authors used for numerical simulations, in that paper, the HDD_{strike} defined in such a way that the 90% of *simulated* HDDs are lower than HDD_{strike}. Other hints in this regard can be found in [8, 14] and in [11]. We opted for the definition of payoff in [11], that is $\text{HDD}_{strike} = 0$ in (17). The mean over 10^4 payoff simulations along the whole year, computing the HDD over each month is represented in Fig. 13, on the left using formula (13), on the right using formula (15). The use of these two different indexes produces the global behavior seen in [4] and small differences only during spring and autumn months. As for the index HDD, the discrepancy between the forecasted payoff and the real one is in the range of a few percentage points: look at Fig. 14, referred to HDD_{EU} payoffs in Fig. 13 on the right.

4 Conclusions

We have investigated the modeling of temperatures in the city of Parma aiming at pricing weather derivatives based on HDD/CDD indexes as underlying. We calibrated an Ornstein-Uhlenbeck model with seasonality from real data. The numerical experiments show the accuracy and reliability of the model selected. This research opens the door to considering more complex models like, for instance, the mean-

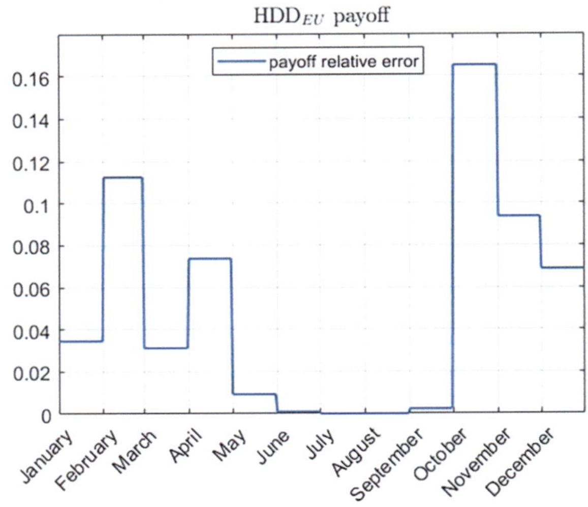

Fig. 14 Relative error of HDD$_{EU}$ payoff simulated w.r.t. HDD$_{EU}$ payoff observed

reverting model for $\tilde{T}_t$ with stochastic volatility considered in [4], as well as modern techniques of forecasting temperature based on machine learning [12, 22].

Funding

This research is part of the project *Financial instruments to offset climate risk* financed by Bank of Italy in 2024. We also acknowledge support from the Departament de Recerca i Universitats, the Departament d'Acció Climàtica, Alimentació i Agenda Rural, and the Fons Climàtic of the Generalitat de Catalunya (2023 CLIMA 00012).

References

1. Financial Stability Report - EIOPA-BoS-23/209, European Insurance and Occupational Pensions Authority (22 June 2023)
2. Alaton, P., Djehiche, B., Stillberger, D.: On modelling and pricing weather derivatives. Appl. Math. Financ. **9**(1), 1–20 (2002)
3. Albano, G., Giorno, V.: Inferring time non-homogeneous Ornstein Uhlenbeck type stochastic process. Comput. Stat. Data Anal. **150**, 1–17 (2020)
4. Alfonsi, A., Vadillo, N.: A stochastic volatility model for the valuation of temperature derivatives (2023). arXiv:2209.05918 [q-fin.RM]
5. Benth, F.E., Saltyte-Benth, J.: Stochastic modelling of temperature variations with a view towards weather derivatives. Appl. Math. Financ. **12**(1), 53–85 (2005)
6. Benth, F.E., Saltyte-Benth, J.: The volatility of temperature and pricing of weather derivatives. Quant. Financ. **7**(5), 553–561 (2007)
7. Bressan, G.M., Romagnoli, S.: Climate risks and weather derivatives: a copula-based pricing model. J. Financ. Stab. **54**, 100877 (2021)
8. Brody, D.C., Syroka, J., Zervos, M.: Dynamical pricing of weather derivatives. Quant. Financ. **2**(3), 189–198 (2002)
9. Campbell, S.D., Diebold, F.X.: Weather forecasting for weather derivatives, Technical report, University of Pennsylvania, Applied Mathematical Finance (2002)

10. CIDEA: Interdepartmental Center for Energy and Environment (2002). https://www.centritecnopolo.unipr.it/cidea. Accessed 10 June 2024
11. D'Aversa, M., Mainini, A., Moretto, E., Stefani, S., Uberti, P.: Minimizing the impact of geographical basis risk on weather derivatives. Ann. Oper. Res. (2023)
12. Fister, D., Pérez-Aracil, J., Peláez-Rodríguez, C., Del Ser, J., Salcedo-Sanz, S.: Accurate long-term air temperature prediction with machine learning models and data reduction techniques. Appl. Soft Comput. **136**, 110118 (2023)
13. ENEL (Italian National Agency for Electrical Energy) https://globaltrading.enel.com/it/prodotti-strumenti-finanziari/weather-derivati. Accessed 14 July 2024
14. Huang, H.H., Shiu, Y.M., Lin, P.S.: HDD and CDD option pricing with market price of weather risk for Taiwan. J. Futur. Mark. **28**(8), 790–814 (2008)
15. Patel, K.B.: Financial Stability Report - Number 1. Technical report, Bank of Italy (2024)
16. Patel, K.B.: Managing climate risk in mortgage markets: a role for derivatives. Letter, Chicago FED (2021)
17. Piovani, G., Torluccio, G., Cotugno, M.: Weather derivatives market and the Italian case. Eur. J. Soc. Sci. **30**, 352–365 (2012)
18. Saltyte-Benth, J., Benth, F.E.: A critical view on temperature modelling for application in weather derivatives markets. Energy Econ. **34**, 592–602 (2012)
19. Eurostat: the Statistical Office of the European Union. https://ec.europa.eu/eurostat/cache/metadata/en/nrg_chdd_esms.htm. Accessed 14 July 2024
20. Working Group I to the Sixth Assessment Report, Climate change: the physical science basis, Technical report, p. 2021. United Nations, Intergovernmental Panel on Climate Change (2021)
21. Tong, K.Z., Liu, A.: Modeling temperature and pricing weather derivatives based on subordinate Ornstein-Uhlenbeck processes. Green Financ. **2**(1), 1–19 (2020)
22. Ustaoglu, B., Cigizoglu, H.K., Karaca, M.: Forecast of daily mean, maximum and minimum temperature time series by three artificial neural network methods. Phys. A **566** (2021)
23. Witt, J., Gordon, S.: Gloomy forecast: the economy costs of extreme weather, Technical report, Barclays Research, 02 Mar 2023
24. Zanotti, G., Laboratore, D., Gabbi, G.: Climate variables and weather derivatives. Gas demand, temperature and seasonality effects in the Italian case (2003). https://papers.ssrn.com/sol3/papers.cfm?abstract_id=488745. Accessed 26 July 2024

Approximation of Singular Functions

Francesc Aràndiga

Abstract This paper is concerned with the approximation order of ENO-SR piecewise polynomial interpolatory techniques for piecewise smooth functions with isolated singularities. When the interpolatory pieces are of degree r, we propose a detection procedure based on differences of order r that allows us to prove that the global order of accuracy of the interpolatory reconstruction is improved by a factor of h relative to the linear methods, where h is the sampling rate, for isolated discontinuities in $f^{(q)}$ and $q \leq r$. In addition, we show that full accuracy is recovered for h below a critical threshold, that depends on the jump in $f^{(q)}$ at the isolated singularities.

1 Introduction

This paper is concerned with the approximation of scalar functions continuously differentiable except for an unknown singular points. There are many applications where we can find functions of this kind like finite volume methods for shock computations (see [10]). Another example is image representation where the edges correspond to discontinuities of the underlying function. In this case, the adaptive treatment of singularities results in more accurate reconstructions and therefore in sparser decompositions than when using standard wavelet basis. In recent years, Essentially Non-Oscillatory with Subcell Resolution (ENO-SR) techniques have been extended to 2D image data, either by tensor product [1, 2, 6] or by intrinsically 2D reconstructions [3, 14]. Other related nonlinear multiscale representations have been introduced in [7] in the context of the lifting scheme.

There are many results studying singular functions and detection and localization of singularities. One of the approaches is based in Fourier analysis [18]. Note that most nonlinear approximation methods that deal with local singularities are based

F. Aràndiga (✉)
Departament de Matemàtica Aplicada, Universitat de València, Burjassot, Spain
e-mail: arandiga@uv.es

© The Author(s), under exclusive license to Springer Nature Switzerland AG 2025
M. J. Ibáñez-Pérez et al. (eds.), *Mathematical And Computational Modelling, Approximation and Simulation*, SEMA SIMAI Springer Series 41,
https://doi.org/10.1007/978-3-032-01114-5_6

on either adaptive mesh refinement or wavelet thresholding (see e.g. [8] for surveys on such nonlinear approximation and [15–17]).

In [4] the authors present an algorithm such that if the functions are globally continuous and that their first-order derivatives are discontinuous it is possible to obtain a reconstruction from a finite number of function evaluations on an equispaced grid of size h that recovers the optimal approximation order, as for uniformly smooth functions (the error is bounded by $C||f^{(r)}||_{L^\infty}h^r$), for h below a critical value h_c.

The estimate given in [4] requires nontrivial discontinuity of f' and the result is of no interest for singular functions with continuous f and f' since in this case h_c would have to be zero.

In [15] the authors extend this result and they construct and nonadaptative algorithm such that for continuous and piecewise r-smooth ($r \geq 2$) functions they can obtain the best possible convergence rate h^r.

The main objective of this paper is to extend the results of [4] and to prove that with the techniques presented in [4] it is also possible to recover the convergence rate h^r when $f^{(q)}$, $2 \leq q < r$, is discontinuous.

A specific feature of the proposed approach of this paper, and also of [4], is that it does not rely on any local refinement of the sampled data: the function is accurately reconstructed from a given uniform sampling, by a locally defined data dependent operator ([5, 11, 12]).

Consider now the following situation: from a set of uniformly sampled data $(f(x_k))_{k\in\mathbb{Z}}$, where $x_k = kh$, we are interested in building an interpolant $\mathcal{I}_h f$ i.e. a function such that $\mathcal{I}_h(x_k) = f(x_k)$ for all $k \in \mathbb{Z}$. There are many ways to build an interpolant $\mathcal{I}_h f$ of a prescribed order $r > 0$, i.e. such that if $f \in \mathcal{C}^r(\mathbb{R})$ one has

$$\| \mathcal{I}_h f - f \|_{L^\infty} \leq Ch^r \sup_{x\in\mathbb{R}} |f^{(r)}(x)|.$$

Basically, one can do it with a linear operator $\mathcal{I}_h$ which is local, exact for polynomials of degree $r-1$ and stable. We are interested in the interpolation of functions $f \in \mathcal{C}^{(q-1)}(\mathbb{R})$ such that $f^{(q)}$ have jumps at isolated points. For such functions, we can only expect an error bound of order $\mathcal{O}(h^q)$ with a linear method, independently of its order ($q \leq r$).

In order to explain quickly the ENO and ENO-SR techniques, first consider a standard piecewise polynomial interpolation of the data $(f(x_k))_{k\in\mathbb{Z}}$: to each interval $I_k := [x_k, x_{k+1}]$, $k \in \mathbb{Z}$, we take the stencil S_k of size r around I_k i.e. $S_k := \{x_{k-r_1}, \ldots, x_{k+r_2}\}$, where $r_1 \geq 0$ and $r_2 > 0$ are fixed integers such that $r_1 + r_2 = r - 1$. We define a unique polynomial $p_k \in \Pi_{r-1}$ which agrees with f on S_k. A linear interpolation operator is then defined by $\mathcal{I}_h f(x) = p_k(x)$, $x \in I_k$. Since

$$f(x) - \mathcal{I}_h f(x) = f[x_{k-r_1}, \ldots, x_{k+r_2}, x](x - x_{k-r_1})\cdots(x - x_{k+r_2}),$$

if f is $\mathcal{C}^r([x_{k-r_1}, x_{k+r_2}])$, there exists $\xi \in \{x_{k-r_1}, \ldots, x_{k+r_2}, x\}$ such that $f[x_{k-r_1}, \ldots, x_{k+r_2}, x] = f^{(r)}(\xi)/r!$ and we have the estimate

$$\| f - \mathcal{I}_h f \|_{L^\infty(I_k)} \leq C h^r \| f^{(r)} \|_{L^\infty([x_{k-r_1}, x_{k+r_2}])}.$$

If $f \in \mathcal{C}^{q-1}(\mathbb{R})$ and $f^{(q)}$, $q < r$, has an isolated discontinuity, s, situated in the interval I_k we have

$$f[x_{k-r_1}, \ldots, x_{k+r_2}, x] = [f^{(q)}]_s \mathcal{O}\Big(\frac{1}{h^{r-q}}\Big)$$

and

$$f(x) - \mathcal{I}_h f(x) = [f^{(q)}]_s \mathcal{O}(h^q),$$

where $[f^{(q)}]_s = f^{(q)}(s^+) - f^{(q)}(s^-)$.

Clearly, for these functions the order of accuracy is reduced to $\mathcal{O}(h^q)$ on all the intervals I_{k+l} for $l = -r_2 + 1, \ldots, r_1$, due to the systematic use of a fixed stencil.

The principle of ENO (Essentially Non-Oscillatory) interpolation is to allow for data dependent stencils in order to reduce the influence of the singularity on the approximation. For this purpose, one introduces a measure of the oscillation of f on the stencil S_k. Since we are interested in detecting jump discontinuities in the q-derivative, this measure is based on the evaluation of the $(q+1)$-order undivided differences $\Delta_h^{q+1} f(x)$ for $x = x_{k-r_1}, \ldots, x_{k+r_2-2}$, being

$$\Delta_h^{l+1} f(x) := \Delta_h^l f(x) - \Delta_h^l f(x+h), \quad l = 1, \ldots, q,$$

and $\Delta_h^0 f(x) = f(x)$.

For each k, we select among all the stencils containing I_k, $\{S_{k-r_2+1}, \ldots, S_{k+r_1}\}$, the stencil $\tilde{S}_k$ which minimizes a chosen measure. The ENO interpolant is then given by $\mathcal{I}_h f(x) = \tilde{p}_k(x)$, $x \in I_k$, where $\tilde{p}_k$ is the polynomial which agrees with f on the stencil $\tilde{S}_k$. In comparison with the linear interpolation based on a fixed stencil, ENO interpolation has the same order of accuracy r and it reduces the effect of an isolated singularity, since the selected stencil will tend to avoid it. We therefore expect that the precision only deteriorates on the interval which contains the singularity.

The goal of the ENO-SR technique is to improve the approximation properties of the interpolant even on this interval. It is based on a detection mechanism which labels as B (bad) an interval I_k which is suspected to contain a singularity, in the sense that the selected stencils for its immediate neighbors tend to avoid it. Thus I_k is B if $\tilde{S}_{k-1} = S_{k-r_2}$ and $\tilde{S}_{k+1} = S_{k+r_1+1}$. Other intervals are labeled as G (good). On a G interval I_k, we use the above described ENO interpolation to define $\mathcal{I}_h f$. On a B interval I_k, we use the polynomials $\tilde{p}_{k-1}$ and $\tilde{p}_{k+1}$ to predict the location of the singularity. If these polynomials intersect at a single point $\tilde{s}$ of I_k, we define for $x \in I_k$ the interpolant by

$$\mathcal{I}_h f(x) = \begin{cases} \tilde{p}_{k-1}(x), \ x \leq \tilde{s}, \\ \tilde{p}_{k+1}(x), \ x > \tilde{s}. \end{cases}$$

In the case where these polynomials do not intersect at a single point of I_k, the interval is relabeled as G and the ENO interpolation is used.

An intuitive statement is that ENO-SR interpolation has accuracy of order $\mathcal{O}(h^r)$ for functions $f \in \mathcal{C}^{(q-1)}(\mathbb{R})$. Some first results suggesting the validity of this statement were given in [13] and in [9, 10] (for ENO and ENO-SR respectively) in the context of building and analyzing high order schemes for conservation laws, as well as in [6] in the context of signal and image approximation and compression. In [4] we see that this is true for $q = 1$. Our goal here is to investigate this statement in a rigorous way for any $q \geq 2$. For simplicity we consider functions which are smooth except at one unknown point s, but are globally q-continuous. We also assume that $f \in \mathcal{C}^m(\mathbb{R} \setminus \{s\})$ in the sense that its derivatives up to order m are uniformly bounded on $\mathbb{R} \setminus \{s\}$. Thus the derivatives of f have jumps $([f']_s, [f'']_s, \ldots)$ at the point s. Ideally we could hope for an estimate of the form

$$\|f - \mathcal{I}_h f\|_{L^\infty} \leq Ch^r \sup_{x \in \mathbb{R}\setminus\{s\}} |f^{(r)}(x)|, \tag{1}$$

for all $h > 0$. Unfortunately, we see in [4], with a simple example ($q = 1$), that we cannot hope for such a result for $m > 2$. In fact (1) holds for h smaller than a fixed fraction of a critical scale h_c depending itself on the function f according to

$$h_c := \frac{[f^{(q)}]_s}{2^{q+1} \sup_{x \in \mathbb{R}\setminus\{s\}} |f^{(q+1)}(x)|}.$$

This critical scale h_c corresponds to the minimal level of resolution which ensures the detection of the singularity.

Therefore we can only achieve

$$\|f - \mathcal{I}_h f\|_{L^\infty} \leq Ch^r \sup_{x \in \mathbb{R}\setminus\{s\}} |f^{(r)}(x)|, \;\; h \leq K h_c(f).$$

We shall yet prove that we have for all $h > 0$, an estimate of the form

$$\|f - \mathcal{I}_h f\|_{L^\infty} \leq Ch^{q+1} \sup_{x \in \mathbb{R}\setminus\{s\}} |f^{(q+1)}(x)|,$$

i.e. at least a gain of one order of accuracy relative to any linear method. Note that since the interpolation process is local, our analysis applies also to the case of several isolated singularities which are sufficiently separated relative to the sampling scale (typically by rh).

Our paper is organized as follows. We present in Sect. 2 a specific singularity detection mechanism together with an ENO-SR interpolation process, which slightly differs from the original ENO-SR, yet with the same basic principles, and we discuss the organization of the intervals which are detected by this mechanism. We prove in Sect. 3 that detection always occurs for $h < h_c$ and that the position of the singularity is accurately estimated. We then use these results in Sect. 4 to prove that our version of the ENO-SR interpolation technique has accuracy of order $\mathcal{O}(h^r)$ for h smaller than Kh_c where $0 < K < 1$ is a fixed constant, and that the approximation order is improved by a factor of h relative to the linear methods, which is the best that we can hope for according to the example of [4]. These findings are confirmed in Sect. 5 by numerical examples.

2 A Modified ENO-SR Detection and Interpolation Mechanism

For a given approximation order r, our detection mechanism defines a set of intervals labeled as B, which potentially contain the singularity, according to the following rules.

Assume that $f \in C^{(q-1)}(\mathbb{R})$, $[f^{(q)}]_s = f^{(q)}(s^+) - f^{(q)}(s^-) \neq 0, q < r$, and $f \in C^r(\mathbb{R} \setminus \{s\})$.

1. If
$$\min_{i=-q,\dots,-1} |\Delta_h^{q+1} f(x_{k+i})| > |\Delta_h^{q+1} f(x_{k+n})|,$$
for $n = 0, \dots, r-1$ and $n = -r-q, \dots, -q-1$ then I_k and I_{k-1} are labeled as B.
2. If
$$\min_{i=-q,\dots,0} |\Delta_h^{q+1} f(x_{k+i})| > |\Delta_h^{q+1} f(x_{k+n})|,$$
for $n = 1, \dots, r$ and $n = -r-q, \dots, -q-1$, then I_k is labeled as B.
3. All other intervals are labeled as G.

This detection mechanism is designed in such a way that for h sufficiently small, the interval I_k containing the singularity s is labeled as B, while all intervals labeled as G are in smooth regions of f. On the other hand it is also possible that an interval I_k might be labeled as B in a smooth region at an arbitrarily small scale. In case of such false alarms, it is crucial that the polynomials which are used to construct the interpolation are built from stencils which only contain G intervals, i.e. from smooth regions. This is ensured by the following lemma which describes the organization of the B and G intervals.

Lemma 2.1 The groups of adjacent B intervals are at most of size 2. They are separated by groups of adjacent G intervals which are at least of size $r-1$. □

Proof Assume that I_0 and I_k are B with $1 < k < r$. We have three cases:

1. I_0 and I_k have been labeled B by the second rule. Then it follows that $|\Delta_h^q f(0)| > |\Delta_h^q f(x_{k-1})|$ and $|\Delta_h^q f(0)| < |\Delta_h^q f(x_{k-1})|$ which is a contradiction.
2. I_0 has been labeled B by the second rule and I_k has been labeled B by the first rule. Then either I_{k-1} or I_{k+1} is also a B interval. Hence we obtain that both $|\Delta_h^q f(0)| > |\Delta_h^q f(x_u)|$ and $|\Delta_h^q f(0)| < |\Delta_h^q f(x_u)|$ for some $u \in \{k-1, k\}$, which is a contradiction. The case where I_0 has been labeled B by the first rule and I_k has been labeled B by the second rule is treated in a similar way.

3. I_0 and I_k have been labeled B by the first rule, hence each one is a member of a B-pair (two adjacent B intervals). Hence we obtain that both $|\Delta_h^q f(x_v)| > |\Delta_h^q f(x_u)|$ and $|\Delta_h^q f(x_v)| < |\Delta_h^q f(x_u)|$ for some $v \in \{-1, 0\}$ and $u \in \{k-1, k\}$, which is a contradiction.

We therefore obtain that no two B intervals can have a difference of indices strictly between 1 and r, which concludes the proof. □

Based on the above described detection mechanism, we propose the following interpolation procedure:

1. If I_k is a G interval, define $\mathcal{I}_h f$ on I_k as a polynomial p_k of degree $r-1$ obtained by interpolation of f on a stencil $\{x_v, \ldots, x_{v+r-1}\}$ such that $v \le k < k+1 \le v+r-1$ and such that this stencil only contains G intervals. Such a stencil always exists according to Lemma 2.1, yet is not unique. In practice, we may choose the stencil which is the most centered around the interval I_k or we may use the standard ENO procedure.
2. If I_k is an isolated B interval, we obtain polynomials p_k^- and p_k^+ by interpolation of f on the stencils $\{x_{k-r+1}, \ldots, x_k\}$ and $\{x_{k+1}, \ldots, x_{k+r}\}$ and use them to predict the location of the singularity. If the $q-1$ derivatives of these polynomials intersect at a single point $\tilde{s}$ of I_k, we define for $x \in I_k$ the interpolant by

$$\mathcal{I}_h f(x) = \begin{cases} p_k^-(x), & x \le \tilde{s}, \\ p_k^+(x), & x > \tilde{s}. \end{cases}$$

   In the case where these polynomials do not intersect at a single point of I_k, the interval is relabeled as G and we return to the previous case.
3. If (I_{k-1}, I_k) is a B-pair, we treat $I_{k-1} \cup I_k$ as I_k in the previous case, i.e. we obtain polynomials p_{k-1}^- and p_k^+ by interpolation of f at stencils $\{x_{k-r}, \ldots, x_{k-1}\}$ and $\{x_{k+1}, \ldots, x_{k+r}\}$ and use them to predict the location of the singularity. If the $q-1$ derivatives of these polynomials intersect at a single point $\tilde{s}$ of $I_{k-1} \cup I_k$, we define for $x \in I_{k-1} \cup I_k$ the interpolant by

$$\mathcal{I}_h f(x) = \begin{cases} p_k^-(x), & x \le \tilde{s}, \\ p_{k+1}^+(x), & x > \tilde{s}. \end{cases}$$

   In the case where these polynomials do not intersect at a single point of $I_{k-1} \cup I_k$, both intervals are relabeled as G and we return to the first case.

Note that the interpolation operator $\mathcal{I}_h f$ described above does not make use of the data at mid-points of B-pairs. Hence $\mathcal{I}_h f$ does not interpolate f at these points. This is a specific feature of the modified ENO-SR interpolation which facilitates the proof of our main approximation result in Sect. 4.

3 Properties of the Detection Mechanism.

The goal of this section is to establish some properties of the detection mechanism which shall be used in Sect. 4 for proving the improved approximation order of $\mathcal{I}_h f$ announced in the introduction.

The properties are expressed by two lemmas. The first one ensures that the singularity is always detected under some critical scale.

Lemma 3.1 Let $f \in \mathcal{C}^{(q-1)}(\mathbb{R})$ such that $[f^{(q)}]_s = f^{(q)}(s^+) - f^{(q)}(s^-) \neq 0$, $q < r$, and with a bounded $(q+1)$-derivative on $\mathbb{R} \setminus \{s\}$. We define the critical scale

$$h_c := \frac{|[f^{(q)}]_s|}{2^{q+1} \sup_{x\in\mathbb{R}\setminus\{s\}} |f^{(q+1)}(x)|}. \tag{2}$$

Then for $h < h_c$, the interval I_k which contains s is labeled as B. Moreover, there exist $0 < \hat{s}_q < h/4$ such that if $x_k \leq s < x_k + \hat{s}_q$ then the interval adjacent to this edge is also labeled as B. □

Proof Without loss of generality, we can assume that s is located on the first half of the interval I_0, i.e. $0 \leq s \leq h/2$. For $k \neq 0, \ldots, -q$, we find that

$$|\Delta_h^{q+1} f(x_k)| \leq h^{q+1} \sup_{x\in\mathbb{R}\setminus\{s\}} |f^{(q+1)}(x)|. \tag{3}$$

For $k = -q, \ldots, 0$, the $(q+1)$-order finite differences can be estimated by decomposing f into

$$f(x) = f_1(x) + f_2(x),$$

with $f_1(x) = [f^{(q)}]_s (x-s)_+^q$ and $f_2(x)$ is a $\mathcal{C}^q$ function with a bounded $q+1$ derivative on $\mathbb{R} \setminus \{s\}$, such that

$$\sup_{x\in\mathbb{R}\setminus\{s\}} |f_2^{(q+1)}(x)| = \sup_{x\in\mathbb{R}\setminus\{s\}} |f^{(q+1)}(x)|.$$

We therefore have for all $k \in \mathbb{Z}$

$$|\Delta_h^{q+1} f_2(x_k)| \leq h^{q+1} \sup_{x\in\mathbb{R}\setminus\{s\}} |f^{(q+1)}(x)|.$$

Since

$$|\Delta_h^{q+1} f_1(-x_q)| = |(h-s)^q [f^{(q)}]_s|,$$

it follows that

$$|\Delta_h^{q+1} f(-x_q)| \geq |(h-s)^r [f^{(q)}]_s| - h^{q+1} \sup_{x\in\mathbb{R}\setminus\{s\}} |f^{(q+1)}(x)|, \tag{4}$$

and therefore

$$|\Delta_h^{q+1} f(-x_q)| \geq \frac{h^q}{2^q}|[f^{(q)}]_s| - h^{q+1} \sup_{x\in\mathbb{R}\setminus\{s\}} |f^{(q+1)}(x)|.$$

So if $h < h_c$ we get by (2)

$$|\Delta_h^{q+1} f(-x_q)| > h^{q+1} \sup_{x\in\mathbb{R}\setminus\{s\}} |f^{(q+1)}(x)|.$$

Combining this with (3), we find that if $h < h_c$

$$|\Delta_h^{q+1} f(-x_q)| > |\Delta_h^{q+1} f(x_k)|,$$

for $k \neq -q, \ldots, 0$.

For $l = 1, \ldots, q-1$ we also have $|\Delta_h^{q+1} f(-x_l)| \geq \frac{h^q}{2^q}|[f^{(q)}]_s|$ and

$$|\Delta_h^{q+1} f(-x_l)| > |\Delta_h^{q+1} f(x_k)|,$$

for $k \neq -q, \ldots, 0$ if $h < h_c$.

In the case where $\min_{l=1,\ldots,q} |\Delta_h^{q+1} f(-x_l)| > |\Delta_h^{q+1} f(0)|$, we find that I_{-1} and I_0 are a B-pair according to the first detection rule. Otherwise, if $\min_{l=1,\ldots,q} |\Delta_h^{q+1} f(-x_l)| < |\Delta_h^{q+1} f(0)|$, we find that I_0 must be labeled as B according to the first detection rule.

We finally notice that

$$|\Delta_h^{q+1} f_1(0)| = s^r|[f^{(q)}]_s|,$$

so that

$$|\Delta_h^{q+1} f(0)| \leq s^q|[f^{(q)}]_s| + h^{q+1} \sup_{x\in\mathbb{R}\setminus\{s\}} |f^{(q+1)}(x)|. \tag{5}$$

Therefore, by (5) and (4), we have

$$|\Delta_h^{q+1} f(0)| \leq |\Delta_h^{q+1} f(-x_q)| \tag{6}$$

if

$$2h^{q+1} \sup_{x\in\mathbb{R}\setminus\{s\}} |f^{(q+1)}(x)| \leq ((h-s)^q - s^q)|[f^{(q)}]_s|. \tag{7}$$

Since $h < h_c$ we have

$$2h^{q+1} \sup_{x\in\mathbb{R}\setminus\{s\}} |f^{(q+1)}(x)| \leq \frac{h^q}{2^q}|[f^{(q)}]_s|.$$

Let $\hat{s}_q$ the value such that if $s < \hat{s}_q$ we have

$$\frac{h^q}{2^q} \leq ((h-s)^q - s^q). \tag{8}$$

Therefore, when $h < h_c$ and $s < \hat{s}_q$, we have (8), (7) and (6). Hence I_{-1} and I_0 is B-pair. □

The next lemma expresses the fact that the location of the singularity is accurately estimated when h is less than a fixed fraction of the critical scale.

Lemma 3.2 There exist constants $C > 0$ and $0 < K < 1$ such that for all $f \in \mathcal{C}^{(q-1)}(\mathbb{R})$ such that $[f^{(q)}]_s \neq 0$ and with a bounded $q+1$ derivative on $\mathbb{R} \setminus \{s\}$ and for $h < Kh_c$ with h_c defined by (2), the following holds:

1. The singularity s is contained in an isolated B interval I_k (case 1) or in a B-pair (I_k, I_{k+1}) (case 2).
2. The two polynomials (p_k^-, p_k^+) (case 1) or (p_k^-, p_{k+1}^+) (case 2) which are used in the definition of $\mathcal{I}_h f$ have only one intersection point $\tilde{s}$ inside I_k (case 1) or inside $I_k \cup I_{k+1}$ (case 2).
3. The distance between s and $\tilde{s}$ is bounded by

$$|s - \tilde{s}| \leq C \frac{h^{r-q+1} \sup_{x \in \mathbb{R} \setminus \{s\}} |f^{(r)}(x)|}{|[f^{(q)}]_s|} \tag{9}$$

Proof Assume that s is located on the first half of I_0. Since $K < 1$, the first statement has already been proved in Lemma 3.1. Without loss of generality, we assume that $0 \leq s \leq \hat{s}_q$. In this case we know by Lemma 3.1 that I_0 is B for $h < h_c$. For the sake of notational simplicity we denote by $I = [b, c]$ the interval where we do the subcell resolution process, which is either I_0 (case 1) or $I_{-1} \cup I_0$ (case 2) or $I_0 \cup I_1$ (case 2). By Lemma 3.1, we are ensured that $I = I_{-1} \cup I_0$ when $s < \hat{s}_q$, and therefore

$$\min\{|s - b|, |s - c|\} \geq \hat{s}_q \tag{10}$$

We also denote by (p_-, p_+) the polynomials which are used in the subcell resolution of I. Finally, we note that for any $q+1 \leq k \leq r$ we can write $f(x) = f_-(x) * (x \leq s) + f_+(x) * (x > s)$, where f_- and f_+ are functions which are globally $\mathcal{C}^k$ over $\mathbb{R}$ and such that

$$\sup_{x \in \mathbb{R}} |f_\pm^{(k)}(x)| \leq \sup_{x \in \mathbb{R} \setminus \{s\}} |f^{(k)}(x)|.$$

For example, we can define these functions by extension of f using its left or right Taylor expansion of order k at the point s. That is

$$f_-(x) = \begin{cases} f(x), & x \leq s, \\ f(s^-) + f'(s^-)(x-s) + \ldots + \frac{f^{k)}(s^-)}{k!}(x-s)^k, & x > s, \end{cases}$$

$$f_+(x) = \begin{cases} f(x), & x \geq s, \\ f(s^+) + f'(s^+)(x-s) + \cdots + \frac{f^{k)}(s^+)}{k!}(x-s)^k, & x < s. \end{cases}$$

In order to prove the second statement of the theorem, we choose $k = q + 1$. We note that p_- and p_+ can also be viewed as Lagrange interpolation of f_- and f_+. It then follows from classical results on Lagrange interpolation that there exists a constant D independent of f such that for all $t \in I$,

$$|f_\pm(t) - p_\pm(t)| \leq Dh^{q+1} \sup_{x\in\mathbb{R}} |f_\pm^{(q+1)}(x)| = Dh^{q+1} \sup_{x\in\mathbb{R}\setminus\{s\}} |f^{(q+1)}(x)|, \tag{11}$$

$$|f_\pm'(t) - p_\pm'(t)| \leq Dh^{q} \sup_{x\in\mathbb{R}} |f_\pm^{(q+1)}(x)| = Dh^{q} \sup_{x\in\mathbb{R}\setminus\{s\}} |f^{(q+1)}(x)|, \tag{12}$$

and

$$|f_\pm^{(q)}(t) - p_\pm^{(q)}(t)| \leq Dh \sup_{x\in\mathbb{R}} |f_\pm^{(q+1)}(x)| = Dh \sup_{x\in\mathbb{R}\setminus\{s\}} |f^{(q+1)}(x)|. \tag{13}$$

Since $|t - s| \leq 2h$ when $t \in I$, we also have

$$|f_\pm^{(q)}(t) - f_\pm^{(q)}(s)| \leq 2h \sup_{x\in\mathbb{R}\setminus\{s\}} |f^{(q+1)}(x)|,$$

therefore we get from (12)

$$|f_\pm^{(q)}(s) - p_\pm^{(q)}(t)| \leq (D+2)h \sup_{x\in\mathbb{R}\setminus\{s\}} |f^{(q+1)}(x)|, \quad t \in I. \tag{14}$$

It follows that for all $t \in I$,

$$|p_+^{(q)}(t) - p_-^{(q)}(t)| \geq |[f^{(q)}]_s| - 2(D+2)h \sup_{x\in\mathbb{R}\setminus\{s\}} |f^{(q+1)}(x)|. \tag{15}$$

Thus, for $h < \frac{4}{D+2}h_c$ the function $p_+^{(q-1)} - p_-^{(q-1)}$ is strictly monotone on I and has at most one root. Therefore, we are ensured that $p_+^{(q-1)}$ and $p_-^{(q-1)}$ intersect at most at a single point inside I. In order to prove that this point $\tilde{s}$ exists, we need to show that $p_+^{(q-1)} - p_-^{(q-1)}$ has a sign change inside I. Without loss of generality, assume here that $[f^{(q)}]_s > 0$. By Taylor expansion at the point s, we find that

$$(f_+^{(q-1)} - f_-^{(q-1)})(b) \leq -(s-b)[f^{(q)}]_s + (s-b)^2 \sup_{x\in\mathbb{R}\setminus\{s\}} |f^{(q+1)}(x)|,$$

and

$$(f_+^{(q-1)} - f_-^{(q-1)})(c) \geq (c-s)[f^{(q)}]_s - (c-s)^2 \sup_{x\in\mathbb{R}\setminus\{s\}} |f^{(q-1)}(x)|.$$

Combining with (11), we thus obtain

$$(p_+^{(q-1)} - p_-^{(q-1)})(b) \le -(s-b)[f^{(q)}]_s + ((s-b)^2 + 2Dh^2) \sup_{x\in\mathbb{R}\setminus\{s\}} |f^{(q+1)}(x)|,$$

and

$$(p_+^{(q-1)} - p_-^{(q-1)})(c) \ge (c-s)[f^{(q)}]_s - ((c-s)^2 + 2Dh^2) \sup_{x\in\mathbb{R}\setminus\{s\}} |f^{(q+1)}(x)|.$$

Using (10), we therefore obtain,

$$(p_+^{(q-1)} - p_-^{(q-1)})(b) \le -\hat{s}_q[f^{(q)}]_s + (2D+4)h^2 \sup_{x\in\mathbb{R}\setminus\{s\}} |f^{(q+1)}(x)|,$$

and

$$(p_+^{(q-1)} - p_-^{(q-1)})(c) \ge \hat{s}_q[f^{(q)}]_s - (2D+4)h^2 \sup_{x\in\mathbb{R}\setminus\{s\}} |f^{(q+1)}(x)|.$$

It follows that for $h < k_q h_c$, where $k_1 = \frac{1}{2D+4}$ and $k_r = \frac{1}{D+2}$, $q > 1$, we have

$$(p_+^{(q-1)} - p_-^{(q-1)})(b) \le -\hat{b}_r[f^{(q)}]_s < 0,$$

and

$$(p_+^{(q-1)} - p_-^{(q-1)})(c) \ge \hat{b}_q[f^{(q)}]_s > 0,$$

where $\hat{b}_q = \hat{s}_q - h/2^q$ if $q > 1$ and $\hat{b}_1 = h/8$ so that there exists a single intersection point $\tilde{s} \in I$. So defining $K := K_q$, we have proved the two first statements of the theorem.

In order to prove the third statement (9), we now choose $k = r$ in the definition of the extensions f_+ and f_-. It again follows from classical results on Lagrange interpolation that there exists a constant $\tilde{D}$ such that for all $t \in I$,

$$|f_\pm^{(q-1)}(t) - p_\pm^{(q-1)}(t)| \le \tilde{D}h^{r-q+1} \sup_{x\in\mathbb{R}\setminus\{s\}} |f^{(r)}(x)|,$$

and therefore, if we define $g = f_+ - f_-$ and $q = p_+ - p_-$, we obtain for all $t \in I$,

$$|g^{(q-1)}(t) - q^{(q-1)}(t)| \le 2\tilde{D}h^{r-q+1} \sup_{x\in\mathbb{R}\setminus\{s\}} |f^{(r)}(x)|, \qquad (16)$$

so that for $t = s$, we have $g^{(q-1)}(s) = 0$ and

$$|q^{(q-1)}(s)| \le 2\tilde{D}h^{r-q+1} \sup_{x\in\mathbb{R}\setminus\{s\}} |f^{(r)}(x)|. \qquad (17)$$

Now, note that for $t \in I$ and $h < Kh_c$, by (15)

$$|q^{(q)}(t)| = \frac{|[f^{(q)}]_s|}{2},$$

and therefore

$$|q^{(q-1)}(s)| = |q^{(q-1)}(\tilde{s}) - q^{(q-1)}(a)| \geq |\tilde{s} - s||[f^{(q)}]_s|/2.$$

Combining this with (17), we therefore obtain (9) with $C = 4\tilde{D}$. □

4 Approximation Properties of $\mathcal{I}_h$

We are now ready to derive our main approximation result.

Theorem 4.1 For all $f \in \mathcal{C}^{(q-1)}(\mathbb{R})$ such that $[f^{(q)}]_s \neq 0$ and with a bounded $q+1$ derivative on $\mathbb{R} \setminus \{s\}$ the nonlinear interpolant $\mathcal{I}_h f$ satisfies

$$\|f - \mathcal{I}_h f\|_{L^\infty} \leq Ch^{q+1} \sup_{x \in \mathbb{R}\setminus\{s\}} |f^{(q+1)}(x)|,$$

for all $h > 0$, with $C > 0$ independent of f. Moreover there exists $0 < K < 1$ independent of f such that for $h < Kh_c$ with h_c defined by (2), we have

$$\|f - \mathcal{I}_h f\|_{L^\infty} \leq Ch^r \sup_{x \in \mathbb{R}\setminus\{s\}} |f^{(r)}(x)|.$$

Proof We choose for K the constant in Lemma 3.2. Note first that for $h < Kh_c$, according to Lemmas 2.1 and 3.1, all the polynomials which are used in the construction of $\mathcal{I}_h$ are built from stencils over which the function is smooth. It follows from classical results on Lagrange interpolation that the estimate

$$|f(x) - \mathcal{I}_h f(x)| \leq Ch^r \sup_{x \in \mathbb{R}\setminus\{s\}} |f^{(r)}(x)|,$$

holds whenever x belongs to a G interval or to an isolated B interval or B-pair which does not contain s (i.e. false alarms do not deteriorate the convergence rate). Let us now assume that x belongs to the group of adjacent B intervals which contains s. Here, we shall assume again without loss of generality that $0 \leq s \leq h/2$ and use the notations $I = [b, c]$, p_+, p_-, f_+, f_- that were introduced in the proof of Lemma 3.2. We also assume that $s \leq \tilde{s}$, the case $\tilde{s} \leq s$ being treated in a similar way. For $x \in [b, s]$, we have the estimate

$$|f(x) - \mathcal{I}_h f(x)| = |f_-(x) - p_-(x)| \leq Ch^r \sup_{x \in \mathbb{R}\setminus\{s\}} |f^{(r)}(x)|.$$

and for $x \in [\tilde{s}, c]$, we have the estimate

$$|f(x) - \mathcal{I}_h f(x)| = |f_+(x) - p_+(x)| \leq Ch^r \sup_{x \in \mathbb{R} \setminus \{s\}} |f^{(r)}(x)|.$$

It remains to consider the case $s < x < \tilde{s}$. In this case, we have

$$|f(x) - \mathcal{I}_h f(x)| = |f_+(x) - p_-(x)| \leq |f_+(x) - f_-(x)| + |f_-(x) - p_-(x)|.$$

The second term is again bounded by $Ch^r \sup_{x \in \mathbb{R} \setminus \{s\}} |f^{(r)}(x)|$. For the first term, we use $(q+1)$-order Taylor expansion to derive

$$\begin{aligned}|f_+(x) - f_-(x)| &\leq \tfrac{1}{q!}|[f^{(q)}]_s|(\tilde{s} - s)^q + \tfrac{2}{(q+1)!}(\tilde{s} - s)^{(q+1)} \sup_{x \in \mathbb{R} \setminus \{s\}} |f^{(q+1)}(x)| \\ &\leq (\tilde{s} - s)h^{q-1}\Big(\tfrac{1}{q!}|[f^{(q)}]_s| + \tfrac{2}{(q+1)!}h \sup_{x \in \mathbb{R} \setminus \{s\}} |f^{(r+1)}(x)|\Big).\end{aligned}$$

Since $h < h_c$, this gives

$$|f_+(x) - f_-(x)| \leq \Big(\frac{1}{q!} + \frac{1}{(q+1)!\,2^q}\Big)h^{q-1}|[f^{(q)}]_s|(\tilde{s} - s).$$

Combining this with the estimate (9) of Lemma 3.2, we also obtain the bound $Ch^r \sup_{x \in \mathbb{R} \setminus \{s\}} |f^{(r)}(x)|$ for $|f_+(x) - f_-(x)|$ which concludes the proof in the case $h < Kh_c$.

In the case $h \geq Kh_c$, we are ensured of the estimate

$$|f(x) - \mathcal{I}_h f(x)| \leq Ch^r \sup_{x \in \mathbb{R} \setminus \{s\}} |f^{(r)}(x)|,$$

only when x is at distance at least $(r+1)h$ from s. We also have the lower order estimate

$$|f(x) - \mathcal{I}_h f(x)| \leq Ch^{q+1} \sup_{x \in \mathbb{R} \setminus \{s\}} |f^{(q+1)}(x)|.$$

Let us now prove that this estimate remains valid if $|x - s| \leq (r+1)h$. For this purpose we consider the decomposition $f = f_1 + f_2$ used in the proof of Lemma 3.1. The errors of polynomial interpolation of f_1 and f_2 are respectively bounded by $Ch^q|[f^{(q)}]_s|$ and $Ch^{q+1}\sup_{x \in \mathbb{R} \setminus \{s\}} |f^{(q+1)}(x)|$. Since $h \geq Kh_c$ the second bound dominates the first one so that the above estimate is valid. The proof of the theorem is now complete. □

5 Numerical Examples

We consider the following function

$$f_{r,\varepsilon}(x) = \begin{cases} (x - \frac{\pi}{6})^q (x - \frac{\pi}{6} - \varepsilon(q+1)2^{q+1}) + \frac{\sin(\frac{\pi x}{8})}{8}, & 0 \le x < \frac{\pi}{6}, \\ \frac{\sin(\frac{\pi x}{8})}{8}, & \frac{\pi}{6} \le x \le 1, \end{cases}$$

for different values of ε $(2^{-8}, 2^{-9}, 2^{-10}, 2^{-11})$ and $q = 1, 2$. Each of these functions is globally continuous with a jump of ε in its q-derivative at the point $\pi/6$, while its higher derivatives are uniformly bounded independently of ε. The functions $f_{q,\varepsilon}(x)$ are piecewise polynomials to which we add the smooth function $\sin(\pi x/8)/8$ in order to avoid a 0 approximation error for h less than a critical scale.

For different values of $h = 2^{-n}$ we compute the lineal reconstruction with $m = 4$, $\mathcal{I}_h^{LIN} f_{q,\varepsilon}$, for the different values of ε. In Table 1 we give the corresponding values of $\| f_{q,\varepsilon} - \mathcal{I}_h^{LIN} f_{q,\varepsilon} \|_{L^\infty}$.

Now, we apply our technique with $r = 4$ and we compute $\mathcal{I}_h^{SR} f_{1,\varepsilon}$. We apply the detection mechanism of Sect. 2 for $q = 1, 2, \ldots, r-1$ until the discontinuity is detected (we are assuming that q is unknown). In Table 2 we give the corresponding values of $\| f_{1,\varepsilon} - \mathcal{I}_h^{SR} f_{1,\varepsilon} \|_{L^\infty}$ and in Fig. 1, $log(\| f_{1,\varepsilon} - I_h^{SR} f_{1,\varepsilon} \|_{L^\infty})$ versus $-log(h)$.

We observe that we have accuracy of order $\mathcal{O}(h^2)$ when $h >> \varepsilon$ and that for h smaller than some h_ε the discontinuity has been found and we recover the optimal accuracy of order $\mathcal{O}(h^4)$. This is ensured by Theorem 4.1 for $h < K h_c$ where

$$h_c := \frac{|[f^{(q)}]_s|}{2^{q+1} \sup_{x \in \mathbb{R} \setminus \{s\}} |f^{(q+1)}(x)|} \approx \varepsilon \tag{18}$$

and $0 < K < 1$. However we observe that $h_\varepsilon > \varepsilon$, which means that the order of accuracy $\mathcal{O}(h^4)$ is attained even for some $h > K h_c$.

Table 1 $\| f_{1,\varepsilon} - I_h^{LIN} f_{1,\varepsilon} \|_{L^\infty}$ for different values of ε and different values of h

$\varepsilon \setminus h$	2^{-4}	2^{-5}	2^{-6}	2^{-7}	2^{-8}	2^{-9}
2^{-9}	4.09e-4	1.01e-4	4.54e-5	1.03e-5	3.57e-6	1.55e-6
2^{-10}	1.25e-4	7.83e-5	2.32e-5	7.14e-6	2.30e-6	6.74e-7

Table 2 $\| f_{1,\varepsilon} - I_h^{SR} f_{1,\varepsilon} \|_{L^\infty}$ for different values of ε and different values of h

$\varepsilon \setminus h$	2^{-4}	2^{-5}	2^{-6}	2^{-7}	2^{-8}	2^{-9}
2^{-9}	4.09e-4	1.01e-4	2.27e-5	2.19e-12	1.39e-13	8.90e-15
2^{-10}	3.37e-4	7.83e-5	2.32e-5	4.04e-6	1.39e-13	8.92e-15

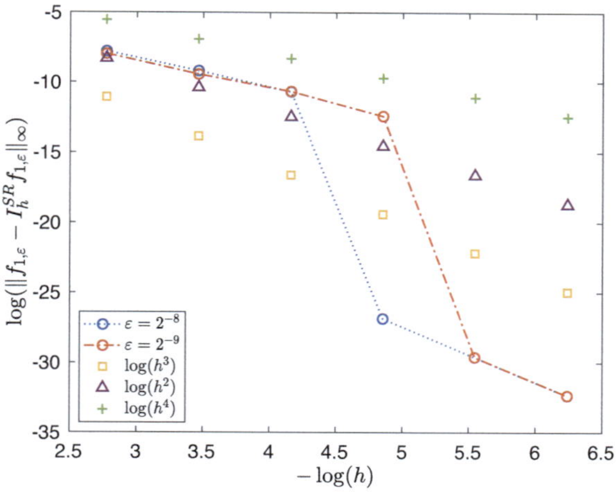

Fig. 1 Dot-dot and dash-dot lines: $log(\| f_{1,\varepsilon} - I_h^{SR} f_{1,\varepsilon} \|_\infty)$ versus $-log(h)$ for the different values of ε (2^{-9} and 2^{-10}). Diamonts: $log(h^4)$ versus $-log(h)$; Triangle: $log(h^3)$ versus $-log(h)$ and Plus: $log(h^2)$ versus $-log(h)$

Table 3 $\| f_{2,\varepsilon} - I_h^{LIN} f_{2,\varepsilon} \|_{L^\infty}$ for different values of ε and different values of h

$\varepsilon \setminus h$	2^{-4}	2^{-5}	2^{-6}	2^{-7}	2^{-8}	2^{-9}
2^{-8}	4.68e-5	4.19e-6	8.33e-7	4.09e-7	9.98e-8	2.54e-8
2^{-9}	3.95e-5	2.52e-6	6.47e-7	2.21e-7	5.20e-8	1.30e-8

Table 4 $\| f_{2,\varepsilon} - I_h^{SR} f_{2,\varepsilon} \|_{L^\infty}$ for different values of ε and different values of h

$\varepsilon \setminus h$	2^{-4}	2^{-5}	2^{-6}	2^{-7}	2^{-8}	2^{-9}
2^{-8}	4.68e-5	4.19e-6	8.27e-11	2.19e-12	1.39e-13	8.40e-15
2^{-9}	3.95e-5	5.04e-6	6.47e-7	2.19e-12	1.39e-13	8.38e-15

Now, we repeat the experiments for a fixed but unknown $q = 2$ (see Tables 3, 4 and Fig. 2).

We can see (Table 2) that we have accuracy of order $\mathcal{O}(h^3)$ when $h >> \varepsilon$ and that when h is smaller than some h_ε the discontinuity has been found and we recover the optimal accuracy of order $\mathcal{O}(h^4)$.

6 Conclusions

In this paper we have presented a uniform approximation of continuous an piecewise r-smooth ($r \geq 2$) functions with a finite number of isolated singular points. We have used a class of nonlinear interpolation procedures for uniformly sampled univariate data based on Essentially Non-Oscillatory with Subcell Resolution (ENO-SR) reconstruction techniques.

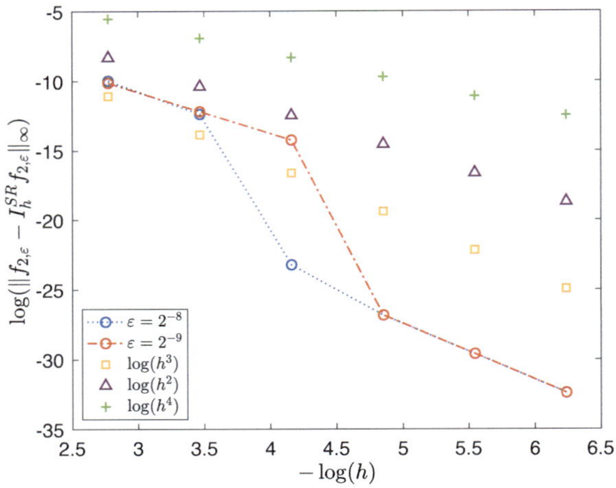

Fig. 2 Dot-dot and dash-dot lines: $log(\| f_{1,\varepsilon} - I_h^{SR} f_{1,\varepsilon} \|_\infty)$ versus $-log(h)$ for the different values of ε (2^{-8} and 2^{-9}). Diamonts: $log(h^4)$ versus $-log(h)$; Triangle: $log(h^3)$ versus $-log(h)$ and Plus: $log(h^2)$ versus $-log(h)$

We have proved that detection always occurs for $h < h_c$ and that the position of the singularity is accurately estimated. We also have seen that our version of the ENO-SR interpolation technique has accuracy of order $\mathcal{O}(h^r)$ for h smaller than Kh_c where $0 < K < 1$ is a fixed constant, and that the approximation order is improved by a factor of h relative to the linear methods, which is the best that we can hope for according to the example of [4]. These findings have been confirmed in Sect. 5 by numerical examples.

References

1. Amat, S., Aràndiga, F., Cohen, A., Donat, R.: Tensor product multiresolution analysis with error control for compact image representation. Signal Process. **4**, 587–608 (2002)
2. Amat, S., Aràndiga, F., Cohen, A., Donat, R., Garcia, G., von Oehsen, M.: Data compression with ENO schemes: a case study. Appl. Comput. Harmon. Anal. **11**(2), 273–288 (2001)
3. Aràndiga, F., Cohen, A., Doblas, M., Matei, B.: Edge adapted nonlinear multiscale transform for compact image representation. In: Proceedings of the 2003 IEEE International Conference Image Processing, vol. 1, pp. 701–704 (2003)
4. Aràndiga, F., Cohen, A., Donat, R., Dyn, N.: Interpolation and approximation of piecewise smooth functions. SIAM J. Numer. Anal. **43**(1), 41–57 (2005)
5. Aràndiga, F., Donat, R.: Nonlinear multiscale decompositions: the approach of A. Harten. Numer. Algorithms **23**(2–3), 175–216 (2000)
6. Chan, T., Zhou, H.: ENO-wavelet transforms for piecewise smooth functions. SIAM J. Numer. Anal. **40**(4), 1369–1404 (2002)
7. Claypoole, R.L., Davis, G., Sweldens, W., Baraniuk, R.: Nonlinear wavelet transforms for image coding via lifting scheme. IEEE Trans. Image Process. **12**(12), 1449–1459 (2003)
8. DeVore, R.: Nonlinear approximation. Acta Numer. **7**, 51–150 (1998)
9. Donat, R.: Studies on error propagation for certain nonlinear approximations to hyperbolic equations: discontinuities in derivative. SIAM J. Numer. Anal. **31**, 655–679 (1994)
10. Harten, A.: ENO schemes with sub cell resolution. J. Comput. Phys. **83**, 148–184 (1989)

11. Harten, A.: Discrete multiresolution analysis and generalized wavelets. J. Appl. Num. Math. **12**, 153–193 (1993)
12. Harten, A.: Multiresolution representation of data: a general framework. SIAM J. Numer. Anal. **33**, 1205–1256 (1996)
13. Harten, A., Engquist, B., Osher, S., Chakravarthy, S.R.: Uniformly high-order accurate essentially nonoscillatory schemes III. J. Comput. Phys. **71**(2), 231–303 (1987)
14. Matei, B.: Métodes multirésolution non-linéaires- applications au traitement d'image. Ph.D. thesis, Université Pierre et Marie Curie, Paris (2002)
15. Plaskota, L., Wasilkowski, G.W.: Uniform approximation of piecewise r-smooth and globally continuous functions. SIAM J. Numer. Anal. **47**(1), 762–785 (2009)
16. Plaskota, L., Wasilkowski, G.W.: The power of adaptive algorithms for functions with singularities. J. Fixed Point Theory Appl. **6**(2), 227–248 (2009)
17. Plaskota, L., Wasilkowski, G.W., Zhao, Y.: The power of adaption for approximating functions with singularities. Math. Comp. **77**(264), 2309–2338 (2008)
18. Vetterli, M., Marziliano, P., Blu, T.: Sampling signals with finite rate of innovation. IEEE Trans. Signal Proc. **50**(6), 1417–1428 (2002)

Powell-Sabin Finite Element Point Interpolation via the Constrained Mock-Chebyshev Least Squares Operator

Domingo Barrera, Francesco Dell'Accio, Salah Eddargani, and Federico Nudo

Abstract Splines over triangulation are a fundamental tool in finite element analysis (FEM), due to their ability to perform local refinements and accurately represent complex geometries. An important example is the C^1 quadratic splines on the Powell-Sabin (PS) 6-split triangulations. This spline space is characterized by specifying discrete values and first derivative values at the vertices of the triangulation on which it is defined. However, a common challenge arises in many applications where only functional evaluations are known at equally spaced nodes, making the direct application of the Powell-Sabin finite element method impractical. To address this limitation in this setting, we present a novel technique for approximating derivative values at the set of vertices, essential for effectively defining the Powell-Sabin finite element. The idea of this technique lies in leveraging the provided functional evaluations and approximating the partial derivatives at the data points using an interpolation-regression operator.

D. Barrera · S. Eddargani
Department of Applied Mathematics and Institute of Mathematics (IMAG), University of Granada, 18071 Granada, Spain
e-mail: dbarrera@uge.es

S. Eddargani
e-mail: seddargani@ugr.es

F. Dell'Accio (✉) · F. Nudo
Department of Mathematics and Computer Science, University of Calabria, Rende, Italy
e-mail: francesco.dellaccio@unical.it

F. Nudo
e-mail: federico.nudo@unical.it

F. Dell'Accio
Istituto per le Applicazioni del Calcolo "Mauro Picone", National Research Council of Italy, Napoli, Italy

© The Author(s), under exclusive license to Springer Nature Switzerland AG 2025
M. J. Ibáñez-Pérez et al. (eds.), *Mathematical And Computational Modelling, Approximation and Simulation*, SEMA SIMAI Springer Series 41,
https://doi.org/10.1007/978-3-032-01114-5_7

1 Introduction

The finite element method is a widely used tool for numerically solving partial differential equations encountered in engineering and mathematical modeling, over domains $\Omega \subset \mathbb{R}^d$, where $d \geq 1$ [1, 2]. Its widespread adoption can be attributed, in part, to its adaptability to different geometries. In this method, the domain $\bar{\Omega}$ is partitioned into polytopes, with each polytope providing a local approximation of the differential solution. The global approximation is then defined as a piecewise function, composed of these local approximations [3]. This partitioning of the domain into polytopes aligns with the fundamental approach of splines over triangulation, which are renowned for their ability to adeptly represent intricate geometries, thereby facilitating precise numerical simulations. A well-known technique used to improve the accuracy of the finite element method is the Enriched Finite Element Method (EFEM). This technique enhances the standard finite element approach by incorporating additional functions that capture complex features of the solution, such as singularities or discontinuities, thereby improving both accuracy and convergence in numerical simulations [4–8]. Among the most prominent finite elements is the Powell-Sabin finite element, introduced in [9]. It aims to construct quadratic splines that are C^1 continuous on a general triangulation, thus reducing the gap between spline degree and smoothness. In particular, achieving C^r smooth splines on such triangulations requires using degrees greater than or equal to $4r + 1$ or $3r + 2$ [10, 11]. However, this gap can be mitigated through the Powell-Sabin macro-element split, which consists of dividing each triangle into six small micro-triangles, enabling the attainment of C^r smooth splines of degree $3r - 1$. Since its inception, the C^1 quadratic spline space of the Powell-Sabin 6-split has attracted significant interest. For instance, Paul Dierckx proposed an elegant geometric approach in [12] to construct a normalized B-spline basis, which later became a primary method for obtaining B-spline-like bases for high-degree splines on the Powell-Sabin split. Various quasi-interpolating schemes were also proposed in [13–15]; and a Gaussian quadrature formula was studied in [16]. Additionally, blossoming was employed to construct quadratic quasi-interpolants on the Powell-Sabin split [17]. Extensive discussions on high-degree splines on the Powell-Sabin split are found in the literature, including studies on C^1-smooth cubic splines [18–23], almost C^2 smooth splines of degree four [24], geometric approaches characterizing triangulations for constructing full C^2 smooth splines of degree four [25], C^2 smooth splines of degree five [26–28], and degree six [29]. General studies are also discussed [30–32].

In this work, we focus on the C^1 continuous quadratic space, characterized by both the value and the first derivative values at each vertex of the initial triangulation. This defines the space as a solution to a Hermite interpolation problem. However, in many application areas, information on the derivative values is often missing. This necessitates the use of new techniques to accurately approximate the derivative values needed for the full prescription of the Powell-Sabin quadratic space. To address this issue, we consider the use of the mock-Chebyshev least squares operator. This approximation operator, presented in [33] for the univariate case and subsequently generalized to the multivariate case in [34], is a polynomial approximation operator based on equally spaced points on $[-1, 1]^d$ introduced to mitigate the Runge phenomenon. The idea involves interpolating the unknown function only on the subset of nodes near the well-known Chebyshev-Lobatto nodes, to leverage the optimality of the Lebesgue constant on these nodes. The remaining nodes are then employed for simultaneous regression. Hence, this operator is commonly referred to as the interpolation-regression operator. Notably, this approximation operator has recently found applications in various contexts. It has been used to determine stable quadrature formulas with a high degree of exactness [36–38], solve Fredholm integral equations [39], and derive numerical differentiation formulas [40]. Due to its outstanding accuracy characteristics, this paper aims to leverage the constrained mock-Chebyshev least squares operator for approximating the partial derivatives of the unknown function. Subsequently, we intend to employ this derivative approximation to approximate the Powell-Sabin finite element method on a regular Cartesian grid of points.

The organization of the paper is outlined below. In Sect. 2, we begin by providing a review of fundamental definitions and properties concerning bivariate polynomials on triangles, as well as the C^1 quadratic spline space on Powell-Sabin 6-split triangulations. Section 3 briefly revisits the properties of the constrained mock-Chebyshev least squares approximation. Additionally, we use this approximation to derive a numerical differentiation formula for partial derivatives, which is then employed to introduce an approximation of the Powell-Sabin finite element method. In Sect. 4, we conduct numerical experiments to assess the accuracy of the proposed method using various examples based on data derived from continuous functions.

2 Powell-Sabin Finite Element

2.1 *Bivariate Quadratic Polynomials on Triangles*

In the analysis of a spline over triangulations, it is often essential to understand its polynomial behaviour within a single triangle. An intuitive geometric method of expressing this polynomial is through the Bernstein-Bézier representation, using the Bernstein polynomials.

Let T be a triangle with vertices $\mathbf{p}_1, \mathbf{p}_2, \mathbf{p}_3$ in $\mathbb{R}^2$ with a non-zero signed area, that is $T = \langle \mathbf{p}_1, \mathbf{p}_2, \mathbf{p}_3 \rangle$. Every point $\mathbf{p}$ in T can be expressed in terms of its barycentric coordinates with respect to T, as $\mathbf{p} = \tau_1 \mathbf{p}_1 + \tau_2 \mathbf{p}_2 + \tau_3 \mathbf{p}_3$, where $\tau_1 + \tau_2 + \tau_3 = 1$. The quadratic Bernstein polynomials $\mathcal{B}_{i,j,k}$, $i, j, k \in \mathbb{N}_0$, $i + j + k = 2$, on T are defined as

$$\mathcal{B}_{i,j,k}(\mathbf{p}) := \frac{2!}{i!j!k!} \tau_1^i \, \tau_2^j \, \tau_3^k. \tag{1}$$

The quadratic Bernstein polynomials $\left\{\mathcal{B}_{i,j,k}, \quad i + j + k = 2\right\}$ form a basis of the linear space of quadratic polynomials $\mathbb{P}_2(\mathbb{R}^2)$. They form a convex partition of unity when restricted to T. For more details, readers are directed to [11]. Additionally, for each polynomial $q \in \mathbb{P}_2$ (e.g., the space of quadratic polynomials) defined on T, there exist coefficients $b_{i,j,k} \in \mathbb{R}$, where $i + j + k = 2$, such that

$$q(\mathbf{p}) = \sum_{i+j+k=2} b_{i,j,k} \mathcal{B}_{i,j,k}(\tau). \tag{2}$$

Since the barycentric coordinates of a point in T are non-negative, the restriction of q to T can be interpreted as a convex combination of the coefficients $b_{i,j,k}$. These coefficients, termed the BB-coefficients of q with respect to T, are associated with the domain points in

$$\left\{ \mathbf{p}_1, \frac{1}{2}(\mathbf{p}_1 + \mathbf{p}_2), \mathbf{p}_2, \frac{1}{2}(\mathbf{p}_2 + \mathbf{p}_3), \mathbf{p}_3, \frac{1}{2}(\mathbf{p}_3 + \mathbf{p}_1) \right\}.$$

The quadratic Bernstein-Bézier representation can be illustrated schematically, as depicted in Fig. 1.

Now, we define the C^1 smoothness between two pieces in terms of the BB-coefficients. Let $\widehat{T} = \langle \mathbf{p}_4, \mathbf{p}_2, \mathbf{p}_3 \rangle$ be an adjacent triangle of T sharing the common edge $\langle \mathbf{p}_2, \mathbf{p}_3 \rangle$. Let p and $\widehat{p}$ be polynomials defined on T and $\widehat{T}$, respectively. We aim to ensure C^1 continuity across the shared edge. To achieve this, we examine the BB-coefficients of $\widehat{p}$ with respect to $\widehat{T}$. We denote these coefficients as $\hat{b}_{i,j,k}$. The C^1 smoothness between p and $\widehat{p}$ across $\langle \mathbf{p}_2, \mathbf{p}_3 \rangle$ is established if the following conditions hold

$$\begin{aligned} \hat{b}_{0,j,k} &= b_{0,j,k}, \quad j + k = 2, \\ \hat{b}_{1,j,k} &= \hat{\tau}_1 b_{1,j,k} + \hat{\tau}_2 b_{0,j+1,k} + \hat{\tau}_3 b_{0,j,k+1}, \quad j + k = 1, \end{aligned}$$

where $(\hat{\tau}_1, \hat{\tau}_2, \hat{\tau}_3)$ are the barycentric coordinates of $\mathbf{p}_4$ w.r.t. T.

Let Δ be a triangulation of a polygonal domain $\Omega \subset \mathbb{R}^2$, and Δ_{PS} be the Powell-Sabin 6-split of Δ, which divides each triangle of Δ into six small micro-triangles [9]. The space of C^1 continuous quadratic polynomial splines is defined as

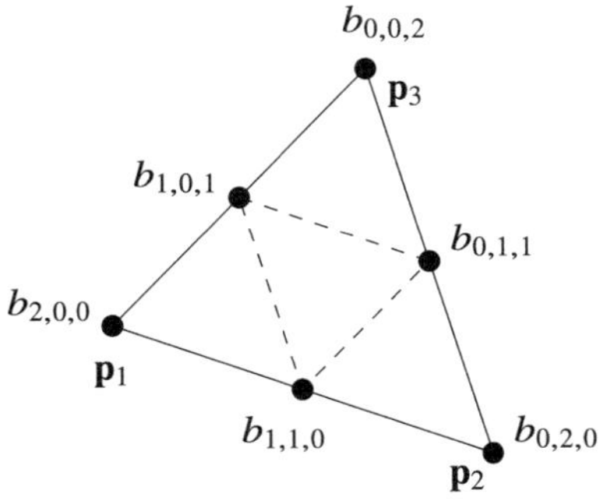

Fig. 1 Schematic representation of the BB-coefficients of a quadratic bivariate polynomial

$$\mathbb{S}_2^1(\Delta_{\text{PS}}) := \left\{ s \in C^1(\Omega) : s_{|T} \in \mathbb{P}_2, \quad T \in \Delta \right\}.$$

2.2 Uniform Triangulation with the Powell-Sabin 6-Split

Let $\mathbf{n} := (n_x, n_y) \in \mathbb{N}^2$, and let $\mathcal{V}_\mathbf{n}$ be the uniform grid of nodes in the square $S = [-1, 1]^2$, defined as follows:

$$\mathcal{V}_\mathbf{n} := \left\{ \mathbf{p}_{i,j} := \left(-1 + \frac{2}{n_x} i, -1 + \frac{2}{n_y} j \right) : \quad i = 0, \ldots, n_x, \ j = 0, \ldots, n_y \right\}. \tag{3}$$

The vertices $\mathbf{p}_{i,j}$ define the faces of the lattice $\mathcal{V}_\mathbf{n}$, that can be decomposed into the triangles $T_{i,j,1} := \langle \mathbf{p}_{i,j}, \mathbf{p}_{i+1,j}, \mathbf{p}_{i,j+1} \rangle$ and $T_{i,j,2} := \langle \mathbf{p}_{i+1,j}, \mathbf{p}_{i,j+1}, \mathbf{p}_{i+1,j+1} \rangle$. This decomposition generates a uniform triangulation $\Delta := \bigcup_{i,j} \left(T_{i,j,1} \cup T_{i,j,2} \right)$.

We denote the Powell-Sabin 6-split of Δ by Δ_{PS}, which is obtained by connecting the vertices of each triangle in Δ to the midpoint of the opposite edge. Figure 2 illustrates an example of the triangulation Δ (blue), the Δ_{PS} split of Δ (dashed lines), the Powell-Sabin triangles (triangles in black) and the Powell-Sabin points (points in black).

For each vertex $\mathbf{p}_{i,j}$ of Δ, we associate a set of points known as the Powell-Sabin points, depicted in black in Fig. 2. These points include the vertex itself and the points

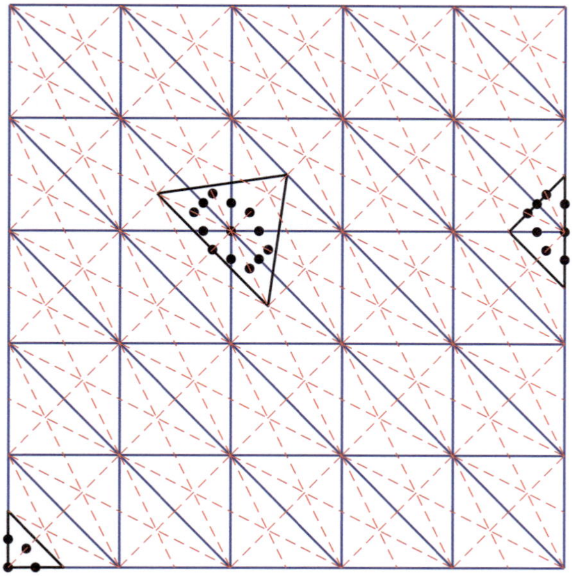

Fig. 2 Example of a triangulation Δ (thick blue lines), Powell-Sabin 6-split Δ_{PS} (dashed red lines) of Δ, Powell-Sabin triangles (black) and Powell-Sabin points (black)

situated at the middle of the edges emanating from the vertex within Δ_{PS}. A Powell-Sabin triangle is depicted in black in Fig. 2, which contains all the Powell-Sabin points. One can employ the algorithms in [29] to find the minimum area triangle that encloses the convex polygon defined by the Powell-Sabin points. From these triangles, a normalized B-spline basis $\left\{\mathcal{B}_{i,j}^{0,0}, \mathcal{B}_{i,j}^{1,0}, \mathcal{B}_{i,j}^{0,1}, \quad i = 1, \ldots, n_x,\ j = 1, \ldots, n_y\right\}$ is constructed for the space $\mathbb{S}_2^1(\Delta_{\mathrm{PS}})$.

It is well known that there exists a unique spline $s \in \mathbb{S}_2^1(\Delta_{\mathrm{PS}})$ such that [9]

$$\partial_x^a\, \partial_y^b\, s\left(\mathbf{p}_{i,j}\right) = f_{i,j}^{a,b}, \quad i = 1, \ldots, n_x,\ i = 1, \ldots, n_y, \quad a, b \geq 0 \text{ and } a + b \leq 1, \tag{4}$$

for given f-values. That is, given function values and partial derivatives at each vertex of the original triangulation Δ, the Hermite interpolation problem (4) has a unique solution in $\mathbb{S}_2^1(\Delta_{\mathrm{PS}})$. Furthermore,

$$s(\mathbf{p}) := \sum_{i,j}\left(c_{i,j}^{0,0}\mathcal{B}_{i,j}^{0,0} + c_{i,j}^{1,0}\mathcal{B}_{i,j}^{1,0} + c_{i,j}^{0,1}\mathcal{B}_{i,j}^{0,1}\right). \tag{5}$$

For each vertex $\mathbf{p}_{i,j}$ in Δ, three normalized B-splines $\mathcal{B}_{i,j}^{0,0}, \mathcal{B}_{i,j}^{1,0}, \mathcal{B}_{i,j}^{0,1}$ are defined, i.e. one for each corner of the Powell-Sabin triangle associated with $\mathbf{p}_{i,j}$. They are solution of (4), for any $\mathbf{p} \neq \mathbf{p}_{i,j}$ we have:

$$\mathcal{B}_{i,j}^{a,b}(\mathbf{p}) = 0, \quad \partial_x \mathcal{B}_{i,j}^{a,b}(\mathbf{p}) = 0, \quad \partial_y \mathcal{B}_{i,j}^{a,b}(\mathbf{p}) = 0, \quad a + b \leq 1.$$

And otherwise

$$\mathcal{B}_{i,j}^{a,b}\left(\mathbf{p}_{i,j}\right) = \alpha_{i,j}^{a,b}, \quad \partial_x \mathcal{B}_{i,j}^{a,b}\left(\mathbf{p}_{i,j}\right) = \beta_{i,j}^{a,b}, \quad \partial_y \mathcal{B}_{i,j}^{a,b}\left(\mathbf{p}_{i,j}\right) = \gamma_{i,j}^{a,b}, \quad a + b \leq 1,$$

with

$$\begin{aligned} 1 &= \alpha_{i,j}^{0,0} + \alpha_{i,j}^{1,0} + \alpha_{i,j}^{0,1}, \\ 0 &= \beta_{i,j}^{0,0} + \beta_{i,j}^{1,0} + \beta_{i,j}^{0,1}, \\ 0 &= \gamma_{i,j}^{0,0} + \gamma_{i,j}^{1,0} + \gamma_{i,j}^{0,1}. \end{aligned}$$

These coefficients are obtained by solving the linear system

$$\begin{pmatrix} \alpha_{i,j}^{0,0} & \alpha_{i,j}^{1,0} & \alpha_{i,j}^{0,1} \\ \beta_{i,j}^{0,0} & \beta_{i,j}^{1,0} & \beta_{i,j}^{0,1} \\ \gamma_{i,j}^{0,0} & \gamma_{i,j}^{1,0} & \gamma_{i,j}^{0,1} \end{pmatrix} \begin{pmatrix} x_{i,j}^1 & y_{i,j}^1 & 1 \\ x_{i,j}^2 & y_{i,j}^2 & 1 \\ x_{i,j}^3 & y_{i,j}^3 & 1 \end{pmatrix} = \begin{pmatrix} -1 + \frac{2}{n_x} i & -1 + \frac{2}{n_x} j & 1 \\ 1 & 0 & 0 \\ 0 & 1 & 0 \end{pmatrix}$$

where $\mathbf{v}_{i,j}^{\ell} := \left(x_{i,j}^{\ell}, y_{i,j}^{\ell}\right)$, $\ell = 1, 2, 3$ stand for the corner coordinates of the Powell-Sabin triangle associated with the vertex $\mathbf{p}_{i,j}$. This nice geometric construction was first introduced by P. Dierckx [12], and further used in the literature by several authors [30, 31]. It yields to B-spline functions with nice properties, i.e.,

$$\sum_{i,j}\left(\mathcal{B}_{i,j}^{0,0}+\mathcal{B}_{i,j}^{1,0}+\mathcal{B}_{i,j}^{0,1}\right)=1,\quad \mathcal{B}_{i,j}^{0,0},\mathcal{B}_{i,j}^{1,0},\mathcal{B}_{i,j}^{0,1}\geq 0.$$

Since the triangulation Δ is uniform, we can distinguish between three types of Powell-Sabin triangles. These types correspond to triangles associated with interior vertices, corner vertices, and boundary vertices that are not corners. For each type, specific values of the coefficients $\alpha_{i,j}^{a,b}$, $\beta_{i,j}^{a,b}$, and $\gamma_{i,j}^{a,b}$ are assigned. Namely, the matrix

$$\begin{pmatrix}\alpha_{i,j}^{0,0} & \alpha_{i,j}^{1,0} & \alpha_{i,j}^{0,1}\\ \beta_{i,j}^{0,0} & \beta_{i,j}^{1,0} & \beta_{i,j}^{0,1}\\ \gamma_{i,j}^{0,0} & \gamma_{i,j}^{1,0} & \gamma_{i,j}^{0,1}\end{pmatrix}$$

is equal to:

- interior vertices

$$\begin{pmatrix}\frac{3}{8} & \frac{3}{8} & \frac{1}{4}\\ -\frac{7}{16}n_x & \frac{1}{16}n_x & \frac{3}{8}n_x\\ \frac{1}{16}n_y & -\frac{7}{16}n_y & \frac{3}{8}n_y\end{pmatrix}$$

- corner vertices: for $\mathbf{p}_{0,0}$, $\mathbf{p}_{n_x,0}$, $\mathbf{p}_{0,n_y}$ and $\mathbf{p}_{n_x,n_y}$, respectively

$$\begin{pmatrix}1 & 0 & 0\\ -n_x & n_x & 0\\ -n_y & 0 & n_y\end{pmatrix},\begin{pmatrix}1 & 0 & 0\\ n_x & -n_x & 0\\ -n_y & 0 & n_y\end{pmatrix},\begin{pmatrix}1 & 0 & 0\\ -n_x & n_x & 0\\ n_y & 0 & -n_y\end{pmatrix},\begin{pmatrix}1 & 0 & 0\\ n_x & -n_x & 0\\ n_y & 0 & -n_y\end{pmatrix}$$

- boundary vertices:

$$\begin{pmatrix}0 & \frac{1}{2} & \frac{1}{2}\\ -n_x & \frac{1}{2}n_x & \frac{1}{2}n_x\\ 0 & \frac{1}{2}n_y & \frac{-1}{2}n_y\end{pmatrix}.$$

As shown in (4), defining a spline in the Powell-Sabin spline space $\mathbb{S}_2^1(\Delta_{\mathrm{PS}})$ requires specifying both the value and the first derivative values at each vertex in Δ. However, in numerous engineering applications, such as Digital Elevation Models [41], the derivative information might be unavailable. In the subsequent section, we present a novel technique based on the constrained mock-Chebyshev least squares operator to approximate the first derivative values from the discrete values at the vertices.

3 Constrained Mock-Chebyshev Least Squares Tensor Product Operator

Let $f(x, y)$ be an unknown function in the square S, and we assume knowledge of this function exclusively on the grid $\mathcal{V}_{\mathbf{n}}$. Let $\mathbf{m} := \left(m_x, m_y\right)$, where

$$m_x = \left\lfloor \pi \sqrt{\frac{n_x}{2}} \right\rfloor, \quad m_y = \left\lfloor \pi \sqrt{\frac{n_y}{2}} \right\rfloor,$$

and consider the Cartesian grid

$$\mathcal{V}_{\mathbf{m}} := \{(x_i', y_j') : i = 0, \ldots, m_x, \ j = 0, \ldots, m_y\}, \tag{6}$$

in which $\mathcal{V}_{\mathbf{m}} \subset \mathcal{V}_{\mathbf{n}}$, is the subset of *mock-Chebyshev nodes* of $\mathcal{V}_{\mathbf{n}}$. For more details, we refer the reader to [42]. Then, we have

$$x_i' \approx -\cos\left(\frac{\pi}{m_x} i\right), \quad y_j' \approx -\cos\left(\frac{\pi}{m_y} j\right), \quad i = 0, \ldots, m_x, \ j = 0, \ldots, m_y.$$

We set

$$p_x = \left\lfloor \pi \sqrt{\frac{n_x}{12}} \right\rfloor, \qquad p_y = \left\lfloor \pi \sqrt{\frac{n_y}{12}} \right\rfloor$$

and we denote by

$$\boldsymbol{r}_{xy} = (r_x, r_y) := (m_x + p_x + 1, m_y + p_y + 1).$$

We also denote by $\mathbb{P}_{r_x} \otimes \mathbb{P}_{r_y}$ the tensor product of polynomial spaces of degree r_x and r_y in the variables x and y, respectively. This is achieved with the basis given by

$$\mathcal{U}_{r_x} \otimes \mathcal{U}_{r_y} = \{\boldsymbol{u}_{ij}(x, y) := u_i(x) u_j(y) : i = 0, \ldots, r_x, \ j = 0, \ldots, r_y\}, \tag{7}$$

such that

$$\operatorname{span}\{u_i(x) \ : \ i = 0, \ldots, m_x\} = \mathbb{P}_{m_x}, \qquad \operatorname{span}\{u_j(y) \ : \ j = 0, \ldots, m_y\} = \mathbb{P}_{m_y}.$$

In this section, we recall the constrained mock-Chebyshev least squares tensor product operator [34]. To this aim some rearrangements of the set $X_{n_x} \times Y_{n_y}$ and of the basis $\mathcal{U}_{r_x} \otimes \mathcal{U}_{r_y}$ are necessary.

Denoted by

$$\mathcal{N}_{\ell,q} := \{1, \ldots, (\ell + 1)(q + 1)\}, \quad \ell, q \in \mathbb{N},$$

we consider the maps

$$\alpha : \mathcal{N}_{n_x,n_y} \to \{0,\dots,n_x\} \times \{0,\dots,n_y\}, \quad \beta : \mathcal{N}_{n_x,n_y} \to \{0,\dots,n_x\} \times \{0,\dots,n_y\}$$

such that

$$X'_{m_x} \times Y'_{m_y} = \left\{ \boldsymbol{x}_{\alpha(k)} \ : \ k = 1,\dots,(m_x+1)(m_y+1) \right\},$$

and

$$\operatorname{span}\left\{ \boldsymbol{u}_{\beta(k)} \ : \ k = 1,\dots,(m_x+1)(m_y+1) \right\} = \mathbb{P}_{m_x} \otimes \mathbb{P}_{m_y}.$$

After these rearrangements, we denote by

$$V = \left[\boldsymbol{u}_{\beta(j)}\left(\boldsymbol{x}_{\alpha(i)} \right) \right]_{\substack{i \in \mathcal{N}_{n_x,n_y} \\ j \in \mathcal{N}_{r_x,r_y}}},$$

and by C the submatrix of V formed by its first $(m_x+1)(m_y+1)$ rows. We also denote by

$$\boldsymbol{b} = \left[f\left(\boldsymbol{x}_{\alpha(1)} \right), \dots, f\left(\boldsymbol{x}_{\alpha((n_x+1)(n_y+1))} \right) \right]^T, \tag{8}$$

$$\boldsymbol{d} = \left[f\left(\boldsymbol{x}_{\alpha(1)} \right), \dots, f\left(\boldsymbol{x}_{\alpha((m_x+1)(m_y+1))} \right) \right]^T. \tag{9}$$

The constrained mock-Chebyshev least squares tensor product operator is defined as [34]

$$\begin{aligned} \hat{P}_{\boldsymbol{r}_{xy},\boldsymbol{n}_{xy}} &: C(S) \to C(S) \\ f(x,y) &\mapsto \hat{P}_{\boldsymbol{r}_{xy},\boldsymbol{n}_{xy}}[f](x,y) = \sum_{j=0}^{r_y} \sum_{i=0}^{r_x} \hat{a}_{ij} u_i(x) u_j(y), \end{aligned} \tag{10}$$

$(x, y) \in S$, where $\hat{\boldsymbol{a}} = [\hat{a}_{00}, \hat{a}_{10}, \hat{a}_{01}, \hat{a}_{20}, \dots, \hat{a}_{r_x r_y}]^T$ is the solution of the KKT linear equations [34]

$$\begin{bmatrix} 2\,V^T V & C^T \\ C & 0 \end{bmatrix} \begin{bmatrix} \hat{\boldsymbol{a}} \\ \hat{z} \end{bmatrix} = \begin{bmatrix} 2\,V^T \boldsymbol{b} \\ \boldsymbol{d} \end{bmatrix}, \tag{11}$$

being $\hat{z}$ the Lagrange multipliers vector. In [34], it is proved that the matrix of the linear system (11), also called KKT matrix, is non singular [34, 35]. Similarly to the univariate case, the approximation operator $\hat{P}_{\boldsymbol{r}_{xy},\boldsymbol{n}_{xy}}$ is a linear operator and reproduces polynomials belonging to $\mathbb{P}_{r_x} \otimes \mathbb{P}_{r_y}$ [36]. Moreover, by definition, we get

$$\hat{P}_{\boldsymbol{r}_{xy},\boldsymbol{n}_{xy}}[f](\boldsymbol{x}_{\alpha(k)}) = f(\boldsymbol{x}_{\alpha(k)}), \qquad k = 1,\dots,(m_x+1)(m_y+1).$$

Additionally, this approximation is entirely determined by the evaluations of f on the set $X_{n_x} \times Y_{n_y}$.

3.1 *Estimating Partial Derivatives from the Constrained Mock-Chebyshev Least Squares Tensor Product Operator*

The aim of this section is to generalize the differentiation formula introduced in [36] to the bivariate case through the tensor product interpolation. Hereafter, we assume that $f \in C^1(S)$, and our knowledge is confined only to the evaluations of f on the regular Cartesian grid (3). Within this framework, the constrained mock-Chebyshev least squares tensor product operator (10) furnishes a comprehensive polynomial approximation of the function f at any point $(x, y) \in S$.

In order to approximate the partial derivatives of f with respect to both x and y, we can use the partial derivatives of the approximation $\hat{P}_{\boldsymbol{r}_{xy},\boldsymbol{n}_{xy}}[f]$ with respect to x and y, that is

$$\partial_x f(x, y) \approx \partial_x \hat{P}_{\boldsymbol{r}_{xy},\boldsymbol{n}_{xy}}[f](x, y) = \sum_{j=0}^{r_y} \sum_{i=0}^{r_x} \hat{a}_{ij} u_i'(x) u_j(y), \quad (x, y) \in S \tag{12}$$

$$\partial_y f(x, y) \approx \partial_y \hat{P}_{\boldsymbol{r}_{xy},\boldsymbol{n}_{xy}}[f](x, y) = \sum_{j=0}^{r_y} \sum_{i=0}^{r_x} \hat{a}_{ij} u_i(x) u_j'(y), \quad (x, y) \in S. \tag{13}$$

These approximations can be represented in the basis (7) as follows

$$\partial_x \hat{P}_{\boldsymbol{r}_{xy},\boldsymbol{n}_{xy}}[f] = \hat{P}_{\boldsymbol{r}_{xy},\boldsymbol{n}_{xy}} \left[\partial_x \hat{P}_{\boldsymbol{r}_{xy},\boldsymbol{n}_{xy}}[f]\right] = \sum_{j=0}^{r_x} \sum_{i=0}^{r_x} \hat{a}_{ij}^x u_i(x) u_j(y)$$

$$\partial_y \hat{P}_{\boldsymbol{r}_{xy},\boldsymbol{n}_{xy}}[f] = \hat{P}_{\boldsymbol{r}_{xy},\boldsymbol{n}_{xy}} \left[\partial_y \hat{P}_{\boldsymbol{r}_{xy},\boldsymbol{n}_{xy}}[f]\right] = \sum_{j=0}^{r_y} \sum_{i=0}^{r_x} \hat{a}_{ij}^y u_i(x) u_j(y)$$

since the operator (10) reproduces polynomials belonging to $\mathbb{P}_{r_x} \otimes \mathbb{P}_{r_y}$ and

$$\partial_x \hat{P}_{\boldsymbol{r}_{xy},\boldsymbol{n}_{xy}}[f], \partial_y \hat{P}_{\boldsymbol{r}_{xy},\boldsymbol{n}_{xy}}[f] \in \mathbb{P}_{r_x} \otimes \mathbb{P}_{r_y}.$$

It is worth noting to emphasize that the differentiation formulas (12) and (13) yield a *global polynomial approximation* by using only functional evaluations on the uniform grid $X_{n_x} \times Y_{n_y}$. Furthermore, these procedures can be iterated to approximate partial derivatives of order $k > 1$, under the assumption that $f \in C^k(S)$.

4 Numerical Results

In this section, we analyze the performance of the proposed approximation approach. To this aim, we consider the following test functions defined on the square S

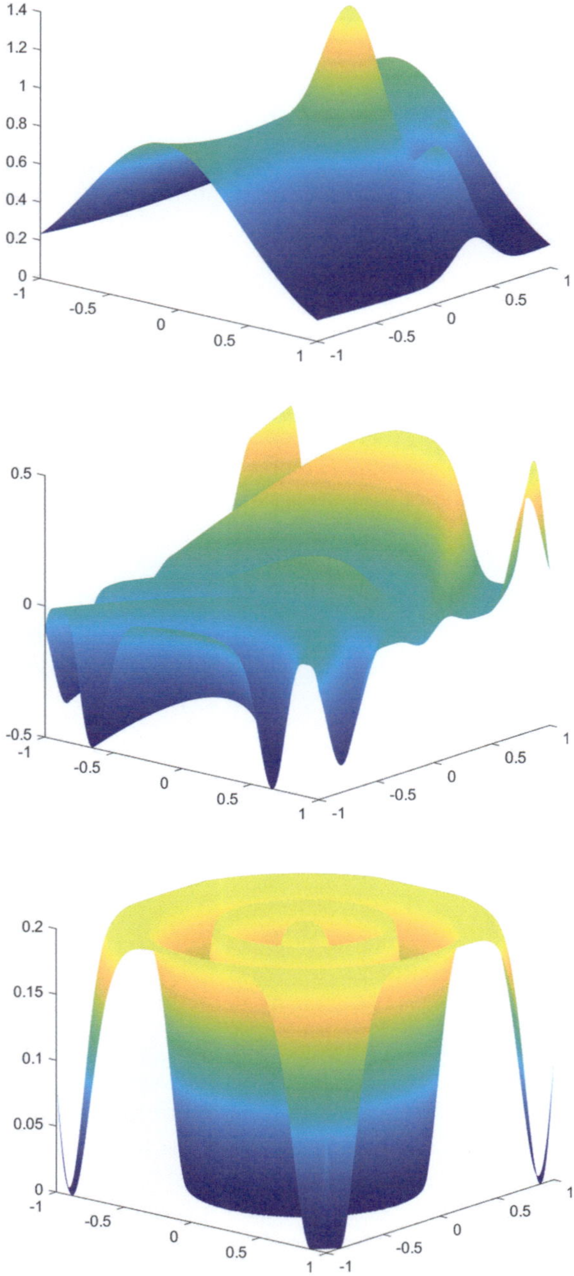

Fig. 3 Plots of the test functions: f_1 (left), f_2 (center) and f_3 (right)

$$f_1(x, y) = 0.75e^{-\frac{1}{4}\left((9x-2)^2+(9y-2)^2\right)} + 0.75e^{\left(-\frac{1}{49}(9x+1)^2-\frac{1}{10}(9y+1)\right)}$$
$$+ 0.5e^{-\frac{1}{4}\left((9x-7)^2+(9y-3)^2\right)} + 0.2e^{-(9x-4)^2-(9y-7)^2},$$

Table 1 Errors and the corresponding NCOs for the functions f_1, f_2 and f_3

Δ_n	f_1	NCO	f_2	NCO	f_3	NCO
20	2.592940e-01	–	6.335118e-01	–	7.048398e-01	–
40	9.312955e-02	1.4773	8.054009e-02	2.9756	1.275937e-01	2.4657
80	1.858789e-02	2.3249	1.184876e-03	6.0869	8.859532e-03	3.8482
160	1.855795e-03	3.3243	8.788211e-05	3.7530	9.411460e-05	6.5567

$$f_2(x, y) = \frac{y}{2} \cos^4 \left(4 \left(x^2 + y - 1\right)\right),$$
$$f_3(x, y) = 0.1 \left(1 + \cos \left(12\pi \cos \left(\pi \sqrt{x^2 + y^2}\right)\right)\right).$$

The functions f_1 and f_2 correspond to the Franke and Nielsen functions [43, 44], respectively, whereas f_3 represents a highly osculated function, see Fig. 3. We determine the accuracy of the approximation by assessing the maximum approximation error

$$E_n := \max_{T \in \Delta_n} \max_{\mathbf{x} \in \mathcal{T}_M} |f(\mathbf{x}) - s(\mathbf{x})|,$$

where $\mathcal{T}_M$ is the set of points generated by dividing the triangle T into $M = 200$ uniformly distributed points. In addition, we calculate the numerical convergence order, defined as

$$\text{NCO} := \log_2 \left(\frac{E_{2n}}{E_n}\right).$$

We estimate the approximation errors for various values of n and determine the corresponding NCOs. The results are presented in Table 1.

5 Conclusion

In this work, we propose using the constrained mock-Chebyshev least squares approach to approximate partial derivatives from discrete data. This approximation is integrated into the definition of the space of quadratic continuous splines on the Powell-Sabin 6-split. This space solves a Hermite interpolation problem that specifies discrete values along the first derivatives at each vertex of the initial triangulation. Such a technique can be particularly valuable in engineering fields where derivatives are often not provided. Our approach is applied within a square domain. The numerical experiments conducted in this study demonstrate the effectiveness of the constrained mock-Chebyshev least squares operator in approximating the Powell-Sabin finite element method. The results validate that the proposed approach achieves high accuracy in the approximation of derivative values, which are crucial for defining the Powell-Sabin finite element. The observed convergence orders align with the theoretical expectations. These convergence rates confirm the reliability of the proposed method, especially in scenarios where derivative information is not directly available and needs to be approximated from function values. Overall, the proposed method provides a robust and accurate alternative for applying the Powell-Sabin finite element method, even when only functional evaluations at equally spaced nodes are available.

In future work, we aim to extend this method to handle more general triangulations.

Acknowledgements This research has been achieved as part of RITA "Research ITalian network on Approximation" and as part of the UMI group "Teoria dell'Approssimazione e Applicazioni". The research was supported by GNCS-INdAM 2025 project "Polinomi, Splines e Funzioni Kernel: dall'Approssimazione Numerica al Software Open-Source". The work of F. Nudo has been funded by the European Union – NextGenerationEU under the Italian National Recovery and Resilience Plan (PNRR), Mission 4, Component 2, Investment 1.2 "Finanziamento di progetti presentati da giovani ricercatori", pursuant to MUR Decree No. 47/2025. The third author is a member of the research group GNCS (Gruppo Nazionale per il Calcolo Scientifico) of INdAM (Istituto Nazionale di Alta Matematica), and acknowledges the MUR Excellence Department Project awarded to the Department of Mathematics, University of Rome Tor Vergata, CUPE83C23000330006.

References

1. Ciarlet, P.G.: The finite element method for elliptic problems. SIAM (2002)
2. Guessab, A.: Sharp Approximations Base on Delaunay Triangulations and Voronoi Diagrams. Publisher: NSU Publishing and Printing Center, Novosibirsk (2022)
3. Pepper, D.W., Heinrich, J.C.: The finite element method: basic concepts and applications. Taylor & Francis (2005)
4. Bachar, M., Guessab, A.: Characterization of the existence of an enriched linear finite element approximation using biorthogonal systems. Res. Math. **70**, 401–413 (2016)
5. Bachar, M., Guessab, A.: A simple necessary and sufficient condition for the enrichment of the Crouzeix-Raviart element. Appl. Anal. Discret. Math. **10**, 378–393 (2016)
6. Dell'Accio, F., Di Tommaso, F., Guessab, A., Nudo, F.: A unified enrichment approach of the standard three-node triangular element. Appl. Numer. Math. **187**, 1–23 (2023)
7. Dell'Accio, F., Di Tommaso, F., Guessab, A., Nudo, F.: Enrichment strategies for the simplicial linear finite elements. Appl. Math.Comput. **451**, 128023 (2023)
8. Dell'Accio, F., Di Tommaso, F., Guessab, A., Nudo, F.: New quadratic and cubic polynomial enrichments of the Crouzeix-Raviart finite element. Comput. Math. Appl. **170**, 204–212 (2024)
9. Powell, M.J.D., Sabin, M.A.: Piecewise quadratic approximations on triangles. ACM Trans. Math. Softw. **3**, 316–325 (1977)
10. Ženíšek, A.: A general theorem on triangular finite-elements, Revue française d'automatique, informatique, recherche opérationnelle. Analyse numérique **8**(R2), 119–127 (1974)
11. Lai, M.J., Schumaker, L.L.: Spline Functions on Triangulations. Cambridge University Press (2007)
12. Dierckx, P.: On calculating normalized Powell-Sabin B-splines. Comput. Aided Geom. Des. **15**, 61–78 (1997)
13. Manni, C., Sablonnière, P.: Quadratic spline quasi-interpolants on Powell-Sabin partitions. Adv. Comput. Math. **26**, 283–304 (2007)
14. Lamnii, M., Mraoui, H., Zidna, A.: Local quasi-interpolants based on special multivariate quadratic spline space over a refined quadrangulation. Appl. Math. Comput. **219**, 10456–10467 (2013)

15. Barrera, D., Eddargani, S., Ibáñez, M.J., Remogna, S.: Spline quasi-interpolation in the Bernstein basis on the Powell-Sabin 6-split of a type-1 triangulation. J. Comput. Appl. Math. **424**, 115011 (2023)
16. Kosinka, J., Bartoň, M.: Gaussian quadrature for C^1 cubic Clough-Tocher macro-triangles. J. Comput. Appl. Math. **351**, 6–13 (2019)
17. Sbibih, D., Serghini, A., Tijini, A.: Polar forms and quadratic spline quasi-interpolants on Powell-Sabin partitions. Appl. Numer. Math. **59**, 938–958 (2009)
18. Lamnii, M., Mraoui, H., Tijini, A., Zidna, A.: A normalized basis for C^1 cubic super spline space on Powell-Sabin triangulation. Math. Comput. Simul. **99**, 108–124 (2014)
19. Lamnii, A., Lamnii, M., Mraoui, H.: Cubic spline quasi-interpolants on Powell-Sabin partitions. BIT Numer. Math. **54**, 1099–1118 (2014)
20. Sbibih, D., Serghini, A., Tijini, A., Zidna, A.: Superconvergent C^1 cubic spline quasi-interpolants on Powell-Sabin partitions. BIT Numer. Math. **55**, 797–821 (2015)
21. Speleers, H.: A new B-spline representation for cubic splines over Powell-Sabin triangulations. Comput. Aided Geom. Des. **37**, 42–56 (2015)
22. Grošelj, J., Krajnc, M.: C^1 cubic splines on Powell-Sabin triangulations. Appl. Math. Comput. **272**, 114–126 (2016)
23. Grošelj, J., Speleers, H.: Construction and analysis of cubic Powell-Sabin B-splines. Comput. Aided Geom. Des. **57**, 1–22 (2017)
24. Grošelj, J., Krajnc, M.: Quartic splines on Powell-Sabin triangulations. Comput. Aided Geom. Des. **49**, 1–16 (2016)
25. Barrera, D., Eddargani, S., Ibáñez, M.J., Lamnii, A.: A geometric characterization of Powell-Sabin triangulations allowing the construction of C^2 quartic splines. Comput. Math. with Appl. **100**, 30–40 (2021)
26. Speleers, H.: A normalized basis for quintic Powell-Sabin splines. Comput. Aided Geom. Des. **27**, 438–457 (2010)
27. Lamnii, M., Mraoui, H., Tijini, A.: Construction of quintic Powell-Sabin spline quasi-interpolants based on blossoming. J. Comput. Appl. Math. **250**, 190–209 (2013)
28. Speleers, H.: Interpolation with quintic Powell-Sabin splines. Appl. Numer. Math. **62**, 620–635 (2012)
29. Eddargani, S., Ibáñez, M.J., Lamnii, A., Lamnii, M., Barrera, D.: Quasi-interpolation in a space of C^2 sextic splines over Powell-Sabin triangulations. Mathematics **9**, 2276 (2021)
30. Speleers, H.: Construction of normalized B-splines for a family of smooth spline spaces over Powell-Sabin triangulations. Constr. Approx. **37**, 41–72 (2013)
31. Groselj, J.: A normalized representation of super splines of arbitrary degree on Powell-Sabin triangles. BIT Numer. Math. **56**, 1257–1280 (2016)
32. Speleers, H.: A family of smooth quasi-interpolants defined over Powell-Sabin triangulations. Constr. Approx. **41**, 297–324 (2015)
33. De Marchi, S., Dell'Accio, F., Mazza, M.: On the constrained mock-Chebyshev least-squares. J. Comput. Appl. Math. **280**, 94–109 (2015)
34. Dell'Accio, F., Di Tommaso, F., Nudo, F.: Generalizations of the constrained mock-Chebyshev least squares in two variables: Tensor product vs total degree polynomial interpolation. Appl. Math. Lett. **125**, 107732 (2022)
35. Dell'Accio, F., Marcellán, F., Nudo, F.: An extension of a mixed interpolation-regression method using zeros of orthogonal polynomials. J. Comput. Appl. Math. **450**, 116010 (2024)
36. Dell'Accio, F., Di Tommaso, F., Nudo, F.: Constrained mock-Chebyshev least squares quadrature. Appl. Math. Lett. **134**, 108328 (2022)
37. Dell'Accio, F., Di Tommaso, F., Francomano, E., Nudo, F.: An adaptive algorithm for determining the optimal degree of regression in constrained mock-Chebyshev least squares quadrature. Dolomites Res. Notes Approx. **15**, 35–44 (2022)
38. Dell'Accio, F., Mezzanotte, D., Nudo, F., Occorsio, D.: Product integration rules by the constrained mock-Chebyshev least squares operator. BIT Numer. Math. **63**, 24 (2023)
39. Dell'Accio, F., Mezzanotte, D., Nudo, F., Occorsio, D.: Numerical approximation of Fredholm integral equation by the constrained mock-Chebyshev least squares operator. J. Comput. Appl. Math. **447**, 115886 (2024)

40. Dell'Accio, F., Nudo, F:, Polynomial approximation of derivatives through a regression-interpolation method. Appl. Math. Lett. 109010 (2024)
41. Ariza-López, F.J., Barrera, D., Eddargani, S., Ibáñez, M.J., Reinoso, J.F.: Spline quasi-interpolation in the Bernstein basis and its application to digital elevation models. Math. Meth. Appl. Sci. **46**, 1687–1698 (2023)
42. Boyd, J.P., Xu, F.: Divergence (Runge phenomenon) for least-squares polynomial approximation on an equispaced grid and Mock-Chebyshev subset interpolation. Appl. Math. Comput. **210**, 158–168 (2009)
43. Franke, R.: Scattered data interpolation: tests of some methods. Math. Comput. **38**, 181–200 (1982)
44. Nielson, G.M.: A first-order blending method for triangles based upon cubic interpolation. Int. J. Numer. Methods Eng. **15**, 308–318 (1980)

Hybrid Fractional Integro-Differential Equations with Nonlocal Boundary Value Conditions

Khaled Ben Amara and María Isabel Berenguer

Abstract We address the problem of existence, uniqueness and approximation of the solution of a large class of fractional integro-differential equations with boundary conditions. Based on fixed point techniques with Boyd-Wong's type conditions, the existence and uniqueness of the solution is studied. This, together with the use of certain types of biorthogonal systems, allows us to propose a method to approximate the solution that is tested with several examples.

1 Introduction

Due to the nonlocal nature of fractional derivative, fractional differential equations play an important role in the practical mathematical modeling of many real-world problems occurring in engineering and scientific disciplines such as control theory, signal and image processing, biophysics, blood flow phenomena, etc. This is the main reason why initial and boundary value problems of fractional order have emerged as an interesting field of research in recent decades. See, for example, [1, 3, 5–7, 14, 21–23, 27, 28, 30] and their references.

In [29], Zhang considered the existence and multiplicity of positive solutions for the following nonlinear fractional boundary value problem:

$$\begin{cases} D^{\alpha}x(t) = g(t, x(t)), 0 < t < 1, \alpha \in (1, 2], \\ x(0) = -x'(0) \text{ and } x(1) = -x'(1), \end{cases} \tag{1}$$

K. Ben Amara
Higher Institute of Computer Science and Multimedia of Gabès (ISIMG), University of Gabès, Gabès, Tunisia
e-mail: khaled.benamara@isimg.tn

M. I. Berenguer (✉)
Department of Applied Mathematics and Institute of Mathematics (IEMath-GR), E.T.S. de Ingenieria de Edificación, University of Granada, Granada, Spain
e-mail: maribel@ugr.es

© The Author(s), under exclusive license to Springer Nature Switzerland AG 2025
M. J. Ibáñez-Pérez et al. (eds.), *Mathematical And Computational Modelling, Approximation and Simulation*, SEMA SIMAI Springer Series 41,
https://doi.org/10.1007/978-3-032-01114-5_8

where $g : J \times [0, +\infty) \to [0, +\infty)$ is continuous.
In [31], Y. Zhao et al. published a study investigating the existing results for

$$\begin{cases} D^{\alpha}\left(\dfrac{x(t)}{f(t, x(t))}\right) = g(t, x(t)), t \in J, \alpha \in (1, 2], \\ \\ x(0) = x(1) = 0, \end{cases} \tag{2}$$

where $J = [0, 1]$, the functions $f \in \mathscr{C}(J \times \mathbb{R}, \mathbb{R} \setminus \{0\})$ is Lipschitzian and $g \in \mathscr{C}(J \times \mathbb{R}, \mathbb{R})$ is Carathéodory.

Later, Ahmad et al. in [4] established an existence result for the nonlinear fractional problem with a nonlocal boundary value:

$$\begin{cases} D^{\alpha}\left(\dfrac{x(t) - \sum_{j=1}^{m} \mathcal{I}^{\beta_j} h_j(t, x(t))}{f(t, x(t))}\right) = g(t, x(t)), t \in J, \\ \\ x(0) = \mu(x) \text{ and } x(1) = A. \end{cases} \tag{3}$$

where $m \geq 1,\ \beta_j > 0,\ g, h_j \in \mathscr{C}(J \times \mathbb{R}, \mathbb{R}),\ j = 1, \ldots, m,\ f \in \mathscr{C}(J \times \mathbb{R}, \mathbb{R} \setminus \{0\})$, $A \in \mathbb{R}$ and $\mu : \mathscr{C}([0, 1], \mathbb{R}) \to \mathbb{R}$. Their results are based on Dhage's hybrid fixed point theorem.

Recently, Zakir Ullah et al. [28] have used also a Dhage's fixed point result to prove the existence results for

$$\begin{cases} D^{\alpha}\left(\dfrac{x(t) - h(t, x(t))}{f(t, x(t))}\right) = g(t, x(t)), t \in J, \alpha \in (1, 2], \\ \\ \left(\frac{x(t)-h(t,x(t))}{f(t,x(t))}\right)\Big|_{t=0} = 0 \text{ and } \left(\frac{x(t)-h(t,x(t))}{f(t,x(t))}\right)\Big|_{t=1} = 0, \end{cases} \tag{4}$$

where $f \in \mathscr{C}(J \times \mathbb{R}, \mathbb{R} \setminus \{0\})$ and $h,\ g \in \mathscr{C}(J \times \mathbb{R}, \mathbb{R})$.

It is well known that in the context of fractional differential equations, the explicit calculation of the solution is only possible in some simple cases. Therefore, the design of numerical methods that allow us to approximate the solution of these equations is considered one of the most interesting topics in this matter and the approximation of the solutions of such equations attracts the attention of a large number of authors. For example, in [16], by virtue of fractional calculus and fixed point methods, existence theorems as well as fixed point iterative method designed to approximate the solution of fractional differential iterative equations have been proved by Deng et al. In [2], by using Dhage iteration Method, Amara et al. have studied the problem of existence for a Caputo conformable hybrid multi-term integro-differential equations. Based on fixed point theorems developed by Dhage, in [26], Somjaiwang et al. have proved the existence and approximation of mild solutions of fractional hybrid differential equations involving the Caputo fractional derivative of order $0 < \alpha < 1$.

Subsequently, Roy et al. in [24] have discussed existence, uniqueness and approximation results for a class of fractional order perturbed differential equations using modified quasi-linearization and monotone iteration methods.

Motivated by the aforementioned works, in this paper we approach the study of the following nonlinear hybrid fractional differential equations with two-point nonlocal boundary conditions

$$\begin{cases} D^{\alpha}\left(\dfrac{x(t)-h_0(t,x(t))-\sum_{j=1}^{m}\mathcal{I}^{\beta_j}h_j(t,x(t))}{f(t,x(t))}\right)=g(t,x(t)),\ t\in J=[0,1], \\ \left(\frac{x(t)-h_0(t,x(t))}{f(t,x(t))}\right)\Big|_{t=0}=\mu_0(x) \text{ and } \left(\frac{x(t)-h_0(t,x(t))-\sum_{j=1}^{m}\mathcal{I}^{\beta_j}h_j(t,x(t))}{f(t,x(t))}\right)\Big|_{t=1}=\mu_1(x), \end{cases} \tag{5}$$

where D^{α} is understood in Caputo sense of order α, $1<\alpha\le 2$, I^{β_j} is the Riemann-Liouville fractional integral of order $\beta_j>0$, $j=1,\dots,m$, $f: J\times\mathbb{R}\to\mathbb{R}\setminus\{0\}$, $h_j: J\times\mathbb{R}\to\mathbb{R}$, $j=0,\dots,m$, and $\mu_0,\mu_1:\mathscr{C}(J,\mathbb{R})\to\mathbb{R}$ are given functions.

Several authors have discussed some existence results for nonlinear integro-differential equations, by using the class of nonlinear contraction mappings and the Boyd-Wong fixed point theorem (see, for example, [8, 9, 17, 18]). In this paper, using these tools, we prove an existence and uniqueness result for (5). We then construct an approximated method, by a combination method of the Picard operators sequence associated to the modeling problem of (5) and the notion of biorthogonal basis in Banach space. This type of ideas has proven to be very effective in problems without fractional derivatives (see [10, 11, 13]) with the advantages of this method over other numerical methods is its simplicity to implement it in a computer and the approximating functions are the sum of integrals of piecewise univariate and bivariate polynomials with coefficients easy to calculate.

The problem (5) significantly extends and generalizes the mentioned works in the literature [4, 28, 29, 31]. For example, if $f\equiv 1$, $h_i=0$, $i=0,\dots,m$, $\mu_0(x)=-x'(0)$, and $\mu_0(x)=-x'(1)$, then the problem (5) reduces to problem (1). In the special case when $h_i=0$, $i=0,\dots,m$ and $\mu_0(x)=\mu_1(x)=0$, the problem (5) reduces to problem (2). If $\mu_0(x)=\dfrac{\mu(x)-h_0(0,x(0))}{f(0,x(0))}$ and $\mu_1(x)=\dfrac{A-h_0(1,x(1))-\sum_{j=1}^{m}\mathcal{I}^{\beta_j}h_j(1,x(1))}{f(1,x(1))}$, the problem (5) reduces to problem (3). In the case when $h_0(t,x)=h(t,x)$, $h_i=0$, $i=1,\dots,m$, $\mu_0(x)=0$ and $\mu_1(x)=0$, the problem (5) reduces to problem (4).

The paper is organized as follows. Section 2 is devoted to recall the necessary definitions and preliminary results that will be used in the other sections of this paper. In Sect. 3, we establish the existence and uniqueness results for fractional differential equations with boundary conditions (5). On the based on the notion of biorthogonal systems combined with the Picard iterated techniques, a numerical method is constructed to approximated the obtained solution in Sect. 4. To illustrate the applicability of this method, we applied our results to some examples in Sect. 5.

2 Basic Tools

In this section, we present the basic definitions of fractional calculus (see [21, 23]), as well as certain questions about notation and some results necessary for the following sections.

In what follows, $\mathscr{C}(J, \mathbb{R})$ will denoted the space of continuous real-valued functions defined on an real interval J, $\mathscr{C}^m(J, \mathbb{R})$ the space of continuously differentiable functions up to order $m \geq 1$ and $\mathscr{AC}(J, \mathbb{R})$ will denoted the space of absolutely continuous mappings. The Gamma function Γ is defined by $\Gamma(n) = \int_0^\infty t^{n-1}e^{-t}dt$, $n > 0$.

Definition 2.1 Let $f : [0, \infty) \to \mathbb{R}$ be an $(n-1)$-times absolutely continuous mapping. The Caputo derivative of fractional order q of the function f is given by

$$D^q f(t) = \frac{1}{\Gamma(n-q)} \int_0^t (t-s)^{n-q-1} f^{(n)}(s)ds, \; n-1 < q < n, \; n = [q] + 1,$$

where $[q]$ denotes the integer part of the real number q. □

Definition 2.2 Let $f : [0, \infty) \to \mathbb{R}$ be a continuous function. The Riemann-Liouville fractional integral of order $q > 0$ is given by

$$I^q f(t) = \frac{1}{\Gamma(q)} \int_0^t \frac{f(s)}{(t-s)^{1-q}} ds, \; t \in [0, \infty)$$

provided the right-hand side is point-wise defined. □

Lemma 2.1 ([21]) Let $y \in \mathscr{AC}([0, 1], \mathbb{R})$ and $x \in \mathscr{C}^m([0, 1], \mathbb{R})$. For $q \in (m-1, m)$ and $t \in [0, 1]$ it is satisfied that:

(a) $I^q D^q x(t) = x(t) - \sum_{k=0}^{m-1} \frac{t^k}{k!} x^k(0)$.

(b) $D^q I^q y(t) = y(t)$.

□

In the following definition, we introduced the class of $\mathcal{D}$-Lipschitzian mappings, which play an efficient tool to solve, generalize and extend many works focussed to the study different nonlinear integro-differential equations under Lipshitzian conditions.

Definition 2.3 ([18, 19]) Let $(X, \|\cdot\|)$ be a normed linear space. A mapping $T : X \to X$ is $\mathcal{D}$-Lipschitzian if

$$\|Tx - Ty\| \leq \phi(\|x - y\|) \text{ for all } x, y \in X$$

for some continuous nondecreasing function $\phi : \mathbb{R}_+ \to \mathbb{R}_+$ with $\phi(0) = 0$. The function ϕ is called the $\mathcal{D}$-function of T. □

In the sequel we need the following technical lemma which ensures that every $\mathcal{D}$-Lipschitzian operator maps bounded subsets into bounded one.

Lemma 2.2 Let $(X, \|\cdot\|)$ be a normed linear space with zero element θ and $T : X \to X$ a $\mathcal{D}$-Lipschitzian mapping with $\mathcal{D}$-function ϕ. Then $T(M)$ is bounded for all bounded subset M of X, with bound $\phi(\|M\|) + \|T(\theta)\|$ where $\|M\| = \sup\{\|x\|, x \in M\}$. □

Proof Let M be a bounded subset of X. For any arbitrary $x \in M$, we have

$$\|T(x)\| \leq \|T(x) - T(\theta)\| + \|T(\theta)\|.$$

Since T is $\mathcal{D}$-Lipschitzian with $\mathcal{D}$-function ϕ, we get

$$\|T(x)\| \leq \phi(\|x\|) + \|T(\theta)\|.$$

Taking into account the fact that ϕ is nondecreasing, the desired result follows by passing to the supremum over $x \in M$. □

As a generalisation of the celebrated Banach's fixed point theorem, Boyd and Wong have established in [15] the following fixed point result.

Theorem 2.1 (Boyd-Wong Theorem) Let $(X, \|\cdot\|)$ be a Banach space and let $T : X \to X$. Assume that there exists a continuous function $\Theta : [0, \infty) \to [0, \infty)$ such that $\Theta(r) < r$ if $r > 0$, and

$$\|T(x) - T(y)\| \leq \Theta(\|x - y\|), \quad \forall x, y \in X.$$

Then T has a unique fixed point $\tilde{x} \in X$. Moreover, for any $x \in X$, the sequence $\{T^n(x)\}_{n\in\mathbb{N}}$ converges to $\tilde{x}$. □

The concept of biorthogonal system in a Banach space X plays a crucial role in the subsequent development. We refer the reader to [12, 20, 25] where the basic definitions and the most common examples of biorthogonal systems to be used in the following can be found.

3 Existence and Uniqueness of the Solution

We consider the following hybrid fractional differential equation:

$$\begin{cases} D^{\alpha}\left(\dfrac{x(t)-h_0(t,x(t))-\sum_{j=1}^{m}\mathcal{I}^{\beta_j}h_j(t,x(t))}{f(t,x(t))}\right)=g(t,x(t)),\, t\in J, \\ \left(\frac{x(t)-h_0(t,x(t))}{f(t,x(t))}\right)\Big|_{t=0}=\mu_0(x) \text{ and } \left(\frac{x(t)-h_0(t,x(t))-\sum_{j=1}^{m}\mathcal{I}^{\beta_j}h_j(t,x(t))}{f(t,x(t))}\right)\Big|_{t=1}=\mu_1(x), \end{cases} \tag{6}$$

where $J=[0,1]$, D^{α} is the Caputo derivative of order α with $1<\alpha\leq 2$, I^{β_j} is the Riemann-Liouville fractional integral of order $\beta_j>0$, $j=1,\ldots,m$, $f:J\times\mathbb{R}\to\mathbb{R}\setminus\{0\}$, $h_j:J\times\mathbb{R}\to\mathbb{R}$, $j=0,\ldots,m$, $\mu_0,\mu_1:\mathscr{C}(J,\mathbb{R})\to\mathbb{R}$ and $x\in\mathscr{C}(J,\mathbb{R})$ is the unknown to be determined. Specially, we discuss the existence of a solution x on a closed ball of the Banach algebra $\mathscr{C}(J,\mathbb{R})$ and problem (6) is considered under the following conditions:

(i) For every $x\in\mathbb{R}$, the partial mapping $t\to f(t,x)$ is continuous on J.

(ii) There exist R, k_1, $k_2>0$, continuous functions $\alpha,\eta,\gamma_i:J\to J$, and continuous nondecreasing functions $\phi,\psi,\varphi_i:\mathbb{R}_+\to\mathbb{R}_+$, $i=\{0,\ldots,m\}$, such that

  (a) $|f(t,x)-f(t,y)|\leq\alpha(t)\phi(|x-y|)$, $t\in J$ and $x,y\in\mathbb{R}$ with $|x|,|y|\leq R$,

  (b) $|g(t,x)-g(t,y)|\leq\eta(t)\psi(|x-y|)$, $t\in J$ and $x,y\in\mathbb{R}$ with $|x|,|y|\leq R$,

  (c) $|h_i(t,x)-h_i(t,y)|\leq\gamma_i(t)\varphi_i(|x-y|)$, $t\in J$ and $x,y\in\mathbb{R}$ with $|x|,|y|\leq R$,

  (d) $\left|\mu_j(u)-\mu_j(v)\right|\leq k_j\|u-v\|$, $j=1,2$, $u,v\in\mathscr{C}(J,\mathbb{R})$ with $\|u\|,\|v\|\leq R$.

(iii) For every $x\in\mathbb{R}$, the partial mappings $t\to g(t,x)$ and $t\to h_i(t,x)$, $i=0,\ldots,m$, are continuous on J.

With a technique similar to that used in [4], we obtain the following result.

Lemma 3.1 Let $y\in\mathscr{AC}(J,\mathbb{R})$, then x is a solution of the hybrid fractional integro-differential problem

$$\begin{cases} D^{\alpha}\left(\dfrac{x(t)-h_0(t,x(t))-\sum_{j=1}^{m}\mathcal{I}^{\beta_j}h_j(t,x(t))}{f(t,x(t))}\right)=y(t),\ t\in J,\\[2ex] \left(\frac{x(t)-h_0(t,x(t))}{f(t,x(t))}\right)\Big|_{t=0}=\mu_0(x) \text{ and } \left(\frac{x(t)-h_0(t,x(t))-\sum_{j=1}^{m}\mathcal{I}^{\beta_j}h_j(t,x(t))}{f(t,x(t))}\right)\Big|_{t=1}=\mu_1(x), \end{cases} \tag{7}$$

if and only if

$$x(t)=f(t,x(t))\left(\int_0^1 G(t,s)y(s)ds+(1-t)\mu_0(x)+t\mu_1(x)\right)+\sum_{j=1}^{m}\mathcal{I}^{\beta_j}h_j(t,x(t)) \\ +h_0(t,x(t)),$$

where

$$G(t,s)=\begin{cases}(t-s)^{\alpha-1}-t(1-s)^{\alpha-1} & \text{if } 0\le s\le t,\\ -t(1-s)^{\alpha-1} & \text{if } t\le s\le 1.\end{cases}$$

Here $G(t,s)$ is called the Green's function of the hybrid fractional integro-differential problem (7).

Proof Suppose that

$$D^{\alpha}\left(\frac{x(t)-\sum_{j=1}^{m}\mathcal{I}^{\beta_j}h_j(t,x(t))}{f(t,x(t))}\right)=y(t),$$

applying the Riemann-Liouville fractional integral operator on the right-hand side, we get

$$\frac{x(t)-\sum_{j=1}^{m}\mathcal{I}^{\beta_j}h_j(t,x(t))}{f(t,x(t))}=I^{\alpha}y(t)+k_1t+k_2. \tag{8}$$

Using the given boundary conditions, we find that $k_1=\mu_1(x)-\mu_0(x)-I^{\alpha}y(1)$ and $k_2=\mu_0(x)$. Substituting the values of k_1, k_2 in (8), we get solution. The converse follows by direct computation. □

Using Lemma 3.1, the problem (6) can be transformed as fixed point problem of hybrid operator equation of the form:

$$x=A(x)\cdot B(x)+C(x), \tag{9}$$

where $A, B, C:\mathscr{C}(J,\mathbb{R})\to\mathscr{C}(J,\mathbb{R})$ and for $x\in\mathscr{C}(J,\mathbb{R})$ and $t\in J$ are defined by

$$\begin{cases} (Ax)(t) = f(t, x(t)), \\ (Bx)(t) = \displaystyle\int_0^1 G(t, s)g(s, x(s))ds + (1-t)\mu_0(x) + t\mu_1(x) \text{ and} \\ (Cx)(t) = \displaystyle\sum_{j=1}^{m} \int_0^t \frac{1}{\Gamma(\beta_j)(t-s)^{1-\beta_j}} h_j(s, x(s))ds + h_0(t, x(t)). \end{cases}$$

In order to establish the main result in this section, we need to prove the following two auxiliary Lemmas.

Lemma 3.2 Let B_R be the closed bull of $\mathscr{C}(J, \mathbb{R})$ centered at zero with radius R. The operators A, B and C map B_R into $\mathscr{C}(J, \mathbb{R})$. □

Proof The claim regarding A follows immediately from assumption (i). On the other hand, let $x \in B_R$ and $t, t' \in J$. Without loss of generality, we may assume that $t' \leq t$. We have

$$|(Bx)(t) - (Bx)(t')| \leq \int_0^1 \left|G(t, s) - G(t', s)\right| |g(s, x(s))| \, ds$$
$$+|t - t'||\mu_1(x) - \mu_0(x)|.$$

From assumptions $(ii.b)$ and $(ii.c)$, it follows that the partial functions $x \to g(t, x)$, $i = 1, \ldots, m$, are $\mathcal{D}$-Lipschitzian with $\mathcal{D}$-function $\eta(t)\, \psi(\cdot)$. Using Lemma 2.2 we obtain

$$\int_0^1 \left|G(t, s) - G(t', s)\right| |g(s, x(s))| \, ds \leq \int_0^1 \left|G(t, s) - G(t', s)\right| \eta(s)\, \psi(|x(s)|) ds +$$
$$\int_0^1 \left|G(t, s) - G(t', s)\right| |g(s, 0)| \, ds.$$

By the continuity of the functions η and the partial mapping $s \to g(s, 0)$ on the compact interval J we deduce that

$$\left|B(x)(t) - B(x)(t')\right| \leq \Delta_{\eta,\psi,g} \left(|t - t'||\mu_1(x) - \mu_0(x)|\right),$$

where

$$\Delta_{\eta,\psi,g} := \psi(R) \, \|\eta(\cdot)\| + \|g(\cdot, 0)\| \, .$$

As $t \to t'$, the right-hand side of the above inequality tends to zero, and consequently we get $B(x) \in \mathscr{C}(J, \mathbb{R})$.

Similarly, we have

$$\begin{aligned}\left|C(x)(t)-C(x)(t')\right| \leq &\sum_{j=1}^{m}\int_{t'}^{t}\frac{(t-s)^{\beta_j-1}}{\Gamma(\beta_j)}\left|h_j(s,x(s))\right|ds+\\ &\sum_{j=1}^{m}\frac{1}{\Gamma(\beta_j)}\int_0^{t'}\left|(t-s)^{\beta_j-1}-(t'-s)^{\beta_j-1}\right|\left|h_j(s,x(s))\right|ds+\\ &\left|h_0(t,x(t))-h_0(t',x(t'))\right|.\end{aligned}$$

By using Lemma 2.2 we get

$$\left|C(x)(t)-C(x)(t')\right| \leq$$

$$\sum_{j=1}^{m}\frac{1}{\Gamma(\beta_j)}\Delta_{1,\varphi_j,h_j}\left(\frac{|t-t'|^{\beta_j}}{\beta_j}+\int_0^{t'}\left|(t-s)^{\beta_j-1}-(t'-s)^{\beta_j-1}\right|ds\right)$$

$$+\left|h_0(t,x(t))-h_0(t',x(t))\right|+|\gamma(t')|\varphi_0(|x(t)-x(t')|)$$

where $\Delta_{\gamma_i,\varphi_i,h_i}=\|\gamma_i(\cdot)\|\ \varphi_i(R)+\ \|h_i(\cdot,0)\|$.
Therefore, as $t\to t'$, the right-hand side of the above inequality tends to zero, and consequently we get $C(x)\in\mathscr{C}(J,\mathbb{R})$. □

Lemma 3.3 The operators A, B and C are $\mathcal{D}$-Lipschitzian. □

Proof Using assumption (ii) together with the continuity of $\alpha(\cdot)$ on J, we can see that A is $\mathcal{D}$-Lipschitzian with $\mathcal{D}$-function Φ given by $\Phi(t)=\|\alpha\|_\infty\phi(t),\ t\in J$.

Now, let us prove that B is $\mathcal{D}$-Lipschitzian. To see this, let $x,y\in B_R$, and $t\in J$. Keeping in mind the assumption (ii), we obtain

$$|B(x)(t)-B(y)(t)| \leq$$

$$\int_0^1|G(t,s)(g(s,x(s))-g(s,y(s)))|\,ds+$$

$$(1-t)|\mu_0(x)-\mu_0(y)|+t|\mu_1(x)-\mu_1(y)| \leq$$

$$\int_0^1|G(t,s)|\eta(s)\psi(|x(s)-y(s)|)ds+((1-t)k_0+tk_1)\,\|x-y\| \leq$$

$$\frac{2}{\Gamma(\alpha+1)}\|\eta(\cdot)\|\psi(\|x-y\|)+((1-t)k_0+tk_1)\,\|x-y\|.$$

Taking the supremum over t, we obtain that B is $\mathcal{D}$-Lipschitzian with $\mathcal{D}$-function Ψ given by $\Psi(t)=\dfrac{2}{\Gamma(\alpha+1)}\,\|\eta(\cdot)\|\ \psi(t)+(k_0+k_1)\,t,\ t\geq 0$.

Proceeding as above, Assumption (ii) ensures that

$$|C(x)(t) - C(y)(t)| \leq \sum_{j=1}^{m} \frac{1}{\Gamma(\beta_j)} \int_0^t \frac{1}{(t-s)^{1-\beta_j}} \left|h_j(s, x(s)) - h_j(s, y(s))\right| ds$$
$$+ |h_0(t, x(t)) - h_0(t, y(t))|$$
$$\leq \sum_{j=1}^{m} \frac{1}{\Gamma(\beta_j + 1)} \|\gamma_j\| \varphi_j(\|x - y\|) + \|\gamma_0(\cdot)\| \varphi_0(\|x - y\|),$$

from where, it follows that

$$\|C(x) - C(y)\| \leq \sum_{j=1}^{m} \frac{1}{\Gamma(\beta_j + 1)} \|\gamma_j\| \varphi_j(\|x - y\|) + \|\gamma_0(\cdot)\| \varphi_0(\|x - y\|).$$

Consequently, C is a $\mathcal{D}$-Lipschitzian mapping on $\mathscr{C}(J, \mathbb{R})$ with $\mathcal{D}$-function Ξ defined by $\Xi(t) = \sum_{j=1}^{m} \frac{1}{\Gamma(\beta_j + 1)} \|\gamma_j\| \varphi_j(t) + \|\gamma_0(\cdot)\| \varphi_0(t), t \geq 0.$ □

Remark 3.1 Notice that the sets $A(B_R)$, $B(B_R)$ and $C(B_R)$ are bounded and in view of Lemma 2.2, we have $\|A(B_R)\| \leq \Phi(R) + \|A(\theta)\|$, $\|B(B_R)\| \leq \Psi(R) + \|B(\theta)\|$, and $\|C(B_R)\| \leq \Xi(R) + \|C(\theta)\|$. □

Therefore, we are in position to present the main result of this section.

Theorem 3.1 Suppose that

$$\begin{aligned} &\|A(B_R)\| \; \|B(B_R)\| + \|C(B_R)\| \leq R \\ &\|A(B_R)\| \; \Psi(t) + \|B(B_R)\| \; \Phi(t) + \Xi(t) < t, \; t > 0. \end{aligned} \tag{10}$$

Then, the nonlinear Caputo fractional differential equation with boundary conditions (6) has a unique solution $\tilde{x}$ in B_R, and

$$\tilde{x} = \lim_{n \to \infty} (A \cdot B + C)^n (x) \text{ for all } x \in B_R. \tag{11}$$

□

Proof On one hand, it can be seen from the first estimate of (10) that $A \cdot B + C$ maps B_R into itself.

On the other hand, let $x, y \in B_R$, using Lemma 3.2 one obtains

$$\|A(x) \cdot B(x) + C(x) - A(y) \cdot B(y) - C(y)\| \leq$$

$$\|A(x) \cdot (B(x) - B(y))\| + \|(A(x) - A(y)) \cdot B(y)\| + \|C(x) - C(y)\| \leq$$

$$\|A(B_R)\| \; \Psi(\|x - y\|) + \|B\,(B_R)\| \; \Phi(\|x - y\|) + \Xi(\|x - y\|).$$

Hence, from the second estimate of (10) and Remark 3.1 we deduce that $A \cdot B + C : B_R \to B_R$ satisfies the hypothesis of Theorem 2.1 with

$$\Theta(r) = \|A(B_R)\| \, \psi(r) + \|B(B_R)\| \, \Phi(r) + \Xi(r), \; r > 0.$$

In consequence we conclude that there exists a unique point $\tilde{x} \in B_R$ such that $A(\tilde{x}) \cdot B(\tilde{x}) + C(\tilde{x}) = \tilde{x}$ and the sequence $\{(A \cdot B + C)^n(x), n \in \mathbb{N}\}$ converges to $\tilde{x}$ for all $x \in B_R$. □

Remark 3.2 Notice that by induction argument we can prove that

$$\|(A \cdot B + C)^n(x) - (A \cdot B + C)^n(y)\| \leq \Theta^n(\|x - y\|), \text{ for all } n = 1, 2, \ldots \tag{12}$$

□

4 Numerical Approximation Method

The above section allows us to express the solution of (6) as the limit of the sequence of functions $\{(A \cdot B + C)^n(x)\}_{n \in \mathbb{N}}$, with $x \in B_R$. Obviously, if it were possible to explicitly calculate, for each iteration, the expression $(A \cdot B + C)^n(x)$, then for each n we would have an approximation of the fixed point $\tilde{x}$. But, as a practical matter, such an explicit calculation is only possible in very particular cases. For this reason, we need to construct another approximation of the fixed point which is simple to calculate in practice.

Specifically, the main idea to construct the approximated scheme is the following: take $x \in B_R$, in the first step, we approximate $B(x)$ by $\widetilde{B}_1(x)$, and $C(x)$ by $\widetilde{C}_1(x)$, then we obtain $N_1(x) := A(x) \cdot \widetilde{B}_1(x) + \widetilde{C}_1(x)$ as an approximation of the first term of the Picard iterate, $A(x) \cdot B(x) + C(x)$. We obtain the second term of our scheme combining $N_1(x)$ with $A \cdot \widetilde{B}_2 + \widetilde{C}_2$ where $\widetilde{B}_2$ and $\widetilde{C}_2$ are approximation of the operator B and C respectively. This way, $N_2 \circ N_1(x) = (A \cdot \widetilde{B}_2 + \widetilde{C}_2)(N_1(x))$ approximates to $(A \cdot B + C)^2(x)$. Repeating the process, we construct two sequences of operators $\{\widetilde{B}_n\}_{n \in \mathbb{N}}$ and $\{\widetilde{C}_n\}_{n \in \mathbb{N}}$ in order to obtain a sequence $\{N_n \circ \ldots \circ N_1(x)\}_{n \in \mathbb{N}}$ of approximations of the fixed point $\tilde{x}$.

In order to construct an approximation scheme for the solution $\tilde{x}$, we have to prove the following technical Lemma.

Lemma 4.1 Let $\tilde{x}$ be the unique solution of the nonlinear Caputo fractional differential equation with boundary conditions (6). Let $x \in B_R$, $\varepsilon > 0$, and $m \in \mathbb{N}$ such that

$$\left\|(A \cdot B + C)^m(x) - \tilde{x}\right\| \leq \varepsilon/2. \tag{13}$$

Let $\varepsilon_1, \ldots, \varepsilon_m > 0$ such that

$$\sum_{p=1}^{m} \Theta^{m-p}\left(\varepsilon_p\right) < \varepsilon/2. \tag{14}$$

and let $N_0, N_1, \ldots, N_m$ be m operators on $\mathscr{C}(J, \mathbb{R})$, satisfying for $p = 1, \ldots, m$ that

$$\left\|(A \cdot B + C) \circ N_{p-1} \circ \ldots \circ N_1(x) - N_p \circ \ldots \circ N_1(x)\right\| \leq \varepsilon_p, \tag{15}$$

□

with N_0 the identity operator on X. Then

$$\|\tilde{x} - N_m \circ \ldots \circ N_1(x)\| \leq \varepsilon.$$

Proof Taking into account the triangular inequality and (12),

$$\left\|(A \cdot B + C)^m(x) - N_m \circ \ldots \circ N_1(x)\right\|$$

$$\left\|(A \cdot B + C)^m(x) - (A \cdot B + C)^{m-1} \circ N_1(x)\right\| +$$

$$\left\|(A \cdot B + C)^{m-1} \circ N_1(x) - (A \cdot B + C)^{m-2} \circ N_2 \circ N_1(x)\right\| + \cdots +$$

$$\left\|(A \cdot B + C)^2 \circ N_{m-2} \circ \ldots \circ N_1(x) - (A \cdot B + C) \circ N_{m-1} \circ \ldots \circ N_1(x)\right\|$$

$$+ \|(A \cdot B + C) \circ N_{m-1} \circ \ldots \circ N_1(x) - N_m \circ \ldots \circ N_1(x)\| \leq$$

$$\sum_{p=1}^{m-1} \Theta^{m-p}\left(\left\|(A \cdot B + C) \circ N_{p-1} \circ \ldots \circ N_1(x) - N_p \circ \ldots \circ N_1(x)\right\|\right) +$$

$$\|(A \cdot B + C) \circ N_{m-1} \circ \ldots \circ N_1(x) - N_m \circ \ldots \circ N_1(x)\|.$$

By using (14) and (15), this implies that

$$\left\|(A \cdot B + C)^m(x) - N_m \circ \ldots \circ N_1(x)\right\| \leq \sum_{p=1}^{m} \Theta^{m-p}\left(\varepsilon_p\right) < \varepsilon/2,$$

which combined with (13) allows us to ensure that

$$\|\tilde{x} - N_m \circ \ldots \circ N_1(x)\| \leq$$

$$\left\|\tilde{x} - (A \cdot B + C)^m(x)\right\| + \left\|(A \cdot B + C)^m(x) - N_m \circ \ldots \circ N_1(x)\right\| \leq$$

$$\left\|\tilde{x} - (A \cdot B + C)^m(x)\right\| + \frac{\varepsilon}{2} < \varepsilon,$$

□and the proof is completed. □

4.1 Construction of the Approximation Scheme and Main Approximation Result

Let us consider a Schauder basis $\{\tau_n\}_{n\geq 1}$ in $\mathscr{C}(J,\mathbb{R})$ and the sequence of associated projections $\{\xi_n\}_{n\geq 1}$, (see [20, 25] for details). Note that, in view of the Baire category theorem, for all $n \geq 1$, ξ_n is continuous. This yields, in particular, that

$$\lim_{n\to\infty} \|\xi_n(x) - x\| = 0. \tag{16}$$

For each $p \in \mathbb{N}$, we consider $n_p \in \mathbb{N}$ and we define $\widetilde{B}_p, \widetilde{C}_p : \mathscr{C}(J,\mathbb{R}) \to \mathscr{C}(J,\mathbb{R})$ by

$$\widetilde{B}_p(x)(t) = (1-t)\mu_0(x) + t\mu_1(x) + \int_0^1 G(t,s)\xi_{n_p}\left(U_0(x)\right)(s)ds,$$

$$\widetilde{C}_p(x)(t) = V_0^0(x)(t) + \sum_{j=1}^{m} \frac{1}{\Gamma(\beta_j)} \int_0^t \frac{1}{(t-s)^{1-\beta_j}} \xi_{n_p}\left(V_0^j(x)\right)(s)ds,$$

where $U_0, V_0^i : \mathscr{C}(J,\mathbb{R}) \to \mathscr{C}(J,\mathbb{R})$ are defined by $U_0(x)(s) = g(s,x(s))$ and $V_0^i(x)(s) = h_i(s,x(s))$.

Let us defined the operator $N_p : \mathscr{C}(J,\mathbb{R}) \to \mathscr{C}(J,\mathbb{R})$ which assigns to each element $x \in \mathscr{C}(J,\mathbb{R})$ the value $N_p(x) \in \mathscr{C}(J,\mathbb{R})$ such that

$$N_p(x)(t) = A(x)(t)\widetilde{B}_p(x)(t) + \widetilde{C}_p(x)(t),\ t \in J. \tag{17}$$

Lemma 4.2 For all fixed $p \geq 1$, the operator $N_p \circ \ldots \circ N_1$ maps B_R into itself. □

Proof Let $p \in \mathbb{N}\backslash\{0\}$ be fixed. For all $l = 1,\ldots,p$, and all $x \in B_R$, we have

$$|N_l(x)(t)| \leq |f(t,x(t))| \left(|(1-t)\mu_0(x) + t\mu_1(x)| + \int_0^1 |G(t,s)| \left|\xi_{n_l}\left(U_0(x)\right)(s)\right| ds \right)$$
$$+|h_0(t,x(t))| + \sum_{j=1}^{m} \frac{1}{\Gamma(\beta_j)} \int_0^t \frac{1}{(t-s)^{1-\beta_j}} \left|\xi_{n_l}\left(V_0^i(x)\right)(s)\right| ds,$$

Proceeding essentially as in the above section and using the fact that ξ_{n_l} is a bounded linear operator on $\mathscr{C}(J,\mathbb{R})$, we get

$$\sup_{t\in J} |N_l(x)(t)| \leq \|A(B_R)\|\,\|B(B_R)\| + \|C(B_R)\| \leq R.$$

Consequently, N_l maps B_R into itself, from which we see that $N_p \circ \ldots \circ N_1$ is also maps B_R into itself. □

The objective now is to justify that we can choose $m \in \mathbb{N}$ and $n_1,\ldots,n_m$, so that the operator $N_m \circ N_{m-1} \circ \ldots \circ N_1(x)$ can be used to obtain an approximation of the unique solution to (6).

Theorem 4.1 Let $\tilde{x}$ be the unique solution of the nonlinear hybrid fractional differential equations with boundary conditions (6). Let $x \in B_R$ and $\varepsilon > 0$, then there exists $m \in \mathbb{N}$ and $n_1, n_2, \ldots, n_m$ such that

$$\|\tilde{x} - N_m \circ \ldots \circ N_1(x)\| \leq \varepsilon,$$

Proof Let $x \in B_R$ and let $\varepsilon > 0$. By (11) there is $m \in \mathbb{N}$ such that

$$\|(A \cdot B + C)^m(x) - \tilde{x}\| \leq \varepsilon/2$$

Observe first that from the continuity of the function Θ, we can find $\varepsilon_1, \ldots, \varepsilon_m > 0$ which satisfy (14). By Lemma 4.1 the proof will end when we prove that for $p = 1, \ldots, m$, we can choose $n_p \in \mathbb{N}$ such that the condition (15) is verified.

For all $p = 1, 2, \ldots, m$, using Lemma 4.2, we have $N_p \circ N_{p-1} \circ \ldots \circ N_1(x) \in B_R$ and $N_{p-1} \circ \ldots \circ N_1(x) \in B_R$ and from the triangular inequality, one obtains

$$\left|(A \cdot B + C) \circ N_{p-1} \circ \ldots \circ N_1(x)(t) - N_p \circ N_{p-1} \circ \ldots \circ N_1(x)(t)\right| \leq$$

$$H_{p,1}(x)(t) + H_{p,2}(x)(t),$$

where

$$H_{p,1}(x)(t) = \left|(A \cdot B) \circ N_{p-1} \circ \ldots \circ N_1(x)(t) - (A \cdot \widetilde{B}_p) \circ N_{p-1} \circ \ldots \circ N_1(x)(t)\right|$$

and

$$H_{p,2}(x)(t) = \left|C \circ N_{p-1} \circ \ldots \circ N_1(x)(t) - \widetilde{C}_p \circ N_{p-1} \circ \ldots \circ N_1(x)(t)\right|.$$

Considering $A_p, U_p, V_p^i : \mathscr{C}(J, \mathbb{R}) \to \mathscr{C}(J, \mathbb{R})$ defined by

$$A_p(x)(s) = f\left(s, N_{p-1} \circ \cdots N_1(x)(s)\right),$$

$$U_p(x)(s) = g\left(s, N_{p-1} \circ \cdots N_1(x)(s)\right),$$

and

$$V_p^i(x)(s) = h_i\left(s, N_{p-1} \circ \cdots N_1(x)(s)\right),$$

then, taking the supremum over $t \in J$, we get

$$\left\|H_{p,1}(x)\right\| \leq \left\|A_{p-1}(x)\right\| \int_0^1 |G(t,s)| ds \left\|\xi_{n_p}(U_{p-1})(x) - U_{p-1}(x)\right\| \leq$$

$$\left\|A_{p-1}(x)\right\| \, \|G(t,\cdot)\|_{L^1} \left\|\xi_{n_p}\left(U_{p-1}\right)(x) - U_{p-1}(x)\right\|.$$

Similarly, we can prove that

$$\left\| H_{p,2}(x) \right\| \leq \sum_{j=1}^{m} \frac{1}{\Gamma(\beta_j)} \int_0^t \frac{1}{(t-s)^{1-\beta_j}} ds \, \left\| \xi_{n_p} \left(V_{p-1} \right)(x) - V_{p-1}(x) \right\|$$
$$\leq \sum_{j=1}^{m} \frac{1}{\Gamma(\beta_j + 1)} \left\| \xi_{n_p} \left(V_{p-1} \right)(x) - V_{p-1}(x) \right\|,$$

from where it follows that

$$\left\| (A \cdot B + C) \circ N_{p-1} \circ \ldots \circ N_1(x) - N_p \circ \ldots \circ N_1(x) \right\| \leq$$
$$\left\| A_{p-1}(x) \right\| \, \|G(t, \cdot)\|_{L^1} \left\| \xi_{n_p} \left(U_{p-1} \right)(x) - U_{p-1}(x) \right\|$$
$$+ \sum_{j=1}^{p} \frac{1}{\Gamma(\beta_j + 1)} \left\| \xi_{n_p} \left(V_{p-1} \right)(x) - V_{p-1}(x) \right\|$$

and therefore, since $\{\xi_n\}_{n \geq 1}$ is the sequence of projections associated with a Schauder base, we can find $n_p \geq 1$ and consequently N_p, such that

$$\left\| (A \cdot B + C) \circ N_{p-1} \circ \ldots \circ N_1(x) - N_p \circ \ldots \circ N_1(x) \right\| \leq$$
$$\|A(B_R)\| \sup_{s \in J} \|G(s, \cdot)\|_{L^1} \left\| \xi_{n_p} \left(U_{p-1} \right)(x) - U_{p-1}(x) \right\|$$
$$+ \sum_{j=1}^{m} \frac{1}{\beta_j \Gamma(\beta_j)} \left\| \xi_{n_p} \left(V_{p-1} \right)(x) - V_{p-1}(x) \right\| \leq \varepsilon_p.$$

Using these obtained estimates, together with Lemma 4.1, we conclude that

$$\|\tilde{x} - N_m \circ \ldots \circ N_1(x)\| < \varepsilon,$$

and the proof is completed. □

In the following, we fixed $x \in B_R$ and we denote

$$X_m^{(n_1, \ldots, n_m)} = N_m \circ \ldots \circ N_1(x),$$

which is an approximation of the solution to the problem (6).

Remark 4.1 Let $x \in B_R$, let $\tilde{x}$ be the unique fixed point of $A \cdot B + C$, given $\varepsilon > 0$, if $\overline{m}$ satisfies $\Theta^{\overline{m}-1}(2R) \leq \varepsilon/4$, then,

$$\left\| (A \cdot B + C)^{\overline{m}}(x) - \tilde{x} \right\| \leq \varepsilon/2,$$

it is sufficient to take into account the fact that

$$\begin{aligned}\left\|(A\cdot B+C)^{\overline{m}}(x)-\tilde{x}\right\| &\leq \left\|(A\cdot B+C)^{\overline{m}}(x)-(A\cdot B+C)^{\overline{m}-1}(x)\right\| \\ &\quad+\left\|(A\cdot B+C)^{\overline{m}-1}(x)-(A\cdot B+C)^{\overline{m}-1}(\tilde{x})\right\| \\ &\leq \Theta^{\overline{m}-1}(\|x-(A\cdot B+C)(x)\|)+\Theta^{\overline{m}-1}(\|x-\tilde{x}\|) \\ &\leq 2\Theta^{\overline{m}-1}(2R)\leq \varepsilon/2.\end{aligned}$$

5 Numerical Examples

In this section, we show the numerical results obtained by the proposed method to three specific examples. All numerical examples have been performed using Wolfram Mathematica and considering the Faber-Schauder basis in $\mathscr{C}(J,\mathbb{R})$, $m=4$ and taking $n_p=9$ for $p=1,\ldots,4$. The approximate solution obtained with these conditions will be denoted x^* and we are going to consider the initial function $x(t)=0.16$.

Example 5.1 Example 1 We consider now the following problem:

$$\begin{cases} D^{5/3}\left[\dfrac{x(t)}{f(t,x(t))}\right]=\frac{1}{2}x(t)^2,\, t\in J, \\ \left(\frac{x(t)}{f(t,x(t))}\right)|_{t=0}=\frac{1}{4}x(0)+\frac{3}{40} \text{ and } \left(\frac{x(t)}{f(t,x(t))}\right)|_{t=1}=\frac{1}{5}x(1)+\frac{4}{5}, \end{cases} \tag{18}$$

where $f(t,x)=\dfrac{4400(t+1)\Gamma\left(\frac{5}{3}\right)}{132t^{5/3}+99t^{8/3}+27t^{11/3}-258t+4400(t+1)\Gamma\left(\frac{5}{3}\right)}$.

Notice that this problem is a particular case of (6), where $\alpha=\frac{5}{3}$, $g(t,x)=\frac{1}{2}x^2$, $h_j(t,x)=0$, $j=0,\ldots,m$, $\mu_0(x)=\frac{1}{4}x(0)+\frac{3}{40}$ and $\mu_1(x)=\frac{1}{5}x(1)+\frac{4}{5}$.

We can show that f, g and h_j satisfy (i)-(iii) with $R=\frac{1}{2}$, $k_0=\frac{1}{4}$, $k_1=\frac{3}{40}$, $\alpha(\cdot)\equiv 0, \phi(\cdot)\equiv 0, \eta(t)=1/2$ and $\psi(t)=t, r\geq 0$. Then an application of Theorem 3.1 achieves that the nonlinear fractional integro-differential equation (18) has a unique solution in $B_{1/2}$, in fact, the solution is $\tilde{x}(t)=\frac{1}{10}(t+1)$. The results obtained are collected in Table 1. □

Table 1 Numerical results for (18)

t	0.2	0.4	0.6	0.8	1
$\lvert x^*(t)-\tilde{x}(t)\rvert$	4.9×10^{-2}	3.7×10^{-2}	2.5×10^{-2}	1.3×10^{-2}	3.2×10^{-4}

$\|x^*-\tilde{x}\|=5.53\times10^{-2}$

Table 2 Numerical results for (19)

t	0.2	0.4	0.6	0.8	1
$\lvert x^*(t) - \tilde{x}(t)\rvert$	3.7×10^{-6}	3.4×10^{-5}	2.7×10^{-6}	3.2×10^{-4}	1.4×10^{-3}
$\lVert x^* - \tilde{x}\rVert = 1.52 \times 10^{-3}$					

Example 5.2 Example 2 Consider the following problem

$$\begin{cases} D^{5/3}\left(\dfrac{x(t) - \sum_{j=1}^{2} I^{\frac{2j-1}{2}} h_j(t, x(t))}{f(t, x(t))}\right) = g(t, x(t)), t \in J, \\ \left(\frac{x(t)-\sum_{j=1}^{2} I^{\frac{2j-1}{2}} h_j(t,x(t))}{f(t,x(t))}\right)\Bigg|_{t=0} = \frac{x(0)}{10} + \frac{9}{100} \text{ and} \\ \left(\frac{x(t)-\sum_{j=1}^{2} I^{\frac{2j-1}{2}} h_j(t,x(t))}{f(t,x(t))}\right)\Bigg|_{t=1} = \frac{22}{625}(5x(1) + 4). \end{cases} \tag{19}$$

with $f(t, x) = \dfrac{680\left(51975\sqrt{\pi}t^2 - 128t^{9/2}(3t + 11)\right)\Gamma\left(\frac{5}{3}\right)}{243\sqrt{\pi}\left(405t^{17/3} - 405t + (99484t + 130900)\Gamma\left(\frac{5}{3}\right)\right)}$, $g(t, x) = \dfrac{1}{2}x^2$, and $h_0 = 0$, $h_1(t, x) = \dfrac{1}{3}x^2$, $h_2(t, x) = \dfrac{1}{2}x^2$. Notice that this problem is a particular case of (6), where $m = 2$, $\alpha = 5/3$, $\beta_j = \frac{2j-1}{2}$, $j = 1, 2$, $\mu_0(x) = \frac{x(0)}{10} + \frac{9}{100}$ and $\mu_1(x) = \frac{22}{625}(5x(1) + 4)$.

It is clear that the mappings f, g and h_j satisfy (i)-(iii) with $R = \frac{1}{2}$, $k_0 = \frac{1}{10}$, $k_1 = \frac{22}{125}$, $\alpha(t) = 0$, $\Phi(t) = t$, $\eta(t) = \frac{1}{2}$, $\gamma_0(t) = 0$, $\gamma_1(t) = \frac{1}{3}$ and $\gamma_2(t) = \frac{1}{2}$. Then the conditions (10) are satisfied.

An application of Theorem 3.1 achieves that the nonlinear fractional integro-differential equation (19) has a unique solution in B_R, with $R = 1/2$. The solution is $\tilde{x}(t) = \frac{1}{10}t^2$. The results obtained are shown in Table 2.

□

Acknowledgements The research of M. I. Berenguer has been partially supported by Junta de Andalucia (Spain), Project FQM359 "Convex and Numerical Analysis".

References

1. Abdon, A.: Modelling the spread of COVID-19 with new fractal-fractional operators: can the lockdown save mankind before vaccination? Chaos, Solitons and Fract. **136**, 38 pp (2020)
2. Amara, A., Etemad, S., Rezapour, S.: Approximate solutions for a fractional hybrid initial value problem via the Caputo conformable derivative. Adv. Differ. Equ. **608**, 19 pp (2020)
3. Ahmad, B., Losada, J., Nieto, J.J.: On antiperiodic nonlocal three-point boundary value problems for nonlinear fractional differential equations. Discret. Dyn. Nat. Soc. Art. ID 973783, 7 pp (2015)
4. Ahmad, B., Ntouyas, S.K., Tariboon,J.: A nonlocal hybrid boundary value problems of Caputo fractional integro-differential equations. Acta Math. Sci. Ser. B (Engl. Ed.) **36**, 1631–1640 (2016)
5. Baleanu, D., Etemad,S., Rezapour,S.: A hybrid Caputo fractional modeling for thermostat with hybrid boundary value conditions. Bound. Value. Probl. **2020**, 16pp (2020)
6. Baleanu, D., Luo, A.C.J., Tenreiro Machado, J.A.: Fractional dynamics and control. Fractional Dynamics and Control, pp. 1–313 (2012)
7. Bai, Z., Lu, H.: Positive solutions for boundary value problem of nonlinear fractional differential equation. J. Math. Anal. Appl. **311**, 495–505 (2005)
8. Ben Amara, K., Jeribi, A., Kaddachi, N.: New fixed point theorems for countably condensing maps with an application to functional integral inclusions. Math. Slovaca **71**, 1–24 (2021)
9. Ben Amara, K., Jeribi, A., Kaddachi, N.: Equivalence of some properties in the theory of Banach algebras and applications. J. Math. Anal. Appl. **520**, 126865 (2022)
10. Berenguer, M.I., Gámez, D., López Linares, A.J.: Fixed point techniques and Schauder bases to approximate the solution of the first order nonlinear mixed Fredholm-Volterra integro-differential equation. J. Comput. Appl. Math. **252**, 52–61 (2013)
11. Berenguer, M.I., Gámez, D.: Projected iterations of fixed-point type to solve nonlinear partial Volterra integro-differential equations. Bull. Malays. Math. Sci. Soc. **43**(6), 4431–4442 (2020)
12. Berenguer, M.I., Gámez, D., López Linares, A.J.: Fixed point techniques and Schauder bases to approximate the solution of the first order nonlinear mixed Fredholm-Volterra integro-differential equation. J. Comput. Appl. Math. **252**, 52–61 (2013)
13. Berenguer, M. I., Garralda-Guillem, A.I., Ruiz Galán, M.: Biorthogonal systems approximating the solution of the nonlinear Volterra integro-differential equation. Fixed Point Theory Appl. Art. ID 470149, 9 pp (2010)
14. Boulares, H., Ardjouni, A., Laskri, Y.: Positive solutions for nonlinear fractional differential equations. Positivity **21**, 1201–1212 (2017)
15. Boyd, D.W., Wong, J.S.W.: On nonlinear contractions. Proc. Am. Math. Soc. **20**, 458–464 (1969)
16. Deng, J., Wang, J.: Existence and approximation of solutions of fractional order iterative differential equations. Cent. Eur. J. Phys. **11**, 1377–1386 (2013)
17. Dhage, B.C.: A fixed point theorem in Banach algebras involving three operators with applications. Kyungpook Math. J. **44**, 145–155 (2004)
18. Dhage, B.C.: On a fixed point theorem in Banach algebras with applications. Appl. Math. Lett. **18**, 273–280 (2005)
19. Dhage, B.C., Jadhav, N.S.: Basic results in the theory of hybrid differential equations with linear perturbations of second type. Tamkang J. Math. **44**, 171–186 (2013)
20. Gelbaum, B.R., Gil de Lamadrid, J.: Bases of tensor products of Banach spaces. Pac. J. Math. **11**, 1281–1286 (1961)
21. Kilbas, A.A., Srivastava, H.M., Trujillo, J.J.: Theory and Applications of Fractional Differential Equations, 204. North-Holland Mathematics Studies, Amsterdam. El- sevier Science B.V. (2006)
22. Nieto, J.J., Pimentel, J.: Positive solutions of a fractional thermostat model. Bound. Value Probl. **2013**, 11 pp (2013)
23. Podlubny, I.: Fractional Differential Equations. Academic, San Diego (1999)

24. Roy, R., Kumar, K. H.: An Iterative Scheme for a Class of Fractional Order Perturbed Differential Equations. M. Singh et al. (Eds.): ICACDS 2019, CCIS 1045, pp. 151–163 (2019)
25. Semadeni, Z.: Product Schauder bases and approximation with nodes in spaces of continuous functions. Bull. Acad. Polon. Sci. **11**, 387–391 (1963)
26. Somjaiwang, D., Sa Ngiamsunthorn, P.: Existence and approximation of solutions to fractional order hybrid differential equations. Adv. Differ. Equ. **2016**, 11 pp (2016)
27. Tuan N. H., Mohammadi H., Rezapour S.: A mathematical model for COVID-19 transmission by using the Caputo fractional derivative. Chaos, Solitons and Fract. **140**, 11 pp (2020)
28. Zakir Ullah, M., Ali, A., Ali, K.R., Iqbal, M.: Existence results to a class of hybrid fractional differential equations. Matriks Sains Matematik **2**, 13–17 (2018)
29. Zhang, S.: Positive solutions for boundary-value problems of nonlinear fractional differential equations. Electron. J. Differ. Equ. **36**, 1–12 (2006)
30. Zhao, Y., Sun, S., Han, Z., Li, Q.: Theory of fractional hybrid differential equations. Comput. Math. Appl. **62**, 1312–1324 (2011)
31. Zhao, Y., Sun, S., Han, Z., Li, Y.: The existence of solutions for boundary value problem of fractional hybrid differential equations. Commun. Nonlinear Sci. Numer. Simul. **17**, 4961–4967 (2012)

Superconvergent Results for Solutions of *Volterra* Integral Equations with Weakly Singular Kernels in Weighted Spaces

Abdelmajid El Bouayadi, Chafik Allouch, Ahmed Boujraf, and Mohamed Tahrichi

Abstract For the purpose of approximating the solution of second-kind *Volterra* integral equations with algebraic singularities at the diagonal and endpoints, this work explores a modified *Nyström* approach based on the zeros of the *Jacobi* polynomials. The technique employs the interpolating operator in suitable weighted spaces and the associated product integration formulae. The method is shown to be stable and more accurate than the standard *Nyström* scheme, and the precision of the solution can further be enhanced by using the *Sloan* iteration. The convergence and stability results are confirmed by several numerical experiments.

1 Introduction

Let K be a *Volterra* integral operator, defined as

$$(Kf)(y) = \int_{-1}^{y} \kappa(x, y) f(x)(y - x)^{\alpha}(1 + x)^{\beta} dx, \quad y \in (-1, 1), \tag{1}$$

where κ is the kernel function and $\alpha, \beta > -1$. We are interested in approximating the solution of the integral equation of the second kind

$$(I - K)f = g, \tag{2}$$

A. El Bouayadi · C. Allouch
The Multidisciplinary Faculty of Nador, Team of Modeling and Scientific Computing, Nador, Morocco
e-mail: c.allouch@ump.ac.ma

A. Boujraf
ENSAH, LSA Laboratory, University Abdelmalek Essaadi, Al-Hoceima, Morocco
e-mail: a.boujraf@uae.ac.ma

M. Tahrichi (✉)
EST, ANAA Team, LANO Laboratory, University Mohammed First of Oujda, Oujda, Morocco
e-mail: m.tahrichi@ump.ac.ma

© The Author(s), under exclusive license to Springer Nature Switzerland AG 2025

M. J. Ibáñez-Pérez et al. (eds.), *Mathematical And Computational Modelling, Approximation and Simulation*, SEMA SIMAI Springer Series 41,
https://doi.org/10.1007/978-3-032-01114-5_9

where I is the identity operator, f is the unknown solution, and g is a known right-hand side. Even if g is smooth, the weak singularity along the diagonal $y = x$ and the line $y = -1$ is inherited by the function f when $\alpha < 0$, as is well known (see [6, Chap. 6] and [17]). Many recent works have focused on numerical methods to solve Eq. (1), and "regularization techniques", which modify the original equation to yield a more smooth solution, have been extensively studied.

Those interested in this subject should consult [1, 9, 12, 13, 15, 17] and the references therein. The resultant integral equation becomes more complicated, however, when such smoothing techniques are used. To overcome this drawback, an interesting technique recently used in [7] consists in transforming the Eq. (1) into a *Fredholm* integral equation with a kernel that possibly is not smooth at $y = -1$ and with an identical right-hand side. The elimination of the singular element $(y - x)^\alpha$ from the equation is a significant benefit of this technique. The discrete collocation method then approximates this equation in the *Banach* space C_u, where $u(x) = (1 - x)^\gamma (1 + x)^\delta$ with $\gamma, \delta \geq 0$ is an appropriate *Jacobi* weight that eliminates the singularities at the ends of the functions being considered.

In [10], an analogous method is developed that exactly integrates the singular factor $(y - x)^\alpha$ by using a product integration rule and only requiring the kernel κ being approximated in C_u. In terms of convergence, it is shown that the resulting weighted *Nyström* scheme acts as the best polynomial approximation of the solution. In this context, an important benefit of the method w.r.t. the collocation approach is the fast convergence even if the function g is not so smooth. In other words, the convergence speed depends more on the smoothness of the kernel κ than on the one of g. Thus, the improvement of the order of convergence of such a powerful method is not a simple task.

In this paper, we define a new modified weighted *Nyström* technique, which combines both the approximations proposed in [7, 10]. The method provides a slight initial enhancement of the accuracy of the weighted *Nyström* solution, and enjoys a theoretically superior convergence order. Furthermore, this approach exhibits a superconvergence phenomenon at the interpolation nodes. Next, we apply the *Sloan* iteration to construct our fastest approximate solution. Convergence rates of numerical methods for solving (1) with orders greater than that of the best polynomial approximation have only been proven in [19, 20] as far as we are aware, and we want to address this gap. With an astonishingly similar condition number, our method's linear system retains the same size as the regular weighted *Nyström* method.

It is noteworthy that the obtained results depend heavily on the work in [10] and also on the essential book of [14]. At last, we notice that the proposed approach is similar to the superconvergent *Nyström* method previously developed for *Fredholm* integral equations in unweighted spaces (see [2–4, 11]).

The following is the outline for the paper. Both the interpolation operator and the corresponding product integration rule are defined in Sect. 2, along with some well-established results. Section 3 presents the numerical methods that will be employed to estimate the integral equation's solution, and `Sect.` 4 conducts numerical tests to assess their accuracy. The proof of the important theorems is reported in the "Appendix".

2 Preliminaries

$\mathcal{C}$ is a constant that will appear multiple times in this paper, and it will have a different meaning in each of the formulas. Furthermore, we will express $\mathcal{C} \neq \mathcal{C}(a, b, \ldots)$ to demonstrate that $\mathcal{C}$ is a positive constant that is unaffected by the parameters $a, b, \ldots$ and $\mathcal{C} = \mathcal{C}(a, b, \ldots)$ to denote that $\mathcal{C}$ is reliant on $a, b, \ldots$. We state that $A \sim B$ occurs for numbers A and B that are greater than zero and rely on certain parameters, if there is a positive constant $\mathcal{C}$ that is independent of A and B and such that

$$\frac{B}{\mathcal{C}} \leq A \leq \mathcal{C}B.$$

Additionally, we denote by κ_x (or κ_y) the bivariate function $\kappa(x, y)$ considered as a function of a single variable y (or x).

2.1 Function Spaces

For $\gamma, \delta > -1$, let $v^{\gamma,\delta}$ be the *Jacobi* weight

$$v^{\gamma,\delta}(x) = (1-x)^{\gamma}(1+x)^{\delta}, \quad x \in (-1, 1).$$

For $\gamma, \delta \geq 0$, let $u := v^{\gamma,\delta}$ and consider the space C_u of continuous functions $f : (-1, 1) \to \mathbb{R}$ such that fu is continuous in the whole interval $[-1, 1]$. For the validity of the *Weierstrass* theorem, we assume that if $\gamma > 0$, then $\lim_{x\to 1}(fu) = 0$, whereas if $\delta > 0$, then $\lim_{x\to -1}(fu) = 0$. If $\gamma = \delta = 0$, the set C_u coincides with the space $C^0([-1, 1])$ of all continuous functions in $[-1, 1]$. The space C_u is a *Banach* space if it is endowed with the weighted uniform norm defined as

$$\|f\|_{C_u} = \|fu\|_\infty = \max_{|x|\leq 1} |(fu)(x)|.$$

$\mathbb{P}_m$ is the usual notation for the set of algebraic polynomials whose degrees are less than or equal to m. One might recall that the weighted error, denoted as $E_m(f)_u$, of the best polynomial approximation is equal to

$$E_m(f)_u = \inf_{p\in\mathbb{P}_m} \|(f-p)u\|_\infty \quad \text{with} \quad \lim_{m\to\infty} E_m(f)_u = 0. \tag{3}$$

For brevity, we shall set $\|f\|_A = \max_{x\in A} |f(x)|, \ A \subseteq [-1, 1]$.

Let $\varphi(x) = \sqrt{1-x^2}$. To deal with smoother functions, we take into account the main part of the *φ-modulus of smoothness*, which is defined for every f in C_u and every integer k greater than or equal to 1 (see [8, p.90])

$$\Omega_\varphi^k(f,t)_u = \sup_{0<h\le t} \|u\Delta_{h\varphi}^k f\|_{I_{kh}},$$

with $I_{kh} = \left[-1+(2kh)^2, 1-(2kh)^2\right]$, $0 < t < 1$ and

$$\Delta_{h\varphi}^k f(x) = \sum_{i=0}^{k} (-1)^i \binom{k}{i} f\left(x + \frac{h\varphi(x)}{2}(k-2i)\right).$$

Thus, we define the *Zygmund* space

$$Z_\lambda(u) = \left\{ f \in C_u : \sup_{t>0} \frac{\Omega_\varphi^k(f,t)_u}{t^\lambda} < \infty,\ k > \lambda \right\},$$

where $\lambda \in \mathbb{R}^+$ is the order, and in this space, we introduce the norm

$$\|f\|_{Z_\lambda(u)} = \|fu\|_\infty + \sup_{t>0} \frac{\Omega_\varphi^k(f,t)_u}{t^\lambda}.$$

Then (see [14, p.172]), it holds

$$\sup_{t>0} \frac{\Omega_\varphi^k(f,t)_u}{t^\lambda} \sim \sup_{n\ge 0} (1+n)^\lambda E_n(f)_u,$$

and the constants involved in "$\sim$" vary with λ. Lastly, the quantity $E_m(f)_u$ can be estimated by means of the weak-*Jackson* inequality (see [14, p. 171])

$$E_m(f)_u \le \mathcal{C} \int_0^{\frac{1}{m}} \frac{\Omega_\varphi^k(f,t)_u}{t} dt, \quad \forall f \in C_u, \quad \mathcal{C} \ne \mathcal{C}(m,f) \tag{4}$$

and the following inequality of the *Favard*-type (see [14, p.172])

$$E_m(f)_u \le \frac{\mathcal{C}}{m^\lambda} \|f\|_{Z_\lambda(u)}, \quad \forall f \in Z_\lambda(u), \tag{5}$$

with $\mathcal{C} \ne \mathcal{C}(m,f)$. From here on, the subscript $C_u \to C_u$ in the norm of operators shall be denoted as u, that is to say $\|.\|_{C_u \to C_u} \equiv \|.\|_u$.

2.2 Interpolation Operator

Let $w := v^{\alpha,\beta}$ with $\alpha, \beta > -1$ and let $\{p_m(w)\}_{m=0}^{\infty}$ be the sequence of the associated orthonormal polynomials with positive leading coefficients.
The *Lagrange* polynomial $\mathcal{L}_m(f, w) \in \mathbb{P}_{m-1}$ which interpolates $f \in C_u$ at the zeros $\{x_i\}_{i=1}^{m}$ of $p_m(w)$ is then defined by

$$\mathcal{L}_m(f, w, x) = \sum_{i=1}^{m} \frac{\ell_i(x)}{u(x_i)}(fu)(x_i), \tag{6}$$

where

$$\ell_i(x) = \frac{p_m(w, x)}{p'_m(w, x_i)(x - x_i)} = \lambda_{m,i}(w) \sum_{k=0}^{m-1} p_k(w, x) p_k(w, x_i) \tag{7}$$

and the weights $\lambda_{m,i}(w)$ are the so-called *Christoffel* numbers. Then, the mth *Lebesgue* constant has a logarithmic growth (see [14, p. 272]) i.e.,

$$\|\mathcal{L}_m\|_u := \max_{|x|\le 1} u(x) \sum_{i=1}^{m} \frac{|\ell_i(x)|}{u(x_i)} \sim \log m, \tag{8}$$

if and only if

$$\max\left\{0, \frac{\alpha}{2} + \frac{1}{4}\right\} \le \gamma \le \frac{\alpha}{2} + \frac{5}{4}, \quad \max\left\{0, \frac{\beta}{2} + \frac{1}{4}\right\} \le \delta \le \frac{\beta}{2} + \frac{5}{4}. \tag{9}$$

This means that, under the above conditions, the *Lagrange* polynomial (6) has an optimal behaviour in C_u, and if $f \in Z_\lambda(u)$, then by (5)

$$\|[f - \mathcal{L}_m(f, w)]u\|_\infty \le \mathcal{C} \frac{\log m}{m^\lambda} \|f\|_{Z_\lambda(u)}, \quad \mathcal{C} \ne \mathcal{C}(m, f). \tag{10}$$

The operator $\mathcal{L}_m$ is linear on C_u and possesses the property $\mathcal{L}_m = (\mathcal{L}_m)^2$. It is therefore a projection, and its range is the set $\mathbb{P}_{m-1}$.
The following result is proved in [14, p. 348] (see also [16]).

Theorem 2.1 Assume that $h(x, y)$ is a bivariate function satisfying

$$\sup_{|y|\le 1} \int_{-1}^{1} \frac{|h(x, y)|}{u(x)} \log\left(2 + \frac{|h(x, y)|}{u(x)}\right) dt < \infty.$$

Then, for any $f \in C_u$,

$$\sup_{|y|\le 1} \int_{-1}^{1} |h(x, y)\mathcal{L}_m(f, w, x)|\, dx \le \mathcal{C}\|fu\|_\infty, \quad \mathcal{C} \ne \mathcal{C}(m, f), \tag{11}$$

if and only if

$$\sup_{|y|\le 1}\int_{-1}^{1}\frac{|h(x,y)|}{\sqrt{w(x)\varphi(x)}}dx<\infty \quad \text{and} \quad \int_{-1}^{1}\frac{\sqrt{w(x)\varphi(x)}}{u(x)}dx<\infty.$$

The following result will be vital for our analysis.

Lemma 2.1 Assume that

$$0\le\gamma<\min\left\{\alpha+1,\frac{\alpha}{2}+\frac{5}{4}\right\},\quad 0\le\delta<\min\left\{\beta+1,\frac{\beta}{2}+\frac{5}{4}\right\}$$

and that

$$\sup_{|y|\le 1}\left(v^{0,1+\alpha+\beta}(y)\sup_{|x|\le 1}|\kappa(x,y)|\right)<\infty. \tag{12}$$

Then, for each $f\in Z_\lambda(u)$, one has

$$\|K(f-\mathcal{L}_m(f,w))u\|_\infty\le\frac{\mathcal{C}}{m^\lambda}\|f\|_{Z_\lambda(u)},\quad \mathcal{C}\ne\mathcal{C}(m,f). \tag{13}$$

Proof Let p denote the best polynomial approximation of degree $m-1$ of a given function $f\in C_u$ and let

$$\sigma(x,y)=\begin{cases}(y-x)^\alpha(1+x)^\beta, & y>x,\\ 0, & y<x.\end{cases}$$

Then, applying Theorem 2.1 to the function $h(x,y)=(\kappa\sigma)(x,y)$ and taking into account that $-1<\alpha-\gamma$ and $-1<\beta-\delta$, we obtain in view of the assumptions made on the kernel that

$$\begin{aligned}
|[K(f-\mathcal{L}_m(f,w))](y)u(y)| &\le u(y)\int_{-1}^{1}|f(x)-\mathcal{L}_m(f,w,x)|\,|(\kappa\sigma)(x,y)|dx,\\
&\le u(y)\left[\int_{-1}^{1}|f(x)-p(x)|\,|(\kappa\sigma)(x,y)|dx+\int_{-1}^{1}|(\kappa\sigma)(x,y)\mathcal{L}_m(f-p)|\,dx\right],\\
&\le u(y)\left[\|(f-p)u\|_\infty\sup_{|x|\le 1}|\kappa(x,y)|\int_{-1}^{1}\left|\frac{\sigma(x,y)}{u(x)}\right|dx+\mathcal{C}\,\|(f-p)u\|_\infty\right],\\
&\le \mathcal{C}E_{m-1}(f)_u\sup_{|x|\le 1}|\kappa(x,y)|u(y)\int_{-1}^{y}v^{\alpha-\gamma,\beta-\delta}(x)dx,
\end{aligned}$$

$$\leq \mathcal{C}E_{m-1}(f)_u \sup_{|x|\leq 1} |\kappa(x,y)| u(y) v^{-\gamma,1+\alpha+\beta-\delta}(y) \int_{-1}^{1} v^{\alpha,\beta-\delta}(t)dt,$$

$$\leq \mathcal{C}E_{m-1}(f)_u \sup_{|y|\leq 1} \left(v^{0,1+\alpha+\beta}(y) \sup_{|x|\leq 1} |\kappa(x,y)| \right). \tag{14}$$

Hence, the desired formula follows from the bound (5) and (12). □

2.3 The Product Integration Rule

We now present the product rule that forms the foundation of both the standard weighted *Nyström* and the modified *Nyström* methods.

The rule is obtained (see [10]) by approximating the function $\kappa_y f$ in (1) by the *Lagrange* polynomial $\mathcal{L}_m(\kappa_y f, w)$, and leads to the *Nyström*-type operator

$$(K_m f)(y) = \sum_{i=1}^{m} c_i(y)\kappa(x_i, y) f(x_i), \tag{15}$$

where

$$\begin{aligned} c_i(y) &= \int_{-1}^{y} \ell_i(x)(y-x)^\alpha (1+x)^\beta dx, \\ &= \lambda_{m,i}(w) \sum_{k=0}^{m-1} p_k(w, x_i) \int_{-1}^{y} p_k(w,x)(y-x)^\alpha (1+x)^\beta dx. \end{aligned}$$

Conditions for the stability and convergence of the integration rule in (15) are stated in the following theorem.

Theorem 2.2 ([10]) Assume that $\kappa_y \in C_u$ and that κ satisfies

$$\sup_{y\in[-1,1]} \left(v^{0,1+\alpha+\beta+\delta}(y) \sup_{x\in[-1,1]} |\kappa(x,y)| \right) < \infty.$$

If

$$0 \leq \gamma < \min\left\{\alpha+1, \frac{\alpha}{2}+\frac{5}{4}\right\},$$

$$\max\{0, -(1+\alpha+\beta)\} \leq \delta < \min\left\{\beta+1, \frac{\beta}{2}+\frac{5}{4}\right\},$$

then, we have

$$\sup_{|y|\leq 1} \sup_m u(y) \sum_{i=1}^{m} \frac{|c_i(y)|}{u(x_i)} < \infty,$$

which means that (15) is stable. In addition, for each $f \in C_u$

$$\lim_{m\to\infty} \|(K - K_m) f u\|_\infty = 0. \tag{16}$$

Lemma 2.2 Assume that

$$\begin{aligned} \max\left\{0, \frac{\alpha}{2} + \frac{1}{4}\right\} \le \gamma < \min\left\{\alpha + 1, \frac{\alpha}{2} + \frac{5}{4}\right\}, \\ \max\left\{0, -(1+\alpha+\beta), \frac{\beta}{2} + \frac{1}{4}\right\} \le \delta < \min\left\{\beta + 1, \frac{\beta}{2} + \frac{5}{4}\right\}, \end{aligned} \tag{17}$$

and

$$\tilde{\kappa}(y) := v^{-\gamma, 1+\alpha+\beta}(y) \sup_{x\in[-1,1]} |\kappa(x, y)| \in Z_\lambda(u). \tag{18}$$

Then, for each $f \in Z_\lambda(u)$

$$\|(K - K_m) f\|_{Z_\lambda(u)} \le \frac{\mathcal{C}}{m^\lambda} \|f\|_{Z_\lambda(u)}, \quad \mathcal{C} \ne \mathcal{C}(m, f). \tag{19}$$

Proof For $f \in Z_\lambda(u)$, let us prove that

$$\sup_{t>0} \frac{\Omega_\varphi^k((K - K_m) f, t)_u}{t^\lambda} \le \frac{\mathcal{C}}{m^\lambda} \|f\|_{Z_\lambda(u)}. \tag{20}$$

It is straightforward to follow the analysis leading to (16) in [10] to obtain

$$\|[(K - K_m) f] u\|_\infty \le \mathcal{C}\left[E_{[\frac{m-1}{2}]}(f)_u + \frac{1}{m^\lambda}\|f u\|_\infty\right]. \tag{21}$$

Then, for each $y \in I_{kh} = [-1 + (2hk)^2, 1 - (2hk)^2]$, we have

$$\begin{aligned} |u(y)\Delta_{h\varphi}^k[(K - K_m) f](y)| &= u(y)\left|\sum_{i=0}^k (-1)^i \binom{k}{i} [(K - K_m) f]\left(y + \frac{h\varphi(y)}{2}(k - 2i)\right)\right|, \\ &\le 2^k \max_{1\le i\le k}\left\{\frac{u(y)}{u\left(y + \frac{h\varphi(y)}{2}(k - 2i)\right)}\right\} \sum_{i=0}^k \left|[(K - K_m) f] u\left(y + \frac{h\varphi(y)}{2}(k - 2i)\right)\right|, \\ &\le 2^k (k+1)\mathcal{C}\left[E_{[\frac{m-1}{2}]}(f)_u + \frac{1}{m^\lambda}\|f u\|_\infty\right], \quad \mathcal{C} = \mathcal{C}(h, k), \end{aligned} \tag{22}$$

where $y + \frac{h\varphi(y)}{2}(k - 2i) \ne \pm 1$. Thus, taking norms on y and $0 < h \le t$, we conclude

$$\Omega_\varphi^k((K - K_m) f, t)_u \le \mathcal{C}\left[E_{[\frac{m-1}{2}]}(f)_u + \frac{1}{m^\lambda}\|f u\|_\infty\right]. \tag{23}$$

Now dividing by t^λ and taking the supremum on $|y| \leq 1$ and $0 < h \leq t$, the estimate (20) follows easily from (5). Finally, the required bound is immediate from (20) and (21). □

2.4 Preliminary Results

Regarding the solvability of the integral equation (1), the next theorem shows that the exact solution exists and is unique.

Theorem 2.3 ([10]) Let $0 \leq \gamma < 1 + \alpha$ and $0 \leq \delta < 1 + \beta$ and assume that (18) is fulfilled. Then, for each $g \in C_u$, equation (1) has a unique solution in C_u. □

Note that under the condition (18), the operator $K : C_u \to C_u$ is compact (see [10, Lemma 6.2] and [18, p. 93]).

To analyse the convergence of the proposed methods, the following crucial results are useful.

Under the hypothesis of Theorem 2.3, we have (see [10, Lemma 6.2])

$$\|(Kf)u\|_\infty \leq \mathcal{C}\|fu\|_\infty, \quad \mathcal{C} \neq \mathcal{C}(f). \tag{24}$$

Moreover, from the same reference

$$\Omega_\varphi^k(Kf, t)_u \leq \mathcal{C}t^\lambda \|fu\|_\infty, \quad \forall f \in C_u. \tag{25}$$

Hence, we conclude that

$$\|Kf\|_{Z_\lambda(u)} \leq \mathcal{C}\|f\|_{C_u} \leq \mathcal{C}\|f\|_{Z_\lambda(u)}, \quad \mathcal{C} \neq \mathcal{C}(f). \tag{26}$$

As a result, if $f \in Z_\lambda(u)$, then the function Kf belongs at least to $Z_\lambda(u)$. From the proof of Theorem 6.3 in [10], we also have

$$E_{[\frac{m-1}{2}]}(K_m f)_u \leq \mathcal{C}\frac{\log m}{m^\lambda}\|fu\|_\infty, \quad \forall f \in C_u. \tag{27}$$

To discuss the existence and uniqueness of the approximate solutions, we need first to recall the following result from [5, Theorem 4.1.1].

Lemma 2.3 Let $L : C_u \to C_u$ be a bounded compact operator and $\{L_m\}$ be a sequence of bounded operators defined on C_u. If

(1) $\lim_{m\to\infty} \|(L - L_m)f\|_{C_u} = 0$ for all $f \in C_u$,
(2) $\lim_{m\to\infty} \|(L - L_m)L_m\|_u = 0$,

then $(I - L_m)^{-1}$ exists as a bounded operator from C_u to C_u. □

3 The Numerical Methods

In this section we develop two efficient methods for solving equation (1). Let us first recall the ordinary weighted *Nyström* method defined in [10].

3.1 The Nyström method

To begin, we consider the weighted *Nyström* equation

$$(I - K_m)f_m = g,$$

where the unknown function f_m is given by

$$f_m(y) = g(y) + \sum_{j=1}^{m} \frac{c_j(y)}{u(x_j)} \kappa(x_j, y) a_j, \tag{28}$$

and $a_j = (f_m u)(x_j)$ are the entries of the solution vector $\mathbf{a}$ of the linear system of size m

$$\sum_{j=1}^{n} \left[\delta_{i,j} - c_j(x_i) \frac{u(x_i)}{u(x_j)} \kappa(x_j, x_i) \right] a_j = (gu)(x_i), \quad i = 1, \ldots, m. \tag{29}$$

For each fixed $y \in [-1, 1]$, let γ be the function defined by

$$\gamma(t, y) = \frac{1+t}{2} y + \frac{t-1}{2},$$

which transforms the interval $[-1, 1]$ into the subinterval $[-1, y]$. Then the integrals $c_j(x_i)$ are evaluated exactly in this way

$$\begin{aligned}
c_j(x_i) &= \lambda_{m,j}(w) \sum_{k=0}^{m-1} p_k(w, x_j) \int_{-1}^{x_i} p_k(w, x)(x_i - x)^{\alpha} (1+x)^{\beta} dx, \\
&= \left(\frac{1+x_i}{2} \right)^{1+\alpha+\beta} \lambda_{m,j}(w) \sum_{k=0}^{m-1} p_k(w, x_j) \int_{-1}^{1} p_k(w, \gamma(x_i, x)) v^{\alpha,\beta}(x) dx, \\
&= \left(\frac{1+x_i}{2} \right)^{1+\alpha+\beta} \lambda_{m,j}(w) \sum_{k=0}^{m-1} p_k(w, x_j) \sum_{l=1}^{m} \lambda_{m,l}(w) p_k(w, \gamma(x_i, x_l)).
\end{aligned}$$

In matrix form, the system (29) can be represented as

$$(I_m - D_m A_m D_m^{-1})\mathbf{a} = b, \tag{30}$$

where I_m is the identity matrix, $D_m = diag(u(x_1), \dots, u(x_m))$ and

$$(A_m)_{ij} = c_j(x_i)\kappa(x_j, x_i), \quad b_i = (gu)(x_i).$$

It has been shown that f_m converges to the unique solution of (1) with the same speed as the product integration formulae (15). More precisely, if (17) and (18) are satisfied and $g \in Z_\lambda(u)$, then the approximation error is of the order $\mathcal{O}(\frac{1}{m^\lambda})$. In the following, we describe a modified form of the weighted *Nyström* method that can yield superconvergent results. Note that to shorten the notation, when necessary, we may choose to write $(f^* - \mathcal{L}_m(f^*, w))$ and $K(f^* - \mathcal{L}_m(f^*, w))$ respectively as $(I - \mathcal{L}_m)f^*$ and $K(I - \mathcal{L}_m)f^*$.

3.2 *The Modified Nyström Method and Its Sloan Iteration*

In the ordinary collocation method, the integral operator K is replaced by $\mathcal{L}_m K$, where K is transformed to (see [7])

$$(Kf)(y) = 2^{-(\alpha+\beta+1)} \int_{-1}^{1} \widetilde{\kappa}(t, y) f(\gamma(t, y)) w(t) dt, \quad y \in [-1, 1], \tag{31}$$

where

$$\widetilde{\kappa}(t, y) = (1 + y)^{\alpha+\beta+1} \kappa(\gamma(t, y), y).$$

Using this form, let us introduce the following approximation of K

$$K_m^* = \mathcal{L}_m K + (I - \mathcal{L}_m) K_m = K_m + \mathcal{L}_m(K - K_m), \tag{32}$$

where $\mathcal{L}_m$ is the *Lagrange* interpolatory projection. This operator is chosen so that the corresponding error has the form

$$K - K_m^* = (I - \mathcal{L}_m)(K - K_m).$$

Therefore, the rate of convergence of the ordinary weighted *Nyström* method can be increased by the additional factor $I - \mathcal{L}_m$. The corresponding approximation of (1) is

$$(I - K_m^*) f_m^* = g, \tag{33}$$

while the *Sloan* iteration is defined by

$$\widetilde{f_m^*} = K f_m^* + g. \tag{34}$$

As a next step, we consider the reduction of (33) to a system of linear equations, taking care to incorporate the *Jacobi* weights u in order to build a stable method.

Theorem 3.1 Let d be the vector with the components $d_j = (Kg)(x_j)$, and let B_m, C_m be the matrices with the entries

$$(B_m)_{ij} = (K\ell_j)(x_i), \quad (C_m)_{ij} = (K\sigma_j)(x_i),$$

where $\sigma_j = c_j\kappa(x_j, .)$. Then, the approximate solution is given by

$$f_m^* = g + \sum_{i=1}^{n} \frac{a_i^*}{u(x_i)} c_i \kappa(x_i, .) + \sum_{i=1}^{n} \left(\frac{a_i^*}{u(x_i)} - g(x_i) \right) \ell_i - \sum_{i=1}^{m} \sum_{j=1}^{m} \frac{a_i^*}{u(x_i)} (A_m)_{ij} \ell_j, \tag{35}$$

where $\mathbf{a}^* = \left\{ a_i^* = (f_m^* u)(x_i) \right\}_i$ satisfies the following system of linear equations

$$(I_m - D_m A_m^* D_m^{-1})\mathbf{a}^* = b^* \tag{36}$$

with

$$A_m^* = C_m - B_m(A_m - I_m) \quad \text{and} \quad b^* = D_m(d + (I_m - B_m)D_m^{-1}b).$$

Proof See "Appendix". □

Remark 3.1 When we look at the expressions of the *Nyström* and the modified *Nyström* solutions f_m and f_m^* in (28) and (35) respectively, we observe that the two solutions probably require the same amount of work to evaluate. The only distinction is that the second and third terms of f_m^*, both polynomials of degree m, are present as additional factors. However, the production of the matrices B_m and C_m adds an additional expense to the proposed method. This increase in cost is offset by the enhancement in the order of convergence.

Note that in practice, the integrals involved in the vector d and in the matrices B_m and C_m are computed throughout the form (31) by an appropriate *Gauss* quadrature rule with $n > m$ points. □

3.3 Orders of Convergence

The following lemma will be helpful in the subsequent analysis.

Lemma 3.1 Assume that the conditions (17) are fullfiled and let K be an integral operator with a kernel κ satisfying the hypothesis (18). Then, the operators $(I - K_m^*)^{-1}$ exist and are uniformly bounded for a sufficiently large m, i.e. there exists a constant $\mathcal{C} > 0$ such that

$$\|(I - K_m^*)^{-1}\|_{C_u \to C_u} \leq \mathcal{C} < \infty, \quad \mathcal{C} \neq \mathcal{C}(m). \tag{37}$$

Proof For any $f \in C_u$, we have $(K - K_m)f \in Z_\lambda(u)$. Hence, by (10)

$$\|(K - K_m^*)f\|_{C_u} = \|(I - \mathcal{L}_m)(K - K_m)f\|_{C_u},$$
$$\leq \mathcal{C}\frac{\log m}{m^\lambda}\|(K - K_m)f\|_{Z_\lambda(u)}. \tag{38}$$

Since we are assuming (17) and (18) then, considering the fact that

$$\|(K - K_m)f\|_{Z_\lambda(u)} = \|(K - K_m)fu\|_\infty + \sup_{t>0}\frac{\Omega_\varphi^k((K - K_m)f, t)_u}{t^\lambda}, \tag{39}$$

in virtue of (16) and (23), we clearly have

$$\|(K - K_m)f\|_{Z_\lambda(u)} \longrightarrow 0 \quad \text{as} \quad m \to \infty.$$

As $\lambda > 0$, it follows from (38) that $K_m^* f \to Kf$ pointwise in C_u. This ends the proof of condition (1) in Lemma 2.3. To prove condition (2) let us first remark that using $K_m\mathcal{L}_m \equiv K_m$, we have

$$(K - K_m^*)K_m^* = (I - \mathcal{L}_m)(KK_m - K_mK),$$
$$= (I - \mathcal{L}_m)[(K - K_m)K - K(K - K_m)].$$

Therefore, replacing f by $(K - K_m)Kf$ in (10) we get, in view of (19) and (26)

$$\|(I - \mathcal{L}_m)(K - K_m)Kf\|_{C_u} \leq \mathcal{C}\frac{\log m}{m^\lambda}\|(K - K_m)Kf\|_{Z_\lambda(u)},$$
$$\leq \mathcal{C}\frac{\log m}{m^{2\lambda}}\|Kf\|_{Z_\lambda(u)},$$
$$\leq \mathcal{C}\frac{\log m}{m^{2\lambda}}\|f\|_{C_u}.$$

Similarly, we are able to show that

$$\|(I - \mathcal{L}_m)K(K - K_m)\|_{C_u} \longrightarrow 0 \quad \text{as} \quad m \to 0.$$

As a consequence

$$\|(K - K_m^*)K_m^*\|_u \longrightarrow 0 \quad \text{as} \quad m \to 0$$

and this concludes the proof. □

The next result summarize the error estimates for f_m^* and $\widetilde{f}_m^*$.

Theorem 3.2 Assume that the hypothesis of Lemma 3.1 are satisfied and let f^* be the exact solution of equation (1). Then, for all sufficiently large m, we have

$$\|[f_m^* - f^*]u\|_\infty \leq \mathcal{C}\|(I - \mathcal{L}_m)(K - K_m)f^*\|_{C_u}, \quad \mathcal{C} \neq \mathcal{C}(m, f^*, g) \tag{40}$$

and

$$\begin{aligned}\|[\widetilde{f}_m^* - f^*]u\|_\infty \leq \mathcal{C}\big\{&\|K(I - \mathcal{L}_m)(K - K_m)f^*\|_{C_u} \\ &+\|K(I - \mathcal{L}_m)(K - K_m)\|_u \times \|[f_m^* - f^*]u\|_\infty\big\}.\end{aligned} \tag{41}$$

Proof By the f^* and f_m^* definitions (1) and (33) respectively, we have

$$\begin{aligned} f_m^* - f^* &= \left[(I - K_m^*)^{-1} - (I - K)^{-1}\right]g, \\ &= (I - K_m^*)^{-1}(K_m^* - K)f^*. \end{aligned}$$

Thus, taking norms and using (37), one gets

$$\begin{aligned} \|f_m^* - f^*\|_{C_u} &\leq \|(I - K_m^*)^{-1}\|_u \|(K_m^* - K)f^*\|_{C_u}, \\ &\leq \mathcal{C}\|(I - \mathcal{L}_m)(K - K_m)f^*\|_{C_u}. \end{aligned}$$

That is (40) is proved. In addition, equations (1) and (34) allow us to write

$$\begin{aligned} \widetilde{f}_m^* - f^* &= K(f_m^* - f^*) \\ &= K(I - K)^{-1}(K_m^* - K)(I - K_m^*)^{-1}g \\ &= -(I - K)^{-1}K(I - \mathcal{L}_m)(K - K_m)(f^* + f_m^* - f^*), \end{aligned}$$

and the bound (41) follows with the constant $\mathcal{C} = \|(I - K)^{-1}\|_u$. This completes the proof. □

For any given matrix A, we will refer to its condition number in the uniform norm as $\text{cond}(A)$ and is equal to $\text{cond}(A) = \|A\|\|A^{-1}\|$. Let M_m and M_m^* be the matrices of the linear systems (30) and (36) respectively. The following result states that $\text{cond}(M_m^*) \sim \text{cond}(M_m)$.

Lemma 3.2 Assume that

$$\sup_{|y|\leq 1} \int_{-1}^{1} v^{\alpha-\gamma,\beta-\delta}(x)|\kappa(x, y)| \log\left(2 + v^{\alpha-\gamma,\beta-\delta}(x)|\kappa(x, y)|\right) dx < \infty. \tag{42}$$

Then, for m sufficiently large,

$$\text{cond}(M_m) \leq \mathcal{C}, \quad \mathcal{C} \neq \mathcal{C}(m) \quad \text{and} \quad \lim_{m\to\infty} \frac{\text{cond}(M_m^*)}{\text{cond}(M_m)} = 1. \tag{43}$$

Convergence rates of the suggested methods are now provided in the following theorem.

Proof See "Appendix". □

Theorem 3.3 Suppose that the hypothesis of Lemmas 2.1 and 2.2 are fulfilled. For $g \in Z_\lambda(u)$, let f^* be the exact solution of (1). Then, for m large enough, the following estimates hold

$$\|[f_m^* - f^*]u\|_\infty = \mathcal{O}\left(\frac{\log m}{m^{2\lambda}}\right), \tag{44}$$

$$\|[\widetilde{f}_m^* - f^*]u\|_\infty = \mathcal{O}\left(\frac{1}{m^{2\lambda}}\right). \tag{45}$$

Moreover, we have the following superconvergence result

$$\max_{1\le i\le m} |[(f_m^* - f^*)u](x_i)| = \mathcal{O}\left(\frac{1}{m^{2\lambda}}\right). \tag{46}$$

Proof See "Appendix". □

4 Numerical Examples

In this section, two integral equations in appropriate weighted spaces are designed to test all the methods described in this paper. The associated linear systems (30) and (36) are solved for each test equation and then we evaluate how well our technique and its iterated form perform in comparison to the standard weighted *Nyström* method. Note that all the numerical tests were carried out in Matlab R2019a. In addition, the *Gauss*-quadrature rule is used to compute various integrals involved in (36) and in the approximate solution (34).

Example 1 The first example is given by

$$f(y) + \frac{1}{e^4}\int_{-1}^{y} f(x)(y-x)^{-\frac{1}{3}}(1+x)^{\frac{1}{2}}dx = g(y), \quad y \in (-1, 1],$$

where g is selected so that $f^*(y) = (1+y)^{\frac{1}{3}}$. We study this equation in the space C_u with $u = v^{\frac{1}{4},\frac{3}{4}}$. Thus, $f^* \in Z_{\frac{13}{6}}(u)$ and according to Theorem 3.3, the theoretical orders are respectively, $\mathcal{O}(\frac{1}{m^{13/6}})$, $\mathcal{O}(\frac{\log m}{m^{13/3}})$ and $\mathcal{O}(\frac{1}{m^{13/3}})$. The weighted norms of the errors with respect to the exact solution f^* are reported in Table 1.

We observe that the convergence of the weighted *Nyström* method is relatively slow and the proposed methods provide more precise approximations. For $m = 14$, the differences are respectively: one, and three correct digits. It can also be seen that the iterated modified *Nyström* solution converges the fastest of all.

Table 2 shows that the matrices M_m and M_m^* of the linear systems (30) and (36) are very well-conditioned, as indicated by their nearly identical condition numbers.

To emphasize the difference between the three methods, Fig. 1 shows a comparison of the corresponding approximation errors.

Table 1 Approximation errors for Example 1

m	$\|[f_m - f^*]u\|_\infty$	$\|[f_m^* - f^*]u\|_\infty$	$\|[\widetilde{f_m^*} - f^*]u\|_\infty$
2	5.79×10^{-4}	2.19×10^{-4}	1.72×10^{-5}
4	5.26×10^{-5}	2.43×10^{-5}	2.67×10^{-7}
6	1.56×10^{-5}	5.99×10^{-6}	5.15×10^{-8}
8	5.47×10^{-6}	2.06×10^{-6}	1.22×10^{-8}
10	2.33×10^{-6}	8.60×10^{-7}	5.03×10^{-9}
12	1.66×10^{-6}	5.96×10^{-7}	3.21×10^{-9}
14	6.70×10^{-7}	2.27×10^{-7}	9.63×10^{-10}

Table 2 Condition numbers for Example 1

m	Cond(M_m)	Cond(M_m^*)
2	1.10872	1.10345
3	1.15165	1.14990
4	1.16806	1.16738
5	1.17512	1.17477
6	1.17841	1.17820
7	1.18016	1.18003
8	1.18115	1.18106

We notice that even with just two nodes, our approximations are satisfactory and more precise than the standard *Nyström* solution. Moreover, following what has been said, we observe that for $i = 1, 2$, $f_2^*(x_i) = \widetilde{f_2^*}(x_i)$, where $x_1 = -0.3506$ and $x_2 = 0.7506$.

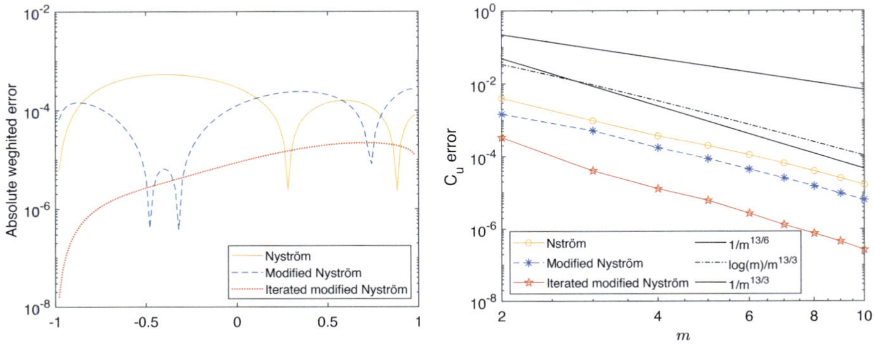

Fig. 1 For $m = 2$, we compare on the left the weighted approximation errors for Example 1 produced by *Nyström* and modified *Nyström* methods as well as its iterated version, while on the right, for varying values of m we provide the error norm for the three methods

Table 3 Approximation errors for Example 2

m	$\|[f_m - f_{128}]u\|_\infty$	$\|[f_m^* - f_{128}]u\|_\infty$	$\|[\widetilde{f}_m^* - f_{128}]u\|_\infty$
8	5.35×10^{-5}	2.22×10^{-5}	6.81×10^{-7}
10	2.01×10^{-5}	7.08×10^{-6}	1.52×10^{-7}
12	9.45×10^{-6}	4.02×10^{-6}	5.97×10^{-8}
14	4.39×10^{-6}	1.76×10^{-6}	1.98×10^{-8}
16	2.35×10^{-6}	9.66×10^{-7}	8.47×10^{-9}
18	1.34×10^{-6}	5.10×10^{-7}	3.97×10^{-9}

Table 4 Condition numbers for Example 2

m	Cond(M_m)	Cond(M_m^*)
2	1.10872	1.10345
3	1.15165	1.14990
4	1.16806	1.16738
5	1.17512	1.17477
6	1.17841	1.17820
7	1.18016	1.18003
8	1.18115	1.18106

The graph on the left shows the fast decreasing of the weighted infinity norm errors for the iterated modified *Nyström* method as compared to the curve $\frac{1}{m^{13/3}}$. A possible explanation for this is the extremely small value of the constants in the error terms.

Example 2 The second test is the following *Volterra* integral equation quoted from [10]

$$f(y) + \int_{-1}^{y} \log(x + y + 2) f(x)(y - x)^{\frac{1}{2}}(1 + x)^{\frac{1}{2}} dx = (y^2 + 1)|\sin y|^{\frac{7}{2}},$$

where the exact solution is not available. We take the *Nyström* solution f_m with polynomial degree $m = 128$ as the 'true' solution. By fixing $u = v^{\frac{3}{4},\frac{3}{4}}$, we see that $f^* \in Z_{7/2}(u)$. The expected orders of convergence are $\mathcal{O}(\frac{1}{m^{7/2}})$, $\mathcal{O}(\frac{\log m}{m^7})$, $\mathcal{O}(\frac{1}{m^7})$. The numerical results are presented in Tables 3 and 4.

The theoretical predictions hold true, with the proposed methods producing satisfying outcomes when compared to the standard approach (Fig. 2). Notably, the iterated method from the above graphs, we observe that the iterated method is more accurate than both of the two other approximations. Furthermore, we speculate that the modified *Nyström* solution is being slightly disturbed, by the big value of the constant in the error terms.

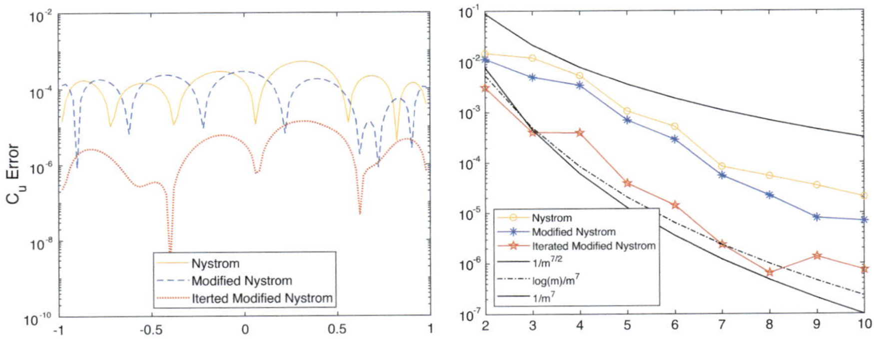

Fig. 2 For $m = 3$, we compare on the left the weighted approximation errors for Example 2 produced by *Nyström* and modified *Nyström* methods as well as its iterated version, while on the right, for varying values of m we provide the error norm for the three methods

5 Appendix

In this section, we report the proofs of Theorems 3.1, 3.3 and Lemma 3.2.

Proof of Theorem 3.1 Remembering that $\mathcal{L}_m$ is a projection, and applying it and $(I - \mathcal{L}_m)$ to Eq. (33), we get

$$\mathcal{L}_m f_m^* - \mathcal{L}_m K f_m^* = \mathcal{L}_m g, \tag{47}$$

$$(I - \mathcal{L}_m) f_m^* - (I - \mathcal{L}_m) K_m f_m^* = (I - \mathcal{L}_m) g. \tag{48}$$

By writing

$$K f_m^* = K(I - \mathcal{L}_m) f_m^* + K \mathcal{L}_m f_m^*, \tag{49}$$

and replacing $(I - \mathcal{L}_m) f_m^*$ by its expression from Eqs. (48) and (49) becomes

$$K f_m^* = K(I - \mathcal{L}_m) K_m f_m^* + K \mathcal{L}_m f_m^* + K(I - \mathcal{L}_m) g.$$

By putting the above form of $K f_m^*$ in Eq. (47), we can figure out that

$$\mathcal{L}_m f_m^* - \mathcal{L}_m K\big((I - \mathcal{L}_m) K_m f_m^* + \mathcal{L}_m f_m^* + (I - \mathcal{L}_m) g\big) = \mathcal{L}_m g.$$

Noting that $K_m f_m^* = K_m \mathcal{L}_m f_m^*$, we can seek $\theta_m^* = \mathcal{L}_m f_m^* \in \mathbb{P}_{m-1}$ such that

$$\theta_m^* - \big(\mathcal{L}_m K + \mathcal{L}_m K(I - \mathcal{L}_m) K_m\big)\theta_m^* = \mathcal{L}_m g + \mathcal{L}_m K(I - \mathcal{L}_m) g, \tag{50}$$

and (48) implies that

$$f_m^* = \theta_m^* + (I - \mathcal{L}_m) K_m \theta_m^* + (I - \mathcal{L}_m) g.$$

By evaluating (50) at the nodes $\{x_i\}_{i=1}^m$, we obtain for $i = 1, \dots m$,

$$f_m^*(x_i) - (K\theta_m^*)(x_i) - [K(I - \mathcal{L}_m)K_m\theta_m^*](x_i) = g(x_i) + [K(I - \mathcal{L}_m)g](x_i). \tag{51}$$

Now by multiplying both sides of the above equation by the weight function u evaluated at x_i, the linear system (36) is derived. □

Proof of Lemma 3.2 The proof of the first bound in (43) is known (see [5, p. 113] and [10]). In fact, it is shown that

$$\max\{\|M_m\|, \|M_m^{-1}\|\} \leq \mathcal{C}.$$

It remains now to prove the second property connecting the condition numbers. In order to achieve this, let us remark first that

$$M_m^* = M_m + D_m(B_mA_m - C_m)D_m^{-1} + D_m(A_m - B_m)D_m^{-1}.$$

Now, we need to invoke the quantity $K(I - \mathcal{L}_m)K_m$ involved in (51) from which the matrix $D_m(B_mA_m - C_m)D_m^{-1}$ has been generated. Let

$$\widetilde{\theta}(y) = \sum_{i=1}^{m} \frac{\ell_i(y)}{u(y)}\theta_i \quad \text{with} \quad \theta_i = (\widetilde{\theta}u)(x_i), \quad i = 1, \dots, m. \tag{52}$$

Then, if we set $\theta = (\theta_1, \dots, \theta_m)$, it is simple to see in view of (57) that

$$\|(B_mA_m - C_m)\theta\| = \|[K(I - \mathcal{L}_m)K_m\widetilde{\theta}]u\|_\infty = \mathcal{O}\left(\frac{\log m}{m^\lambda}\right) \longrightarrow 0 \quad \text{as} \quad m \to \infty.$$

This is true for all vector θ, hence we conclude that $\|B_mA_m - C_m\| \to 0$. In analogous way, by (14) and (16)

$$\begin{aligned}\|(A_m - B_m)\theta\| &= \|[(K\mathcal{L}_m - K_m)\widetilde{\theta}]u\|_\infty, \\ &\leq \|[K(I - \mathcal{L}_m)\widetilde{\theta}]u\|_\infty + \|[(K - K_m)\widetilde{\theta}]u\|_\infty \longrightarrow 0 \quad \text{as} \quad m \to \infty,\end{aligned}$$

and we also obtain

$$\lim_{m\to\infty} \|M_m^* - M_m\| = 0. \tag{53}$$

Now making use of the identities

$$\begin{aligned}M_m^* &= M_m[I_m + M_m^{-1}D_m(B_mA_m - C_m + A_m - B_m)D_m^{-1}], \\ M_m &= M_m^*[I_m - (M_m^*)^{-1}D_m(B_mA_m - C_m + A_m - B_m)D_m^{-1}]\end{aligned}$$

we respectively obtain,

$$\lim_{m\to\infty}\frac{\mathrm{cond}(M_m^*)}{\mathrm{cond}(M_m)}\le 1 \quad\text{and}\quad \lim_{m\to\infty}\frac{\mathrm{cond}(M_m)}{\mathrm{cond}(M_m^*)}\le 1$$

and this reaches the proof. □

Proof of Theorem 3.3 First, to show (44), it suffices to combine (38) with the bound (19) and then use Theorem 5 to obtain the required estimate.
Next, by (13)

$$\|K(I-\mathcal{L}_m)(K-K_m)f^*u\|_\infty \le \frac{\mathcal{C}}{m^\lambda}\|(K-K_m)f^*\|_{Z_\lambda(u)} \tag{54}$$

and hence (19) leads to

$$\|K(I-\mathcal{L}_m)(K-K_m)f^*u\|_\infty = \mathcal{O}\left(\frac{1}{m^{2\lambda}}\right). \tag{55}$$

On the other hand, by applying (13) to the function Kf, where $f\in C_u$ together with (26), gives

$$\begin{aligned}\|K(I-\mathcal{L}_m)Kf\|_{C_u} &\le \frac{\mathcal{C}}{m^\lambda}\|Kf\|_{Z_\lambda(u)},\\ &\le \frac{\mathcal{C}}{m^\lambda}\|f\|_{C_u}\end{aligned}$$

which in turn implies that

$$\|K(I-\mathcal{L}_m)K\|_u = \mathcal{O}\left(\frac{1}{m^\lambda}\right). \tag{56}$$

Similarly, replacing f by $K_m f$ in (14) and using (27), we obtain

$$\begin{aligned}\|K(I-\mathcal{L}_m)K_m f\|_{C_u} &\le \mathcal{C}E_{m-1}(K_m f)_u,\\ &\le \mathcal{C}\frac{\log m}{m^\lambda}\|fu\|_\infty.\end{aligned}$$

Hence

$$\|K(I-\mathcal{L}_m)K_m\|_u = \mathcal{O}\left(\frac{\log m}{m^\lambda}\right) \tag{57}$$

and therefore

$$\|K(I-\mathcal{L}_m)(K-K_m)\|_u = \mathcal{O}\left(\frac{\log m}{m^\lambda}\right). \tag{58}$$

If we combine (41) and (44) with (55) and (58), we obtain

$$\|[f_m^* - f^*]u\|_\infty = \mathcal{O}\left(\frac{1}{m^{2\lambda}}\right) + \mathcal{O}\left(\frac{[\log m]^2}{m^{3\lambda}}\right)$$

then, the remark $\left(\frac{\log m}{m^{\lambda/2}}\right)^2 < 1$ ends the proof of (45).
Lastly, we apply $\mathcal{L}_m$ on both sides of (33) and (34), we find

$$\mathcal{L}_m \widetilde{f}_m^* = \mathcal{L}_m K f_m^* + \mathcal{L}_m g = f_m^*,$$

and this leads to

$$\widetilde{f}_m^*(x_i) = f_m^*(x_i), \quad i = 1, 2, \ldots, m. \tag{59}$$

Now (46) is directly extracted from (44) and (45). This completes the proof. □

6 Conclusions

To approximatively solve problem (1), two efficient numerical methods based on an interpolatory projection have been defined. The approximate solutions are evaluated by solving a linear system of equal size and with an equivalent condition number as the one of the ordinary *Nyström* method. It has been shown that the approach is stable and converges. Further, it exhibits superconvergent results even when the used polynomials are of low degree. Clearly, the iterated modified *Nyström* solution is the most accurate approximation considered in this paper. However, the modified *Nyström* solution possesses the nice property of superconvergence at the collocation points making it especially helpful in certain specific situations, i.e., when we simply need approximations at particular points in the interval.

References

1. Allaei, S., Diogo, T., Rebelo, M.: The Jacobi collocation method for a class of nonlinear Volterra integral equations with weakly singular kernel. J. Sci. Comput. **69**, 673–695 (2016)
2. Allouch, C., Sbibih, D., Tahrichi, M.: Superconvergent Nyström and degenerate kernel methods for integral equations of the second kind. J. Int. Eqns. Appl. **57**, 3–20 (2012)
3. Allouch, C., Sbibih, D., Tahrichi, M.: Superconvergent Nyström method for Urysohn integral equations. BIT Numer. Math. 57 (1), 3–20 (2016)
4. Allouch, C., Tahrichi, M.: Discrete superconvergent Nyström method for integral equations and eigenvalue problems. Math. Comput. Simul. **118**, 17–29 (2015)
5. Atkinson, K.E.: The Numerical Solution of Integral Equations of the Second Kind. Cambridge University Press, Cambridge (1997)
6. Brunner, H.: Collocation Methods for Volterra Integral and Related Functional Differential Equations. Cambridge University Press, Cambridge (2004)

7. Diogo, T., Fermo, L., Occorsio, D.: A projection method for Volterra integral equations in weighted spaces of continuous functions. J Int Eqns Appl (2021)
8. Ditzian, Z., Totik, W.: Moduli of smoothness. SCMG Springer, New York, Berlin, Heidelberg, London, Paris, Tokyo (1987)
9. Fermo, L., Occorsio, D.: A projection method with smoothing transformation for second kind volterra integral equations, Dolomites Res. Notes Approx. **14**, 12–26 (2021)
10. Fermo, L., Occorsio, D.: Weakly singular linear Volterra integral equations: a Nyström method in weighted spaces of continuous functions. J. Comput. Appl. Math. **406** (2022)
11. Hamza, B., Allouch, C., Tahrichi, M.: Legendre Superconvergent Degenerate Kernel and Nyström Methods for Fredholm Integral Equations, Sahand Communications in Mathematical Analysis. https://doi.org/10.22130/scma.2022.550563.1078
12. Kolk, M., Pedas, A., Vainikko, G.: High-order methods for Volterra integral equations with general weak singularities. Numer. Funct. Anal. Optim. **30**(9–10), 1002–1024 (2009)
13. Li, X., Tang, T., Xu, C.: Numerical solutions for weakly singular Volterra integral equations using Chebyshev and Legendre pseudo-spectral Galerkin methods. J. Sci. Comput. **67**, 43–64 (2016)
14. Mastroianni, G., Milovanovic, G.V.: Interpolation Processes: Basic Theory and Applications, Springer Monographs in Mathematics. Springer, Berlin (2008)
15. Mezzanotte, D., Occorsio, D., Russo, M.G.: Combining Nyström Methods for a Fast Solution of Fredholm Integral Equations of the Second Kind, Mathematics (2021)
16. Nevai, P.: Mean convergence of Lagrange interpolation, III. Trans. Am. Math. Soc. **282**(2), 669–698 (1984)
17. Pedas, A., Vainikko, G.: Integral equations with diagonal and boundary singularities of the kernel. Zeitschrift Fur Anal. Und Ihre Anwendung **25**(4), 48–516 (2006)
18. Timan, A.F.: Theory of Approximation of Functions of Real Variable, Dover, New York (1994)
19. Allouch, C.: Collocation and modified collocation methods for solving second kind Fredholm integral equations in weighted spaces. Appl. Numer. Math. **198** 202–216 (2024)
20. Allouch, C.: Fast and accurate solvers for weakly singular Volterra integral equations in weighted spaces. J. Comput. Appl. Math. **438**, 115535 (2024)

Quadratic Spline QI for Approximating Solutions of Love's Integral Equation

Fadila El Mokhtari

Abstract In this paper, we present a simple method based on C^1 quadratic polynomial splines that provides the numerical solution of Love's integral equation. This technique consists in approximating the univariate right section of the kernel of the considered equation by a quadratic quasi-interpolant (QI). The approximate solution is obtained by solving a linear system where the matrix coefficients are computed explicitly. We study the convergence analysis and we show that the method is of order four. Finally, we give some numerical examples that illustrate the theoretical results and the validity of the presented method.

1 Introduction

We provide an explicit numerical solution to the Love's integral equation given by

$$f(x) \pm \frac{1}{\pi} \int_{-1}^{1} \frac{d}{d^2 + (x-t)^2} f(t)\, dt = 1, \quad |x| \leq 1, \quad d \in \mathbb{R}^+ \tag{1}$$

arising in the electrostatic problem of a circular plate capacitor in an unbounded perfect fluid (see e.g. [1]).

The first formulation of the above equation was proposed by E. R. Love [2] in the context of an applied physics study where it was claimed the normalized field created conjointly by two similar circular coaxial conducting disks of radius R, separated by a distance $d \times R$ (see Fig. 1), where d is a positive real parameter, and at equal or opposite potential, with zero potential at infinity, is the solution of the above Love's equation which is a class of second kind Fredholm integral equations [2, 3].

In general, it is not possible to provide a closed form to the exact solution of such an equation involving finitely many terms. It is the case of equation (1), whose

F. El Mokhtari (✉)
LANO Laboratory, FSO, University Mohammed First, Oujda, Morocco

Research Center, High Studies of Engineering School, EHEI, Oujda, Morocco
e-mail: elmokhtari.fadila@gmail.com

© The Author(s), under exclusive license to Springer Nature Switzerland AG 2025

M. J. Ibáñez-Pérez et al. (eds.), *Mathematical And Computational Modelling, Approximation and Simulation*, SEMA SIMAI Springer Series 41,
https://doi.org/10.1007/978-3-032-01114-5_10

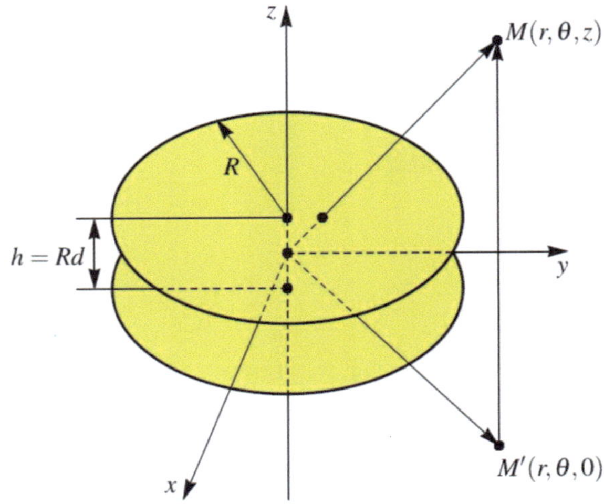

Fig. 1 Love's physical system (cf. [4, p. 81])

solution is given in [2] in the form of infinite series. Hence several methods have been proposed in the literature to approximate its solution. Some of these methods obtain the approximate solution using techniques like Galerkin, collocation, or quadrature approaches such as Nyström (see, e.g., [5–7]). These techniques transform the integral equation into a system of equations. Another class of analytical methods is based on series expansion such as Taylor expansion introduced in [8, 9] and references therein, Chebyshev expansion presented in [10] and Boubaker polynomials expansion studied in [4, 11, 12]. Many other specific methods have been proposed to dealt with this kind of linear and nonlinear equations (see e.g. [13–16] and references therein).

In this paper, we describe an alternative numerical method, based on C^1 quadratic polynomial splines, for approximating the solution of equation (1). This method is different from the ones cited above and it is inspired from the idea developed in [17]. It involves approximating the univariate right section

$$t \to \mathcal{K}(x,t) := \frac{1}{\pi} \frac{d}{d^2 + (x-t)^2}$$

of the kernel $\mathcal{K}$ using a C^1 quadratic spline quasi-interpolant. Such an operator Q is defined by

$$Qg := \sum_{j \in J} g\left(\theta_j\right) \mathcal{L}_j,$$

where $\{\theta_j, \, j \in J\}$ is a sequence of data points in the interval $I := [-1, 1]$ and $\{\mathcal{L}_j, \, j \in J\}$ is the associated sequence of quasi-Lagrange splines, which are finite linear combinations of quadratic B-splines generating the space of C^1 splines of degree 2 defined on I. Substituting this approximate kernel into the integral equation

Table 1 Maximal relative and absolute errors considering (3)

n	$\|(f_n - f^*)/f^*\|_\infty$	NAO	$\|f_n - f^*\|_\infty$	NAO
8	1.89×10^{-5}	–	1.31×10^{-5}	–
16	1.30×10^{-6}	3.86	9.16×10^{-7}	3.83
32	8.51×10^{-8}	3.93	6.05×10^{-8}	3.93
64	5.44×10^{-9}	3.97	3.88×10^{-9}	3.96

Table 2 Maximal relative and absolute errors considering (4)

n	$\|(f_n - f^*)/f^*\|_\infty$	NAO	$\|f_n - f^*\|_\infty$	NAO
8	3.50×10^{-5}	–	6.59×10^{-5}	–
16	2.16×10^{-6}	4.02	4.02×10^{-6}	4.03
32	1.34×10^{-7}	4.015	2.47×10^{-7}	4.02
64	8.29×10^{-9}	4.01	1.53×10^{-8}	4.01

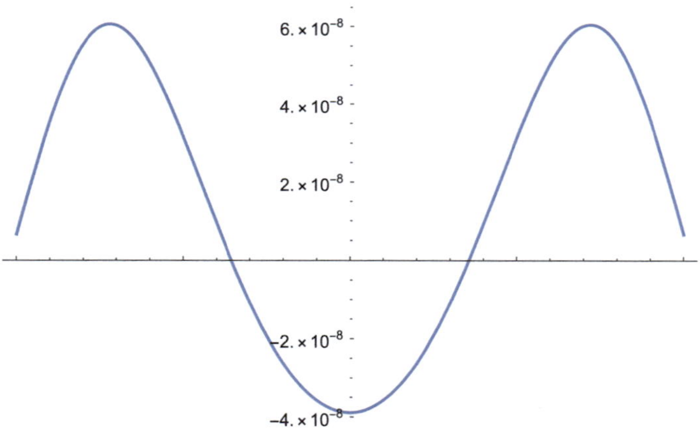

Fig. 2 Graph of the absolute error for $n = 32$

(1) results in a linear system, whose solution is used to construct a numerical solution to (1). We prove that our method achieves a convergence order of four. This result is confirmed by some numerical experiments showed in Tables 1 and 2, and illustrated by Figs. 2 and 3.

The paper is organized as follows. In Sect. 2, we recall the univariate discrete quadratic spline quasi-interpolant on the interval I, which will be used in subsequent sections. Section 3 is dedicated to studying the approximation of the kernel using this spline quasi-interpolant and its associated linear system, as well as providing error estimates for the corresponding solutions. Finally, Sect. 4 contains numerical tests that confirm the theoretical results and method's performance, followed by an appendix with the exact values of the coefficients related to the linear system generated by this method.

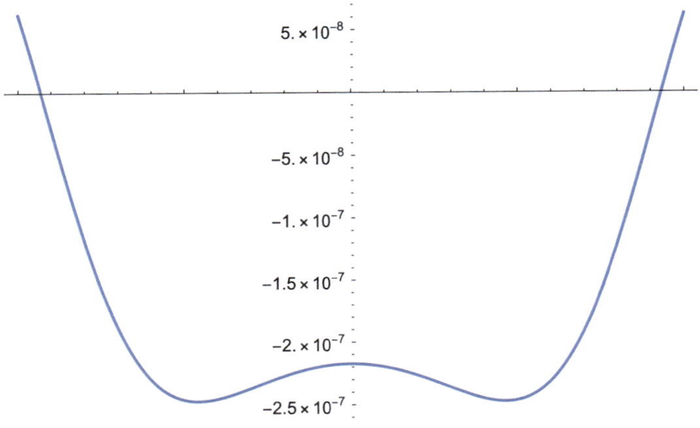

Fig. 3 Graph of the absolute error for $n = 32$

2 Quadratic Spline Quasi-Interpolant

2.1 *Discrete Quasi-Interpolant*

Let $\tau_n := \{-1 + ih, 0 \le i \le n\}$ be the uniform partition of the interval I with meshlength $h := \frac{2}{n}$. The space $S_2(I, \tau_n)$ represents the set of splines of degree 2 (class C^1) on I. A basis of this space is form ed by the $n+2$ quadratic B-splines $\{B_j, 0 \le j \le n+1\}$, with $\text{supp}\left(B_j\right) = [x_{j-2}, x_{j+1}]$ if we add multiple knots at the endpoints: $x_{-2} = x_{-1} = x_0 = -1$ and $1 = x_n = x_{n+1} = x_{n+2}$. We recall the representION of monomials 1, x and x^2 as linear combination of the B-splines B_j (see [18]):

$$1 = \sum_{j=0}^{n+1} B_j, \quad x = \sum_{j=0}^{n+1} \theta_j B_j, \quad \text{and } x^2 = \sum_{j=0}^{n+1} \sigma_j B_j,$$

where $\theta_j := \frac{1}{2}\left(x_{j-1} + x_j\right)$ and $\sigma_j := x_{j-1}x_j$.

A discrete quadratic quasi-interpolant (abbr. dQI) is a spline operator of the form

$$Qg := \sum_{j=0}^{n+1} \mu_j(g) B_j,$$

where the coefficients $\mu_j(g)$ are linear combinations of values $g_j := g\left(\theta_j\right)$ of g on the set $\Theta_n := \{\theta_j, 0 \le j \le n+1\}$.

These coefficients are computed so that the operator Q is exact on quadratic polynomials. Ensuring the exactness of Q on these polynomials requires solving a system of linear equations for interior B-splines and a finite number of specific linear systems for boundary B-splines. Since the determinants of these systems are of Vandermonde type the existence and uniqueness of Q are guaranteed. For more details on the computation of these coefficients, see [19, 20]. The coefficients are given by

$$\mu_0(g) := g_0, \ \mu_1(g) := \frac{1}{6}(-2g_0 + 9g_1 - g_2),$$
$$\mu_n(g) := \frac{1}{6}(-g_{n-1} + 9g_n - 2g_{n+1}), \ \mu_{n+1}(g) := g_{n+1},$$

and

$$\mu_j(g) := \frac{1}{8}\left(-g_{j-1} + 10g_j - g_{j+1}\right), \ 2 \le j \le n-1.$$

It was proved in [20] that $\|Q\|_\infty = 1.4734$. Moreover, for any function $g \in C^3(I)$, we have the error estimate (see [21])

$$\|g - Qg\|_\infty \le C(1 + \|Q\|_\infty)h^3\|g^{(3)}\|_\infty,$$

where $\|\cdot\|_\infty = \|\cdot\|_{\infty,I}$ represents the infinity norm on the interval I, and C is a constant that does not depend on h. Thus,

$$\|g - Qg\|_\infty = O(h^3).$$

For some applications and the rest of this paper, it is useful to write the operator Q under its quasi-Lagrange form

$$Qg = \sum_{j=0}^{n+1} g\left(\theta_j\right)\mathcal{L}_j = \sum_{j=0}^{n+1} g_j\mathcal{L}_j,$$

where, according to [22], the quasi-Lagrange functions $\mathcal{L}_j$ are linear combinations of a maximum of three B-splines. Explicit expressions for $3 \le j \le n-2$ are as follows

$$\mathcal{L}_j := -\frac{1}{8}B_{j-1} + \frac{5}{4}B_j - \frac{1}{8}B_{j+1}.$$

For the extremal functions, we have particular expressions:

$$\mathcal{L}_0 := B_0 - \frac{1}{3}B_1, \ \ \mathcal{L}_1 := \frac{3}{2}B_1 - \frac{1}{8}B_2, \ \ \mathcal{L}_2 := -\frac{1}{6}B_1 + \frac{5}{4}B_2 - \frac{1}{8}B_3,$$
$$\mathcal{L}_{n-1} := -\frac{1}{8}B_{n-2} + \frac{5}{4}B_{n-1} - \frac{1}{6}B_n, \ \ \mathcal{L}_n := -\frac{1}{8}B_{n-1} + \frac{3}{2}B_n, \ \ \mathcal{L}_{n+1} := -\frac{1}{3}B_n - B_{n+1}.$$

2.2 Superconvergence Phenomenon

In this subsection, we show that the quasi-interpolant Q satisfies a superconvergence property, i.e. it is more accurate at the knots x_i and θ_i. More precisely, we demonstrate that the convergence order at these points is $O(h^4)$ instead of $O(h^3)$ on the whole interval I. To do this, we first remark that

$$Qg(-1) = \mu_0(g) = g(-1) \quad \text{and} \quad Qg(1) = \mu_{n+1}(g) = g(1).$$

Moreover, by expressing $Qg(x_i)$ and $Qg(\theta_i)$ as linear combinations of the values $\mu_j(g)$, and by applying the Taylor series of g at these last points, we easily show that for $m_3(x) := x^3$, we have

$$Qm_3(x_i) - m_3(x_i) = 0,\ 2 \le i \le n-2, \quad \text{and} \quad Qm_3(\theta_i) - m_3(\theta_i) = 0,\ 3 \le i \le n-2.$$

The above equalities allow us to prove the following result.

Proposition 2.1 *Assume that $g \in C^4(I)$, then we have:*

$$Qg(x_i) - g(x_i) = -\frac{3}{128} h^4 g^{(4)}(x_{i-1}) + O\left(h^5\right), \qquad i = 2, \ldots, n-2,$$

$$Qg(\theta_i) - g(\theta_i) = -\frac{1}{64} h^4 g^{(4)}(\theta_{i-1}) + O\left(h^5\right), \qquad i = 3, \ldots, n-2.$$

Proof The Taylor expansions of g at the points x_{i-1}, $3 \le i \le n-1$, and θ_{i-1}, $4 \le i \le n-1$ give

$$g(x) = \sum_{j=0}^{4} \frac{g^{(j)}(x_{i-1})}{j!}(x - x_{i-1})^j + O\left((x - x_{i-1})^5\right),$$

$$g(x) = \sum_{j=0}^{4} \frac{g^{(j)}(\theta_{i-1})}{j!}(x - \theta_{i-1})^j + O\left((x - \theta_{i-1})^5\right).$$

If we set $\rho_i = x_i$ or θ_i, then by using the linearity of the operator Q and its exactness on quadratic polynomials, we get

$$Qg(x) = \sum_{j=0}^{2} \frac{g^{(j)}(\rho_{i-1})}{j!}(x - \rho_{i-1})^j + \frac{g^{(3)}(\rho_{i-1})}{3!} Qm_3(x - \rho_{i-1}) + \frac{f^{(4)}(\rho_{i-1})}{4!} Qm_4(x - \rho_{i-1})$$
$$+ O\left((x - \rho_{i-1})^5\right),$$

where $m_4(x) := x^4$.
Therefore,

$$Qg(x) - g(x) = \frac{g^{(3)}(\rho_{i-1})}{3!}\left(Qm_3(x) - m_3(x)\right) + \frac{g^{(4)}(\rho_{i-1})}{4!}\left(Qm_4(x) - m_4(x)\right)$$

$$- 4\rho_{i-1} \left(Qm_3 (x)) - m_3 (x))\right) + O\left((x - \rho_{i-1})^5\right).$$

As $Qm_3 (x_i) = m_3 (x_i)$ for $i = 2, \ldots, n-2$, and $Qm_3 (\theta_i) = m_3 (\theta_i)$ for $i = 3, \ldots, n-2$, we deduce that

$$Qg (x_i) - g (x_i) = \frac{g^{(4)} (x_{i-1})}{4!} (Qm_4 (x_i) - m_4 (x_i)) + O\left(h^5\right),$$

and

$$Qg (\theta_i) - g (\theta_i) = \frac{g^{(4)} (\theta_{i-1})}{4!} (Qm_4 (\theta_i) - m_4 (\theta_i)) + O\left(h^5\right).$$

Hence, thanks to the Taylor series expansions of $Qm_4 (x_i)$ and $Qm_4 (\theta_i)$, the claim follows. □

3 Approximating the Kernel by the Spline Quasi-Interpolant Q

3.1 *Description of the Method*

In this subsection, we approximate the kernel $\mathcal{K}$ of the integral equation (1) with respect to its second variable, the first one remains free [23]. This approximation using the quasi-interpolant Q described above results in a degenerate kernel within the space $S_2 (I, \tau_n)$. More precisely, using the quasi-Lagrange form of Q, allows us to express this approximation as

$$\mathcal{K} (x, t) \approx \mathcal{K}_n (x, t) := Q\mathcal{K} (x, \cdot) (t) = \sum_{i=0}^{n+1} \mathcal{K} (x, \theta_i) \mathcal{L}_i (t) = \sum_{i=0}^{n+1} \mathcal{K}_i (x) \mathcal{L}_i (t).$$

The rational functions $\mathcal{K}_i := \mathcal{K} (\cdot, \theta_i)$, where $0 \le i \le n+1$, are defined as

$$\mathcal{K}_i (x) := \frac{1}{\pi} \frac{d}{d^2 + (x - \theta_i)^2}, \quad -1 \le x \le 1.$$

Then, the approximate solution f_n satisfies the equation

$$f_n (x) = 1 \pm \int_{-1}^{1} \mathcal{K}_n (x, t) f_n (t) \, dt = 1 \pm \sum_{i=0}^{n+1} \mathcal{K}_i (x) \int_{-1}^{1} \mathcal{L}_i (t) f_n (t) \, dt, \qquad (2)$$

and it can be written in the form

$$f_n(x) = 1 \pm \sum_{i=0}^{n+1} Y_i \mathcal{K}_i(x).$$

By substituting into the integral equation (2), we obtain

$$\sum_{i=0}^{n+1} Y_i \mathcal{K}_i(x) = \sum_{i=0}^{n+1} \mathcal{K}_i(x) \int_{-1}^{1} \mathcal{L}_i(t) \left(1 \pm \sum_{j=0}^{n+1} Y_j \mathcal{K}_j(t) \right) dt.$$

It is easy to verify that the functions $\mathcal{K}_i$ are linearly independent. Then,

$$Y_i = b_i \pm \sum_{j=0}^{n+1} \mathcal{M}_{i,j} Y_j, \qquad 0 \leq i \leq n+1.$$

This linear system can be written as

$$(I \pm \mathcal{M})\, Y = b$$

where $Y := (Y_0, Y_1, \ldots, Y_{n+1})^T$. The coefficients of the matrix $\mathcal{M}$ and the vector b are defined as follows:

$$\mathcal{M}_{i,j} := \int_{-1}^{1} \mathcal{L}_i(t)\mathcal{K}_j(t)\, dt \quad \text{and} \quad b_i := \int_{-1}^{1} \mathcal{L}_i(t)\, dt, \qquad 0 \leq i, j \leq n+1.$$

All the above mentioned integrals are computed explicitly, and their exact values are provided in the appendix.

3.2 Order of Convergence

Consider $\mathbf{K}$ and $\mathbf{K}_n$ as integral operators with kernels $\mathcal{K}$ and $\mathcal{K}_n$, respectively, i.e.

$$\mathbf{K} f(x) := \int_{-1}^{1} \mathcal{K}(x,t)\, f(t)\, dt, \quad \mathbf{K}_n f(x) := \int_{-1}^{1} \mathcal{K}_n(x,t)\, f(t)\, dt, \quad x \in [-1, 1].$$

Then, we have

$$\|\mathbf{K} - \mathbf{K}_n\|_\infty = \max_{x \in I} \int_{-1}^{1} |\mathcal{K}(x,t) - \mathcal{K}_n(x,t)|\, dt.$$

To determine the approximation order of $\|\mathbf{K} - \mathbf{K}_n\|_\infty$, we employ the quadrature formula based on the quadratic quasi-interpolant Q for approximating $\int_{-1}^{1} |\mathcal{K}(x,t) - \mathcal{K}_n(x,t)|\, dt$. This quadrature formula, developed in [22], has an approximation order $O(h^4)$ and it is defined by

$$I_Q(g, I) := h\left(\frac{1}{9}(g_0 + g_{n+1}) + \frac{7}{8}(g_1 + g_n) + \frac{73}{72}(g_2 + g_{n-1})h + \sum_{j=3}^{n-2} g_j\right) = h\sum_{j=0}^{n+1} w_j g_j,$$

with

$$w_0 = w_{n+1} = \frac{1}{9},\ w_1 = w_n = \frac{7}{8},\ w_2 = w_{n-1} = \frac{73}{72},\ \text{ and }\ w_j = 1 \text{ for } 3 \leq j \leq n-2.$$

Therefore, we have

$$\int_{-1}^{1} |\mathcal{K}(x,t) - \mathcal{K}_n(x,t)|\, dt = h\sum_{j=0}^{n+1} w_j \left|\mathcal{K}\left(x,\theta_j\right) - \mathcal{K}_n\left(x,\theta_j\right)\right| + O\left(h^4\right).$$

Additionally, since $\mathcal{K}$ is a smooth kernel on $\Omega := [-1, 1]^2$, the following error estimate holds

$$|\mathcal{K}(x,t) - \mathcal{K}_n(x,t)| \leq Ch^3 \left\|D^{0,3}\mathcal{K}\right\|_{\infty,\Omega},$$

and consequently, we get the following inequality:

$$\int_{-1}^{1} |\mathcal{K}(x,t) - \mathcal{K}_n(x,t)| dt \leq h\sum_{j=3}^{n-2} |\mathcal{K}(x,\theta_j) - \mathcal{K}_n(x,\theta_j)| + \tilde{C}h^4 \|D^{0,3}\mathcal{K}\|_{\infty,\Omega} + O(h^4).$$

According to Sect. 2.2, the operator Q is superconvergent at the knots θ_j, $3 \leq j \leq n-2$. Hence, for $3 \leq j \leq n-2$ we obtain

$$\left|\mathcal{K}\left(x,\theta_j\right) - \mathcal{K}_n\left(x,\theta_j\right)\right| \leq \frac{1}{64}h^4 \left\|D^{0,4}\mathcal{K}\right\|_{\infty,\Omega},$$

which yields

$$h\sum_{j=6}^{n-3} \left|\mathcal{K}\left(x,\theta_j\right) - \mathcal{K}_n\left(x,\theta_j\right)\right| \leq \frac{1}{32}h^4 \left\|D^{0,4}\mathcal{K}\right\|_{\infty,\Omega}.$$

Then, finally we deduce that

$$\|\mathbf{K} - \mathbf{K}_n\|_\infty = O(h^4),$$

and based on classical results for kernel approximation (see e.g., [23]) we deduce that

$$\|f - f_n\|_\infty \leq Ch^4 \|f\|_\infty,$$

where C is a constant independent of h.

4 Numerical Illustration

In this section we give some numerical tests for illustrating the theoretical results related to the approximation of the solution of the integral equation (1) by using our method.

As it was proceeded in [4], we use $f^*(x)$, the solution of (1) introduced in [10], as the exact solution. In the next two examples, we present the infinity norms of the relative and absolute errors between f^* and the approximate solutions f_n produced by our method, where n represents the number of subintervals. We also provide the numerical approximation order, denoted as NAO.

For the equation

$$f(x) = 1 - \frac{1}{\pi} \int_{-1}^{1} \frac{d}{d^2 + (x-t)^2} f(t)dt, \tag{3}$$

and $d = 1$, we have the results shown in Table 1.

Figure 2 shows the error $|f_n - f^*|$ for $n = 32$ in the case of the Eq. (3).

For the equation

$$f(x) = 1 + \frac{1}{\pi} \int_{-1}^{1} \frac{d}{d^2 + (x-t)^2} f(t)dt, \tag{4}$$

and $d = 1$, we have the results shown in Table 2.

Figure 3 shows the error $|f_n - f^*|$ for $n = 32$ in the case of the Eq. (4).

When our method is compared with the one in [4], there are significant differences in approach, complexity, and computation time. The Boubaker polynomial-based approach in [4] uses a recursive formula and introduces additional computational layers as the polynomial order increases, which increases the overall computation time. This approach also involves linear system design, which can be difficult to manage for larger systems and generally results in longer processing times due to the recursive computations and associated matrix complexity.

However, our approach is significantly simpler and faster. We use quasi-interpolation in Lagrangian form to create a degenerate kernel approximation by approximating the kernel based on one variable and keeping the other variable free. This leads to a straightforward linear system where each integral is calculated explicitly, significantly reducing both complexity and computation time.

Appendix

The coefficients of the linear system that provides the numerical solution of equation (1) are stored in the matrix $\mathcal{M}$ and the second member vector b.

$$\mathcal{M} := \begin{pmatrix} \mathcal{M}_{0,0} & \cdots & \mathcal{M}_{0,j} & \cdots & \mathcal{M}_{0,n+1} \\ \mathcal{M}_{1,0} & \cdots & \mathcal{M}_{1,j} & \cdots & \mathcal{M}_{1,n+1} \\ \vdots & & \vdots & & \vdots \\ \mathcal{M}_{i,0} & \cdots & \mathcal{M}_{i,j} & \cdots & \mathcal{M}_{i,n+1} \\ \vdots & & \vdots & & \vdots \\ \mathcal{M}_{n,0} & \cdots & \mathcal{M}_{n,j} & \cdots & \mathcal{M}_{n,n+1} \\ \mathcal{M}_{n+1,0} & \cdots & \mathcal{M}_{n+1,j} & \cdots & \mathcal{M}_{n+1,n+1} \end{pmatrix}, \quad b := \begin{pmatrix} b_0 \\ b_1 \\ \vdots \\ b_i \\ \vdots \\ b_n \\ b_{n+1} \end{pmatrix}$$

For the vector b, we get

$$b_0 = b_{n+1} := \frac{1}{9}h, \qquad b_1 = b_n := \frac{7}{8}h, \qquad b_2 = b_{n-1} := \frac{73}{72}h,$$

and

$$b_i := h, \; 3 \leq i \leq n-2.$$

For the computation of the coefficients $\mathcal{M}_{i,j}$, we first write them in the form:

$$\mathcal{M}_{i,j} := -\frac{1}{8}a_{i-1} + \frac{5}{4}a_i - \frac{1}{8}a_{i+1},$$

with

$$\begin{aligned} a_{i-1} &:= \int_{-1}^{1} B_{i-1}(t)\mathcal{K}_j(t)dt = \psi_{i-3}^{i-1} + \psi_{i-2}^{i-1} + \psi_{i-1}^{i-1}, \\ a_i &:= \int_{-1}^{1} B_i(t)\mathcal{K}_j(t)dt = \psi_{i-2}^{i} + \psi_{i-1}^{i} + \psi_{i}^{i}, \\ a_{i+1} &:= \int_{-1}^{1} B_{i+1}(t)\mathcal{K}_j(t)dt = \psi_{i-1}^{i+1} + \psi_{i}^{i+1} + \psi_{i+1}^{i+1}, \end{aligned}$$

where

$$\begin{aligned} \psi_p^{i-1} &:= \int_{-1+ph}^{-1+(p+1)h} B_{i-1}(t)\mathcal{K}_j(t)dt, \qquad i-3 \leq p \leq i-1, \\ \psi_p^{i} &:= \int_{-1+ph}^{-1+(p+1)h} B_{i}(t)\mathcal{K}_j(t)dt, \qquad i-2 \leq p \leq i, \\ \psi_p^{i+1} &:= \int_{-1+ph}^{-1+(p+1)h} B_{i+1}(t)\mathcal{K}_j(t)dt, \qquad i-1 \leq p \leq i+1. \end{aligned}$$

Then, we get

$$\psi_{i-3}^{i-1}\frac{1}{2}\int_{-1+(i-3)h}^{-1+(i-2)h} \left(\frac{t+1}{h} - i + 3\right)^2 \frac{d}{d^2 + \left(t - \theta_j\right)^2}dt$$

$$= \frac{d}{2\pi h} + \frac{d\,(1-(i-3)\,h)+\theta_j d}{2\pi h^2}\log\left(\frac{d^2+\left(-1+(i-2)\,h-\theta_j\right)^2}{d^2+\left(-1+(i-3)\,h-\theta_j\right)^2}\right)$$
$$+\frac{(1-(i-3)\,h)^2-d^2+\theta_j^2+2\theta_j\,(1-(i-3)\,h)}{2\pi h^2}\times$$
$$\left(\arctan\left(\frac{-1+(i-2)\,h-\theta_j}{d}\right)-\arctan\left(\frac{-1+(i-3)\,h-\theta_j}{d}\right)\right),$$
$$\psi_{i-2}^{i-1} = \frac{1}{\pi}\int_{-1+(i-2)h}^{-1+(i-1)h}\left(-\frac{3}{2}-\left(\frac{t+1}{h}-i\right)\left(\frac{t+1}{h}-i+3\right)\right)\frac{d}{d^2+\left(t-\theta_j\right)^2}dt$$
$$= -\frac{d}{\pi h}+\left(\frac{3d}{2\pi h}-\frac{d\,(1-(i-3)\,h)+\theta_j d}{\pi h^2}\right)\log\left(\frac{d^2+\left(-1+(i-1)\,h-\theta_j\right)^2}{d^2+\left(-1+(i-2)\,h-\theta_j\right)^2}\right)$$
$$+\left(-\frac{3}{2}+\frac{3(1-(i-3)\,h)+3\theta_j}{\pi h}+\frac{-\,(1-(i-3)\,h)^2-\theta_j^2+d^2-2\theta_j\,(1-(i-3)\,h)}{\pi h^2}\right)\times$$
$$\left(\arctan\left(\frac{-1+(i-1)\,h-\theta_j}{d}\right)-\arctan\left(\frac{-1+(i-2)\,h-\theta_j}{d}\right)\right).$$

Moreover,

$$\psi_{i-1}^{i-1} = \frac{1}{2}\int_{-1+(i-1)h}^{-1+ih}\left(\frac{t+1}{h}-i\right)^2\frac{d}{d^2+\left(t-\theta_j\right)^2}dt$$
$$= \frac{d}{2\pi h}+\frac{d\,(1-ih)+\theta_j d}{2\pi h^2}\log\left(\frac{d^2+\left(-1+ih-\theta_j\right)^2}{d^2+\left(-1+(i-1)\,h-\theta_j\right)^2}\right)$$
$$+\frac{(1-ih)^2-d^2+t_j^2+2\theta_j\,(1-ih)}{2\pi h^2}\times$$
$$\left(\arctan\left(\frac{-1+ih-\theta_j}{d}\right)-\arctan\left(\frac{-1+(i-1)\,h-\theta_j}{d}\right)\right),$$
$$\psi_{i-2}^{i} = \frac{1}{2}\int_{-1+(i-2)h}^{-1+(i-1)h}\left(\frac{t+1}{h}-i+2\right)^2\frac{d}{d^2+\left(t-\theta_j\right)^2}dt$$
$$= \frac{d}{2\pi h}+\frac{d(1-(i-2)\,h)+\theta_j d}{2\pi h^2}\log\left(\frac{d^2+\left(-1+(i-1)\,h-\theta_j\right)^2}{d^2+\left(-1+(i-2)\,h-\theta_j\right)^2}\right)$$
$$+\frac{(1-(i-2)h)^2-d^2+\theta_j^2+2\theta_j(1-(i-2)h)}{2\pi h^2}\times$$
$$\left(\arctan\left(\frac{-1+(i-1)\,h-\theta_j}{d}\right)-\arctan\left(\frac{-1+(i-2)\,h-\theta_j}{d}\right)\right).$$

The two next values are computed in a similar manner.

$$\psi_{i-1}^{i} = \frac{1}{\pi}\int_{-1+(i-1)h}^{-1+ih}\left(-\frac{3}{2}-\left(\frac{t+1}{h}-i-1\right)\left(\frac{t+1}{h}-i+2\right)\right)\frac{d}{d^2+\left(t-\theta_j\right)^2}dt$$

$$= -\frac{d}{\pi h} + \left(\frac{3d}{2\pi h} - \frac{d\,(1-(i-2)\,h) + \theta_j d}{\pi h^2}\right) \log\left(\frac{d^2 + \left(-1 + ih - \theta_j\right)^2}{d^2 + \left(-1 + (i-1)\,h - \theta_j\right)^2}\right)$$
$$+ \left(-\frac{3}{2} + \frac{3(1-(i-2)\,h) + 3\theta_j}{\pi h} - \frac{(1-(i-2)))^2 + \theta_j^2 - d^2 + 2\theta_j\,(1-(i-2)\,h)}{\pi h^2}\right) \times$$
$$\left(\arctan\left(\frac{-1 + ih - \theta_j}{d}\right) - \arctan\left(\frac{-1 + (i-1)\,h - \theta_j}{d}\right)\right),$$

$$\psi_i^i = \frac{1}{2}\int_{-1+ih}^{-1+(i+1)h} \left(\frac{t+1}{h} - i - 1\right)^2 \frac{d}{d^2 + \left(t - \theta_j\right)^2} dt$$
$$= \frac{d}{2\pi h} + \frac{d\,(1-(i+1h) + \theta_j d}{2\pi h^2} \log\left(\frac{d^2((-1 + (i+1)\,h - \theta_j)^2}{d^2 + \left(-1 + ih - \theta_j\right)^2}\right)$$
$$+ \frac{(1-(i+1)\,h)^2 - d^2 + \theta_j^2 + 2\theta_j\,(1-(i+1)\,h)}{2\pi h^2} \times$$
$$\left(\arctan\left(\frac{-1 + (i+1)\,h - \theta_j}{d}\right) - \arctan\left(\frac{-1 + ih - \theta_j}{d}\right)\right).$$

Finally, we provide the three last values.

$$\psi_{i-1}^{i+1} = \frac{1}{2}\int_{-1+(i-1)h}^{-1+ih} \left(\frac{t+1}{h} - i + 1\right)^2 \frac{d}{d^2 + \left(t - \theta_j\right)^2} dt$$
$$= \frac{d}{2\pi h} + \frac{d\,(1-(i-1)\,h) + \theta_j d}{2\pi h^2} \log\left(\frac{d^2 + \left(-1 + ih - \theta_j\right)^2}{d^2 + \left(-1 + (i-1)\,h - \theta_j\right)^2}\right)$$
$$+ \frac{(1-(i-1)\,h)^2 - d^2 + \theta_j^2 + 2\theta_j\,(1-(i-1)\,h)}{2\pi h^2} \times$$
$$\left(\arctan\left(\frac{-1 + ih - \theta_j}{d}\right) - \arctan\left(\frac{-1 + (i-1)\,h - \theta_j}{d}\right)\right),$$

$$\psi_i^{i+1} = \frac{1}{\pi}\int_{-1+ih}^{-1+(i+1)h} \left(-\frac{3}{2} - \left(\frac{t+1}{h} - i - 2\right)\left(\frac{t+1}{h} - i + 1\right)\right) \frac{d}{d^2 + \left(t - \theta_j\right)^2} dt$$
$$= -\frac{d}{\pi h} + \left(\frac{3d}{2\pi h} - \frac{d(1-(i-1)\,h) + \theta_j d}{\pi h^2}\right) \log\left(\frac{d^2 + \left(-1 + (i+1)\,h - \theta_j\right)^2}{d^2 + \left(-1 + ih - \theta_j\right)^2}\right)$$
$$+ \left(-\frac{3}{2} + \frac{3(1-(i-1)\,h) + 3\theta_j}{\pi h} + \frac{-(1-(i-1)\,h)^2 - \theta_j^2 + d^2 - 2\theta_j\,(1-(i-1)\,h)}{\pi h^2}\right) \times$$
$$\left(\arctan\left(\frac{-1 + (i+1)\,h - \theta_j}{d}\right) - \arctan\left(\frac{-1 + ih - \theta_j}{d}\right)\right),$$

$$\psi_{i+1}^{i+1} = \frac{1}{2}\int_{-1+(i+1)h}^{-1+(i+2)h} \left(\frac{t+1}{h} - i - 2\right)^2 \frac{d}{d^2 + \left(t - \theta_j\right)^2} dt$$
$$= \frac{d}{2\pi h} + \frac{d(1-(i+2)\,h) + \theta_j d}{2\pi h^2} \log\left(\frac{d^2 + \left(-1 + (i+2)\,h - \theta_j\right)^2}{d^2 + \left(-1 + (i+1)\,h - \theta_j\right)^2}\right)$$
$$+ \frac{(1-(i+2)\,h)^2 - d^2 + \theta_j^2 + 2\theta_j\,(1-(i+2)\,h)}{2\pi h^2} \times$$
$$\left(\arctan\left(\frac{-1 + (i+2)\,h - \theta_j}{d}\right) - \arctan\left(\frac{-1 + (i+1)\,h - \theta_j}{d}\right)\right).$$

The remaining coefficients of the matrix A which are generated by the boundary B-splines are computed in a similar way.

References

1. Martin, P.A., Farina, L.: Radiation of water waves by a heaving submerged horizontal disc. J. Fluid. Mech. **337**, 365–379 (1997)
2. Love, E.R.: The electrostatic field of two equal circular co-axial conducting disks. Quart. J. Mech. Appl. Math. **2**, 428–451 (1949)
3. Love, E.R.: The potential due to a circular parallel plate condenser. Mathematika **37**, 217–231 (1990)
4. Milovanović, G.V., Joksimović, D.: Properties of Boubaker polynomials and an application to Love's integral equation. Appl. Math. Comput. **224**, 74–87 (2013)
5. Atkinson, K.E.:The Numerical Solution of Integral Equations of the Second Kind. Cambridge University Press (1997)
6. Delves, L.M., Mohamed, J.L.: Computational Methods for Integral Equations. Combridge University Press, Combridge
7. Pastore, P.: The numerical treatment of Love's integral equation having very small parameter. J. Comput. Appl. Math. **236**, 1267–1281 (2011)
8. Lomthong, P., Huabsomboon, P., Kaneko, H.: A modified Taylor series method to the classical Love's equation. In: Thomas, G., Fleaurant, C., Panagopoulos, T., Chevassus-Lozza, E. (eds.), Recent Researches in Mathematical Methods in Electrical Engineering and Computer Science, Proceedings of the 13th IASME/WSEAS International Conference on Mathematical Methods and Computational Techniques in Electrical Engineering (MMACTEE '11), WSEAS Press, pp. 68–72
9. Ren, Y., Zhang, B., Qiao, H.: A simple Taylor-series expansion method for a class of second kind integral equations. J. Comput. Appl. Math. **110**, 15–24 (1999)
10. Elliott, D.: A Chebyshev series method for the numerical solution of Fredholm integral equations. Comput. J. **6**, 102–111 (1963)
11. Agida, M., Kumar, A.S.: A Boubaker polynomials expansion scheme solution to random Love's equation in the case of a Rational Kernel. Electron. J. Theor. Phys. **7**(24), 319–326 (2010)
12. Kumar, A.S.: An analytical solution to applied mathematics-related Love's equation using the Boubaker polynomials expansion scheme. J. Frank. Inst. **347**, 1755–1761 (2010)
13. Beyrami, H., Lotfi, T., Mahdiani, K.: A new efficient method with error analysis for solving the second kind Fredholm integral equation with Cauchy kernel. J. Comput. Appl. Math. **300**, 385–399 (2016)
14. Borzabadi, A.H., Fard, O.S.: A numerical scheme for a class of nonlinear Fredholm integral equations of the second kind. J. Comput. Appl. Math. **232**, 449–454 (2009)
15. Li, J., Wang, X., Xiao, S., Wang, T.: A rapid solution of a kind of 1D Fredholm oscillatory integral equation. J. Comput. Appl. Math. **236**, 2696–2705 (2012)
16. Maleknejad, K., Mollapourasl, R., Ostadi, A.: Convergence analysis of Sinc-collocation methods for nonlinear Fredholm integral equations with a weakly singular kernel. J. Comput. Appl. Math. **278**, 1–11 (2015)
17. Allouch, C., Sablonnière, P., Sbibih, S.: Solving Fredholm integral equations by approximating kernels by spline quasi-interpolants. Numer. Algorithms **56**, 437–453 (2011)
18. Schumaker, L.L.: Spline Functions, Basic theory. Wiley (1981)
19. Sablonnière, P.: Quadratic spline quasi-interpolants on bounded domains of $\mathbb{R}^d$, $d = 1, 2, 3$. Spline and radial functions. Rend. Sem. Mat. Univ. Pol. Torino **61**, 229–246 (2003)
20. Sablonnière, P.: On Some multivariate quadratic spline quasi–interpolants on bounded domains. In: Modern Developments in Multivariate Approximation: 5th International Conference, Witten-Bommerholz (Germany), Sept. 2002, pp. 263–278. Birkhäuser Basel (2003)

21. DeVore, R.A., Lorentz, G.G.: Constructive Approximation. Springer, Berlin (1993)
22. Sablonnière, P.: A quadrature formula associated with a univariate quadratic spline quasi-interpolant. BIT **47**, 825–837 (2007)
23. Atkinson, K.E., Han, W.: Theoretical Numerical Analysis, 2nd ed. Springer, Berlin (2005)

Curve Network Interpolation by Unequally Smooth Quadratic B-Spline Surfaces with Free Parameters

Paola Lamberti and Sara Remogna

Abstract This paper deals with the study of the interpolation of a bidirectional B-spline curve network by both biquadratic tensor product B-spline surfaces and quadratic B-spline surfaces on criss-cross triangulations, with the aim of simulating discontinuities in the gradient and even in the surface itself. An algorithm for such a construction is proposed and graphical results show the performance of the method.

1 Introduction

The problem of interpolating a curve network by a smooth parametric surface has been addressed in the literature since the second half of last century using various methods ranging from classical ones, involving blending functions [2, 4, 11–13] to piece-by-piece construction, locally imposing the desired smoothness conditions [1, 5, 21], to subdivision schemes [7, 14] and to ribbon surfaces [24, 25]. Moreover, in the context of complex surface modelling, we mention the problem of fitting functions and data on the sphere (see e.g. [18, 20]).

Such fitting methods can be defined *global* if they represent a surface by a directly available expression on the whole parametric domain and *local* when they construct the surface patch-wise, using only local portion of the data for each step and imposing the required smoothness patch by patch.

In this paper we propose a global method which finally encompasses and generalizes previous methods proposed by the same authors [10, 16], where only simple knot B-splines have been considered. In fact here the main contribution is taking into account our results proposed in [9] (see also [8]) on multiple knots in order to simulate unequal smoothness in the surface and/or in its gradient, where it is required. We consider both biquadratic tensor product B-spline surfaces and quadratic

P. Lamberti (✉) · S. Remogna
Department of Mathematics "G. Peano", University of Torino, Torino, Italy
e-mail: paola.lamberti@unito.it

S. Remogna
e-mail: sara.remogna@unito.it

© The Author(s), under exclusive license to Springer Nature Switzerland AG 2025
M. J. Ibáñez-Pérez et al. (eds.), *Mathematical And Computational Modelling, Approximation and Simulation*, SEMA SIMAI Springer Series 41,
https://doi.org/10.1007/978-3-032-01114-5_11

B-spline surfaces on criss-cross triangulations in order to compare them and, as we will demonstrate, their behaviour can be quite different. We remark that in [22, Chapter 10] and [19] the authors construct a tensor product B-spline surface interpolating a curve network. The method is based on Gordon and Coons surfaces and the main difference with respect to the method proposed here is that it is necessary to define three surfaces belonging to different spline spaces to be merged together by the knot refinement algorithm, while the method proposed here is directly based on the knot vectors of the given B-spline curves.

The paper is structured as follows: in Sect. 2 we introduce the B-spline curve network to be interpolated and in Sect. 3 we construct and analyse the interpolating B-spline surfaces, considering both biquadratic tensor product B-spline surfaces and quadratic B-spline surfaces on criss-cross triangulations. Finally, in Sect. 4 we present some graphical results, showing the performances of the method.

2 Curve Network

Let us be given a bidirectional quadratic B-spline curve network, defined by the two sets

$$
\begin{aligned}
C_r(v) &= \sum_{j=0}^{N-1} P_j^{(r)} B_j(v|V) \qquad r = 0, \ldots, R+1, \ v \in [c, d], \\
D_s(u) &= \sum_{i=0}^{M-1} Q_i^{(s)} B_i(u|U) \qquad s = 0, \ldots, S+1, \ u \in [a, b],
\end{aligned}
\tag{1}
$$

with

- control points $\left\{P_j^{(r)}\right\}_{j=0}^{N-1}$, $r = 0, \ldots, R+1$, $\left\{Q_i^{(s)}\right\}_{i=0}^{M-1}$, $s = 0, \ldots, S+1$ and $P_j^{(r)}, Q_i^{(s)} \in \mathbb{R}^3$, $M \geq R+3$, $N \geq S+3$;
- knots

$$U = \{a = u_{-2} \equiv u_{-1} \equiv u_0 \leq u_1 \leq \cdots \leq u_{M-2} \equiv u_{M-1} \equiv u_M = b\},$$

$$V = \{c = v_{-2} \equiv v_{-1} \equiv v_0 \leq v_1 \leq \cdots \leq v_{N-2} \equiv v_{N-1} \equiv v_N = d\},$$

based on the breakpoints $\boldsymbol{\xi} = \{a = \xi_0 < \xi_1 < \ldots < \xi_{m+1} = b\}$ and $\boldsymbol{\eta} = \{c = \eta_0 < \eta_1 < \ldots < \eta_{n+1} = d\}$, respectively. Indeed, given the two smoothness vectors $\bar{\mu}^u = (\mu_i^u)_{i=1}^m$, $\bar{\mu}^v = (\mu_j^v)_{j=1}^n$ whose elements can be 1, 0, -1 and denote C^1, C^0, C^{-1} smoothness, respectively, at knots U and/or V, then $M = 3 + \sum_{i=1}^m (2 - \mu_i^u)$, $N = 3 + \sum_{j=1}^n (2 - \mu_j^v)$ and the break points ξ_i occurs exactly $2 - \mu_i^u$ times in U, while the break points η_j occurs exactly $2 - \mu_j^v$ times in V [3].

We remark that $\text{supp} B_j(v|V) = [v_{j-2}, v_{j+1}]$ and $\text{supp} B_i(u|U) = [u_{i-2}, u_{i+1}]$.

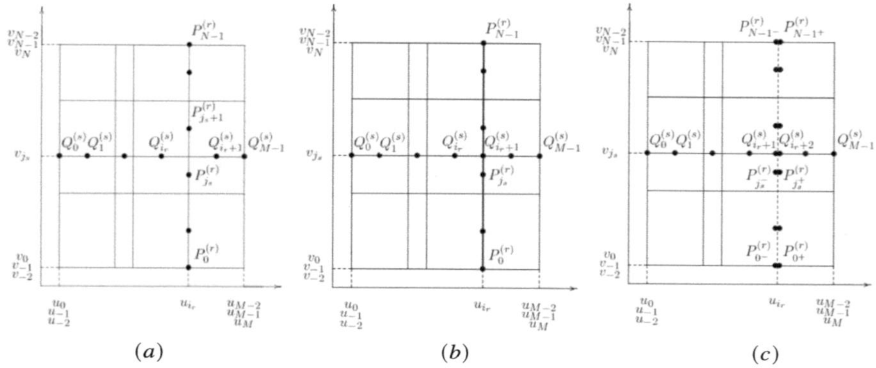

Fig. 1 Preimage locations for **a** simple, **b** double and **c** triple knot partitions

In Fig. 1 we show the control point positions, mapped on the parametric domain in case of simple, double (thick line) and triple (dashed line) knots. In case of triple knots we have two different curves and pairs of different control points, corresponding to each triple knot, i.e. having the same preimage in the parameter domain. We use the sub-indices + and – to denote the values to the right and the left, respectively, due to the presence of a discontinuity (triple knot).

Moreover, there exist parameters

$$u_0 = u_{i_0} < u_{i_1} < \dots < u_{i_R} < u_{i_{R+1}} = u_{M-2},$$
$$v_0 = v_{j_0} < v_{j_1} < \dots < v_{j_S} < v_{j_{S+1}} = v_{N-2},$$

which are subsets of U and V, respectively, and must satisfy

$$C_r(v_{j_s}) = D_s(u_{i_r}), \qquad r = 0, \dots, R+1, \ s = 0, \dots, S+1. \tag{2}$$

We remark that a curve network satisfying such a condition always exists and it can be proven by following the same logic scheme of [16]. In order to verify it, we define the quantities

$$\sigma_i = \frac{h_i}{h_{i-1} + h_i}, \quad \sigma_i' = 1 - \sigma_i = \frac{h_{i-1}}{h_{i-1} + h_i},$$
$$\tau_j = \frac{k_j}{k_{j-1} + k_j}, \quad \tau_j' = 1 - \tau_j = \frac{k_{j-1}}{k_{j-1} + k_j},$$

where $h_i = u_i - u_{i-1}$, $k_j = v_j - v_{j-1}$ and we consider the different cases that can occur:

1. if u_{i_r} and v_{j_s} are simple knots, then the results in [16] are valid;
2. if u_{i_r} is a double knot and v_{j_s} is a simple knot,

$$B_{j_s}(v_{j_s}|V) = \tau_{j_s+1},\ B_{j_s+1}(v_{j_s}|V) = \tau'_{j_s+1},\ B_{i_r}(u_{i_r}|U) = 0,\ B_{i_r+1}(u_{i_r}|U) = 1,$$

hold and consequently

$$\tau_{j_s+1} P^{(r)}_{j_s} + \tau'_{j_s+1} P^{(r)}_{j_s+1} = Q^{(s)}_{i_r+1},\quad r = 1, ..., R,\ s = 1, ..., S; \tag{3}$$

3. (3) is also verified if u_{i_r} is a triple knot and v_{j_s} is simple;
4. if u_{i_r} is a simple knot and v_{j_s} is a double knot,

$$B_{j_s}(v_{j_s}|V) = 0,\ B_{j_s+1}(v_{j_s}|V) = 1,\ B_{i_r}(u_{i_r}|U) = \sigma_{i_r+1},\ B_{i_r+1}(u_{i_r}|U) = \sigma'_{i_r+1},$$

hold and consequently

$$P^{(r)}_{j_s+1} = \sigma_{i_r+1} Q^{(s)}_{i_r} + \sigma'_{i_r+1} Q^{(s)}_{i_r+1},\quad r = 1, ..., R,\ s = 1, ..., S; \tag{4}$$

5. (4) is also verified if v_{j_s} is a triple knot and u_{i_r} is simple;
6. if u_{i_r} and v_{j_s} are both double knots,

$$B_{j_s}(v_{j_s}|V) = 0,\ B_{j_s+1}(v_{j_s}|V) = 1,\ B_{i_r}(u_{i_r}|U) = 0,\ B_{i_r+1}(u_{i_r}|U) = 1,$$

hold and therefore

$$P^{(r)}_{j_s+1} = Q^{(s)}_{i_r+1},\quad r = 1, ..., R,\ s = 1, ..., S; \tag{5}$$

7. (5) is still valid if u_{i_r} and v_{j_s} are both triple, or one double and the other triple knots.

At the boundary of the domain:

1. if u_{i_r} or v_{j_s} are simple knots, then the results in [16] are valid;
2. if u_{i_r} is a double knot, for $s = 0$ we have

$$B_0(v_0|V) = 1,\ B_1(v_0|V) = 0,\ B_{i_r}(u_{i_r}|U) = 0,\ B_{i_r+1}(u_{i_r}|U) = 1,$$

and for $s = S + 1$ we have

$$B_{N-2}(v_{N-2}|V) = 0,\ B_{N-1}(v_{N-2}|V) = 1,\ B_{i_r}(u_{i_r}|U) = 0,\ B_{i_r+1}(u_{i_r}|U) = 1.$$

Therefore

$$\begin{aligned} Q^{(0)}_{i_r+1} &= P^{(r)}_0,\quad s = 0,\ r = 1, ..., R, \\ Q^{(S+1)}_{i_r+1} &= P^{(r)}_{N-1},\quad s = S + 1,\ r = 1, ..., R; \end{aligned} \tag{6}$$

3. (6) is also verified if u_{i_r} is a triple knot;
4. if v_{j_s} is a double knot, for $r = 0$ we have

$$B_0(u_0|U) = 1,\ B_1(u_0|U) = 0,\ B_{j_s}(v_{j_s}|V) = 0,\ B_{j_s+1}(v_{j_s}|V) = 1,$$

and for $r = R + 1$ we have

$$B_{M-2}(u_{M-2}|U) = 0,\ B_{M-1}(u_{M-2}|U) = 1,\ B_{j_s}(v_{j_s}|V) = 0,\ B_{j_s+1}(v_{j_s}|V) = 1.$$

Therefore

$$\begin{aligned} P^{(0)}_{j_s+1} &= Q^{(s)}_0, \quad r = 0,\ s = 1, ..., S, \\ P^{(R+1)}_{j_s} &= Q^{(s)}_{M-1}, \quad r = R + 1,\ s = 1, ..., S; \end{aligned} \tag{7}$$

5. (7) is also verified if v_{j_s} is a triple knot.

   At the four vertices we obtain

$$Q^{(0)}_0 = P^{(0)}_0,\ Q^{(0)}_{M-1} = P^{(R+1)}_0,\ Q^{(S+1)}_0 = P^{(0)}_{N-1},\ Q^{(S+1)}_{M-1} = P^{(R+1)}_{N-1}.$$

Therefore, being $N_d = N(R + 2) + M(S + 2)$ the number of curve network control points, $N_c = (R + 2)(S + 2)$ the number of constraints in (2) and, since $M \geq R + 3$, $N \geq S + 3$, then $N_d \geq N_c$. So a curve network satisfying condition (2) always exists.

3 Construction of the Interpolating Surfaces

Given the above data, we are interested in the construction of B-spline surfaces

$$\mathcal{S}(u, v) = \left(\mathcal{S}_x(u, v), \mathcal{S}_y(u, v), \mathcal{S}_z(u, v)\right)^{\mathrm{T}} = \sum_{i=0}^{M-1} \sum_{j=0}^{N-1} C_{ij} B_{ij}(u, v), \tag{8}$$

with $(u, v) \in \Omega = [a, b] \times [c, d]$, interpolating the two sets of curves (1), where the B_{ij}'s are bivariate B-splines, belonging to the following two different spline spaces:

- $S^{(\overline{\mu}^u, \overline{\mu}^v)}_{2,2}(R_{m,n})$, the space of biquadratic splines on the rectangular partition $R_{m,n} = \boldsymbol{\xi} \times \boldsymbol{\eta}$, i.e. the bivariate piecewise polynomials of coordinate degree 2 in each rectangle of $R_{m,n}$ (Fig. 2 (a)) (see e.g. [3, 22]);
- $S^{(\overline{\mu}^u, \overline{\mu}^v)}_{2}(T_{m,n})$, the space of quadratic splines on the criss-cross triangulation $T_{m,n}$ based on $\boldsymbol{\xi} \times \boldsymbol{\eta}$ (obtained by drawing the two diagonals in each subrectangle of $R_{m,n}$), i.e. the bivariate piecewise polynomials of total degree 2 in each triangle of $T_{m,n}$ (Fig. 2b) (see e.g. [6, 17, 23, 26]).

Let L^0_u (L^0_v) be the number of double knots in U (V), L^{-1}_u (L^{-1}_v) the number of triple knots in U (V). Then

$$M = 3 + m + L^0_u + 2L^{-1}_u, \qquad N = 3 + n + L^0_v + 2L^{-1}_v. \tag{9}$$

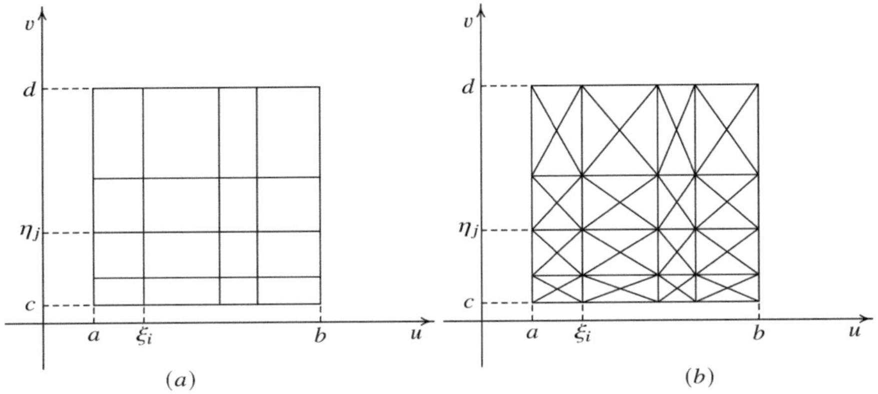

Fig. 2 **a** $R_{m,n}$ and **b** $T_{m,n}$ partitions

Therefore dim $S_{2,2}^{(\overline{\mu}^u,\overline{\mu}^v)}(R_{m,n}) = MN$ (see e.g. [3]) and dim $S_2^{(\overline{\mu}^u,\overline{\mu}^v)}(T_{m,n}) = MN - \gamma$ with

$$\gamma = (1 + L_u^0 + L_u^{-1})(1 + L_v^0 + L_v^{-1}) \tag{10}$$

(see Theorem 1 of [9]). We remark that, although indices in (8) are valid only in the case of a tensor product, where the spline space dimension is MN, while in the case of criss-cross triangulations the corresponding spline space dimension is lower, however for CAGD purpose, where the partition of unity property is important, it is necessary to use the whole set of B-splines.

Let also consider the spanning functions $\{B_{ij}(u, v),\ i = 0, \dots, M-1,\ j = 0, \dots, N-1\}$, being the classical tensor product B-splines $B_{i,j}(u, v) = B_i(u|U) \cdot B_j(v|V)$ in $S_{2,2}^{(\overline{\mu}^u,\overline{\mu}^v)}(R_{m,n})$ [3, 22] and the B-splines with octagonal support, modified in case of multiple knots, $B_{i,j}(u, v)$ in $S_2^{(\overline{\mu}^u,\overline{\mu}^v)}(T_{m,n})$ [6, 8, 9, 26] (some examples of such B-splines are presented in Example 1 of [9], according to possible multiplicity appearing in their support). Since all results proposed in the paper can be conducted similarly, in order to simplify the notation the same symbol for both spline bases is used.

Now our aim is to obtain the control points $\{C_{ij},\ i = 0, \dots, M-1,\ j = 0, \dots, N-1\}$ of the surface (8), such that the initial curves (1) are its *isoparametric curves* at some knots of U and V, i.e.

$$\begin{aligned} \mathcal{S}(u_{i_r}, v) &= C_r(v),\ r = 0, \dots, R+1, \\ \mathcal{S}(u, v_{j_s}) &= D_s(u),\ s = 0, \dots, S+1. \end{aligned} \tag{11}$$

In case of simple knots the conditions that the control points $\{C_{ij}\}$ have to satisfy are given in Theorem 1 of [16]. To deal with unequally smooth surfaces, we have to distinguish double and triple knots, as follows.

Theorem 3.1 If $u_{i_{\overline{r}}}$ is a double knot, then (11) hold if and only if $\{C_{i,j}\}_{i=0,j=0}^{M-1,N-1}$ satisfy the following conditions:

$$\sigma_{i_r+1}C_{i_r,j} + \sigma'_{i_r+1}C_{i_r+1,j} = P_j^{(r)}, \quad j=1,...,N-2,\ r=1,...,R,\ r\neq\overline{r}, \tag{12}$$

$$C_{i_{\overline{r}}+1,j} = P_j^{(\overline{r})}, \quad j=1,...,N-2, \tag{13}$$

$$\tau_{j_s+1}C_{i,j_s} + \tau'_{j_s+1}C_{i,j_s+1} = Q_i^{(s)}, \quad i=1,...,M-2,\ s=1,...,S, \tag{14}$$

$$C_{i,0} = Q_i^{(0)}, \quad C_{i,N-1} = Q_i^{(S+1)}, \quad i=0,...,M-1,$$

$$C_{0,j} = P_j^{(0)}, \quad C_{M-1,j} = P_j^{(R+1)}, \quad j=0,...,N-1.$$

Proof In V all knots are simple and the proof is the one of Theorem 1 of [16], both for inner and boundary control points. The same in U, except when we impose interpolation of curve $C_{\overline{r}}(v)$. So the only relation to be proven is (13). Indeed, since the network curves are isoparametric for $\mathcal{S}$, then for $u = u_{i_{\overline{r}}}$ we get

$$\mathcal{S}(u_{i_{\overline{r}}}, v) = C_{\overline{r}}(v) = \sum_{j=0}^{N-1} P_j^{(\overline{r})} B_j(v|V). \tag{15}$$

From B-spline locality property and since $B_{i_{\overline{r}},j}(u_{i_{\overline{r}}}, v) = 0$ for any $j = 0, ..., N-1$, we obtain

$$\mathcal{S}(u_{i_{\overline{r}}}, v) = \sum_{i=0}^{M-1}\sum_{j=0}^{N-1} C_{i,j}B_{i,j}(u_{i_{\overline{r}}}, v) = \sum_{j=0}^{N-1} C_{i_{\overline{r}}+1,j}B_{i_{\overline{r}}+1,j}(u_{i_{\overline{r}}}, v). \tag{16}$$

Since

$$B_{i_{\overline{r}}+1,j}(u_{i_{\overline{r}}}, v) = B_{i_{\overline{r}}+1}(u_{i_{\overline{r}}}|U)B_j(v|V) = B_j(v|V)$$

holds both in tensor product and criss-cross case, equaling (15) and (16) it results (13). □

Similarly, we can prove that:

- if $v_{j_{\overline{s}}}$ is a double knot, then, instead of (12), (13) and (14) we have

$$\sigma_{i_r+1}C_{i_r,j} + \sigma'_{i_r+1}C_{i_r+1,j} = P_j^{(r)}, \quad j = 1, ..., N-2,\ r = 1, ..., R, \tag{17}$$

$$\tau_{j_s+1}C_{i,j_s} + \tau'_{j_s+1}C_{i,j_s+1} = Q_i^{(s)}, \quad i = 1, ..., M-2,\ s = 1, ..., S,\ s \neq \overline{s}, \tag{18}$$

$$C_{i,j_{\overline{s}}+1} = Q_i^{(\overline{s})}, \quad i = 1, ..., M-2; \tag{19}$$

- if $u_{i_{\overline{r}}}$ and $v_{j_{\overline{s}}}$ are two double inner knots, then

$$\mathcal{S}(u_{i_{\overline{r}}}, v_{j_{\overline{s}}}) = C_{i_{\overline{r}}+1,j_{\overline{s}}+1}$$

  holds.

Theorem 3.2 If an inner knot $u_{i_{\bar{r}}}$ is triple, then (11) hold if and only if $\{C_{i,j}\}_{i=0,j=0}^{M-1,N-1}$ satisfy the following conditions:

$$\sigma_{i_r+1}C_{i_r,j} + \sigma'_{i_r+1}C_{i_r+1,j} = P_j^{(r)}, \quad j=1,\dots,N-2,\ r=1,\dots,R,\ r\neq\bar{r},\bar{r}+1, \tag{20}$$

$$C_{i_{\bar{r}}+1,j^-} = P_{j^-}^{(\bar{r})}, \quad j=1,\dots,N-2, \tag{21}$$

$$C_{i_{\bar{r}}+2,j^+} = P_{j^+}^{(\bar{r})}, \quad j=1,\dots,N-2, \tag{22}$$

$$\tau_{j_s+1}C_{i,j_s} + \tau'_{j_s+1}C_{i,j_s+1} = Q_i^{(s)}, \quad i=1,\dots,M-2,\ i\neq i_{\bar{r}}+1,i_{\bar{r}}+2,\ s=1,\dots,S, \tag{23}$$

$$\tau_{j_s+1}C_{i_{\bar{r}}+1,j_s^-} + \tau'_{j_s+1}C_{i_{\bar{r}}+1,j_s+1^-} = Q_{i_{\bar{r}}+1}^{(s)}, \quad s=1,\dots,S, \tag{24}$$

$$\tau_{j_s+1}C_{i_{\bar{r}}+2,j_s^+} + \tau'_{j_s+1}C_{i_{\bar{r}}+2,j_s+1^+} = Q_{i_{\bar{r}}+2}^{(s)}, \quad s=1,\dots,S, \tag{25}$$

$$C_{i,0} = Q_i^{(0)}, \quad C_{i,N-1} = Q_i^{(S+1)}, \quad i=0,\dots,M-1,$$

$$C_{0,j} = P_j^{(0)}, \quad C_{M-1,j} = P_j^{(R+1)}, \quad j=0,\dots,N-1.$$

Proof As in the case of a double knot in U, we impose that the curves $C_{\bar{r}^-}(v)$ and $C_{\bar{r}^+}(v)$ interpolate the surface. Since the network curves are isoparametric for $\mathcal{S}$, for $u = u_{i_{\bar{r}}}^-$ we have

$$\mathcal{S}(u_{i_{\bar{r}}}^-, v) = C_{\bar{r}^-}(v) = \sum_{j=0}^{N-1} P_{j^-}^{(\bar{r})} B_j(v|V). \tag{26}$$

From the B-spline locality property and since $B_{i_{\bar{r}},j}(u_{i_{\bar{r}}}^-, v) = 0$ for any $j = 0, \dots, N-1$, we get

$$\mathcal{S}(u_{i_{\bar{r}}}^-, v) = \sum_{i=0}^{M-1}\sum_{j=0}^{N-1} C_{i,j}B_{i,j}(u_{i_{\bar{r}}}^-, v) = \sum_{j=0}^{N-1} C_{i_{\bar{r}}+1,j^-}B_{i_{\bar{r}}+1,j}(u_{i_{\bar{r}}}^-, v). \tag{27}$$

Moreover, since

$$B_{i_{\bar{r}}+1,j}(u_{i_{\bar{r}}}^-, v) = B_{i_{\bar{r}}+1}(u_{i_{\bar{r}}}^-|U)B_j(v|V) = B_j(v|V)$$

holds both for tensor product and criss-cross case, equaling (26) and (27) we obtain (21).
Similarly for $u = u_{i_{\bar{r}}}^+$ and taking into account

$$B_{i_{\bar{r}}+2,j}(u_{i_{\bar{r}}}^+, v) = B_{i_{\bar{r}}+2}(u_{i_{\bar{r}}}^+|U)B_j(v|V) = B_j(v|V),$$

we obtain (22). In V we have what has been shown for simple knots in Theorem 1 of [16], except for $i = i_{\bar{r}} + 1$ and $i = i_{\bar{r}} + 2$. In the first case it results

$$\mathcal{S}(u_{i_{\overline{r}}}^-, v_{j_s}) = D_s(u_{i_{\overline{r}}}^-) = Q_{i_{\overline{r}}+1}^{(s)} B_{i_{\overline{r}}+1}(u_{i_{\overline{r}}}^-|U) = Q_{i_{\overline{r}}+1}^{(s)},$$
$$\mathcal{S}(u_{i_{\overline{r}}}^-, v_{j_s}) = \sum_{i=0}^{M-1}\sum_{j=0}^{N-1} C_{i,j} B_{i,j}(u_{i_{\overline{r}}}^-, v_{j_s}) \tag{28}$$
$$= C_{i_{\overline{r}}+1, j_s^-} B_{i_{\overline{r}}+1, j_s}(u_{i_{\overline{r}}}^-, v_{j_s}) + C_{i_{\overline{r}}+1, j_s+1^-} B_{i_{\overline{r}}+1, j_s+1}(u_{i_{\overline{r}}}^-, v_{j_s}).$$

Since

$$B_{i_{\overline{r}}+1, j_s}(u_{i_{\overline{r}}}^-, v_{j_s}) = B_{i_{\overline{r}}+1}(u_{i_{\overline{r}}}^-|U) B_{j_s}(v_{j_s}|V) = B_{j_s}(v_s|V) = \tau_{j_s+1},$$
$$B_{i_{\overline{r}}+1, j_s+1}(u_{i_{\overline{r}}}^-, v_{j_s}) = B_{i_{\overline{r}}+1}(u_{i_{\overline{r}}}^-|U) B_{j_s+1}(v_{j_s}|V) = B_{j_s+1}(v_s|V) = \tau'_{j_s+1},$$

equaling the two expressions in (28), we obtain (24). In the second case we apply the same scheme and we get (25). □

Similarly, we can prove that:

- if $v_{j_{\overline{s}}}$ is a triple knot, then, instead of (20)–(25), we have

$$\sigma_{i_r+1} C_{i_r,j} + \sigma'_{i_r+1} C_{i_r+1,j} = P_j^{(r)}, \quad j=1,\dots,N-2,\ j\neq j_{\overline{s}}+1, j_{\overline{s}}+2,\ r=1,\dots,R, \tag{29}$$
$$\sigma_{i_r+1} C_{i_r^-, j_{\overline{s}}+1} + \sigma'_{i_r+1} C_{i_r+1^-, j_{\overline{s}}+1} = P_{j_{\overline{s}}+1}^{(r)}, \quad r=1,\dots,R, \tag{30}$$
$$\sigma_{i_r+1} C_{i_r^+, j_{\overline{s}}+2} + \sigma'_{i_r+1} C_{i_r+1^+, j_{\overline{s}}+2} = P_{j_{\overline{s}}+2}^{(r)}, \quad r=1,\dots,R, \tag{31}$$
$$\tau_{j_s+1} C_{i,j_s} + \tau'_{j_s+1} C_{i,j_s+1} = Q_i^{(s)}, \quad i=1,\dots,M-2,\ s=1,\dots,S,\ s\neq\overline{s},\overline{s}+1, \tag{32}$$
$$C_{i^-, j_{\overline{s}}+1} = Q_{i^-}^{(\overline{s})}, \quad C_{i^+, j_{\overline{s}}+2} = Q_{i^+}^{(\overline{s})}, \quad i=1,\dots,M-2; \tag{33}$$

- if $u_{i_{\overline{r}}}$ and $v_{j_{\overline{s}}}$ are two triple knots, then

$$\mathcal{S}(u_{i_{\overline{r}}}^-, v_{j_{\overline{s}}}^-) = C_{i_{\overline{r}}+1^-, j_{\overline{s}}+1^-}, \quad \mathcal{S}(u_{i_{\overline{r}}}^+, v_{j_{\overline{s}}}^+) = C_{i_{\overline{r}}+2^+, j_{\overline{s}}+2^+},$$
$$\mathcal{S}(u_{i_{\overline{r}}}^-, v_{j_{\overline{s}}}^+) = C_{i_{\overline{r}}+1^+, j_{\overline{s}}+2^-}, \quad \mathcal{S}(u_{i_{\overline{r}}}^+, v_{j_{\overline{s}}}^-) = C_{i_{\overline{r}}+2^-, j_{\overline{s}}+1^+},$$

  hold;
- if $u_{i_{\overline{r}}}$ is an inner double knot and $v_{j_{\overline{s}}}$ is an inner triple one or $u_{i_{\overline{r}}}$ is triple and $v_{j_{\overline{s}}}$ double, then

$$\mathcal{S}(u_{i_{\overline{r}}}, v_{j_{\overline{s}}}^-) = C_{i_{\overline{r}}+1^-, j_{\overline{s}}+1}, \quad \mathcal{S}(u_{i_{\overline{r}}}, v_{j_{\overline{s}}}^+) = C_{i_{\overline{r}}+1^+, j_{\overline{s}}+2},$$
$$\mathcal{S}(u_{i_{\overline{r}}}^-, v_{j_{\overline{s}}}) = C_{i_{\overline{r}}+1, j_{\overline{s}}+1^-}, \quad \mathcal{S}(u_{i_{\overline{r}}}^+, v_{j_{\overline{s}}}) = C_{i_{\overline{r}}+2, j_{\overline{s}}+1^+},$$

  hold, respectively.

In order to model the surface $\mathcal{S}$, we have to take into account the number of free control points.

Theorem 3.3 The number of unconstrained control points in $\mathcal{S}$ is given by

$$\overline{N} = \begin{cases} \gamma + k\ell + k(1 + L_v^0 + L_v^{-1}) + \ell(1 + L_u^0 + L_u^{-1}) & \text{for } S_{2,2}^{(\overline{\mu}^u,\overline{\mu}^v)}(R_{m,n}) \\ k\ell + k(1 + L_v^0 + L_v^{-1}) + \ell(1 + L_u^0 + L_u^{-1}) & \text{for} S_2^{(\overline{\mu}^u,\overline{\mu}^v)}(T_{m,n}) \end{cases} \tag{34}$$

with γ defined in (10).

□

Proof As in Theorem 2 of [16] for all simple knots, the number of unconstrained parameters ($\overline{N}$) is given by the number of all parameters (N_1), subtracting the ones involved in the interpolation conditions (N_2).

N_1 coincides with the spline space dimension, used to construct the surface, therefore

$$N_1 = \begin{cases} MN & \text{for } S_{2,2}^{(\overline{\mu}^u,\overline{\mu}^v)}(R_{m,n}) \\ MN-\gamma & \text{for } S_2^{(\overline{\mu}^u,\overline{\mu}^v)}(T_{m,n}) \end{cases} .$$

From Theorems 3.1, 3.2 and similar cases, the number of interpolation conditions is $R(N-2)+S(M-2)+2N+2M$, but the following remarks hold: at the four vertices of the domain we have redundant conditions, as seen in Theorem 2 of [16], and at every inner point (u_{i_r}, v_{j_s}), $r = 1, ..., R$, $s = 1, ..., S$, of the domain the RS conditions (2) hold and let us obtain some points from others. Indeed:

- if $u_{i_{\overline{r}}}$ is a double knot, we have $(R-1)S$ redundant equations for simple knots and S conditions (3) for $r=\overline{r}$, that from (13) coincide with (14);
- if $v_{j_{\overline{s}}}$ is a double knot, similarly we have $(S-1)R$ redundant equations for simple knots and R conditions (4) for $s=\overline{s}$, that from (19) coincide with (17);
- if $u_{i_{\overline{r}}}$ and $v_{j_{\overline{s}}}$ are double knots, we have $(R-1)(S-1)$ redundant equations for simple knots, $(S-1)$ conditions (3) $(s\neq\overline{s})$, $(R-1)$ conditions (4) $(r\neq\overline{r})$ and one condition (5) $(s=\overline{s}, r=\overline{r})$;
- if $u_{i_{\overline{r}}}$ is triple, we have $(R-2)S$ redundant equations for simple knots, S conditions (3) of kind
$$Q^{(s)}_{i_{\overline{r}}+1} = \tau_{j_s+1} P^{(\overline{r})}_{j_s^-} + \tau'_{j_s+1} P^{(\overline{r})}_{j_s+1^-}, \tag{35}$$
and further S conditions (3) of kind
$$Q^{(s)}_{i_{\overline{r}}+2} = \tau_{j_s+1} P^{(\overline{r})}_{j_s^+} + \tau'_{j_s+1} P^{(\overline{r})}_{j_s+1^+}, \tag{36}$$
that coincide with (24) and (25), since (21) and (22) hold;
- if $v_{j_{\overline{s}}}$ is triple, we have $(S-2)R$ redundant equations for simple knots, R conditions (4) of kind
$$P^{(r)}_{j_{\overline{s}}+1} = \sigma_{i_r+1} Q^{(\overline{s})}_{i_r^-} + \sigma'_{i_r+1} Q^{(\overline{s})}_{i_r+1^-}, \tag{37}$$
and further R conditions (4) of kind
$$P^{(r)}_{j_{\overline{s}}+2} = \sigma_{i_r+1} Q^{(\overline{s})}_{i_r^+} + \sigma'_{i_r+1} Q^{(\overline{s})}_{i_r+1^+}, \tag{38}$$
that coincide with (30) and (31), since (33) hold;
- if $u_{i_{\overline{r}}}$ and $v_{j_{\overline{s}}}$ are triple knots, we have $(R-2)(S-2)$ redundant equations for simple knots, $2(S-2)$ conditions (35)–(36) $(s\neq\overline{s}^-,\ s\neq\overline{s}^+)$, $2(R-2)$ conditions (37)–(38) $(r\neq\overline{r}^-,\ r\neq\overline{r}^+)$ and $P^{(\overline{r})}_{j_{\overline{s}}+1^-} = Q^{(\overline{s})}_{i_{\overline{r}}+1^-}$, $P^{(\overline{r})}_{j_{\overline{s}}+2^-} = Q^{(\overline{s})}_{i_{\overline{r}}+1^+}$, $P^{(\overline{r})}_{j_{\overline{s}}+1^+} = Q^{(\overline{s})}_{i_{\overline{r}}+2^-}$, $P^{(\overline{r})}_{j_{\overline{s}}+2^+} = Q^{(\overline{s})}_{i_{\overline{r}}+2^+}$;

- if $u_{i_{\overline{r}}}$ is double and $v_{j_{\overline{s}}}$ triple, we have $(R-1)(S-2)$ redundant equations for simple knots, $(S-2)$ conditions (3) $(s \neq \overline{s}^-,\ s \neq \overline{s}^+)$, $2(R-1)$ conditions (37)–(38) $(r \neq \overline{r})$ and $P^{(\overline{r})}_{j_{\overline{s}}+1^-} = Q^{(\overline{s})}_{i_{\overline{r}}+1},\ P^{(\overline{r})}_{j_{\overline{s}}+1^+} = Q^{(\overline{s})}_{i_{\overline{r}}+2}$;
- if $u_{i_{\overline{r}}}$ is triple and $v_{j_{\overline{s}}}$double, we have $(R-2)(S-1)$ redundant equations for simple knots, $2(S-1)$ conditions (35)–(36) $(s \neq \overline{s})$, $(R-2)$ conditions (4) $(r \neq \overline{r}^-,\ \overline{r}^+)$ and $P^{(\overline{r})}_{j_{\overline{s}}+1} = Q^{(\overline{s})}_{i_{\overline{r}}+1^-},\ P^{(\overline{r})}_{j_{\overline{s}}+2} = Q^{(\overline{s})}_{i_{\overline{r}}+1^+}$.

Now, since we consider two different curves corresponding to a triple knot and $m \geq R, n \geq S$, we set $m = R - L_u^{-1} + k$ and $n = S - L_v^{-1} + \ell$, with $k, \ell \geq 0$. Therefore, from (9), by some algebraic manipulations, the number of interpolation conditions becomes

$$
\begin{aligned}
N_2 = {} & R(N-2) + S(M-2) + 2N + 2M - 4 - RS = mn + 3m + 3n + 8 \\
& + nL_u^0 + mL_v^0 + 2L_u^0 + 2L_v^0 + 2nL_u^{-1} + 2mL_v^{-1} + 5L_u^{-1} + 5L_v^{-1} + L_v^0 L_u^{-1} + L_u^0 L_v^{-1} \\
& + 3L_u^{-1}L_v^{-1} - (k\ell + k(1 + L_v^0 + L_v^{-1}) + \ell(1 + L_u^0 + L_u^{-1})).
\end{aligned}
$$

In conclusion, taking into account the dimension of the spline spaces, the number of unconstrained parameters is $\overline{N} = N_1 - N_2$, that is (34). □

In order to show that the constrained control points C_{ij} depend only on the curve network control points $\left\{P_j^{(r)}\right\}_{j=0}^{N-1}$ and $\left\{Q_i^{(s)}\right\}_{i=0}^{M-1}$, we need to consider different possibilities that arise: a minimal configuration (MC) and a non minimal one (NMC). MC means that all isoparametric curves, corresponding to the knots in U and V, are curves of the network, while NMC means that not all isoparametric curves, corresponding to the knots in U and V, are curves of the network.

- **In case of MC**, we note that double and triple knots in U and V give rise to subdomains Ω_r, $r = 1, \dots, \gamma$, of Ω where all knots are simple, so $R = m + L_u^{-1}$ and $S = n + L_v^{-1}$.

  Regarding the interpolation parameters and all knots in U and V, we can list the following situations:

  – if $u_{i_{\overline{r}}}$ is double, $u_{i_r} = u_r,\ r = 0, ..., \overline{r},\ u_{i_r} = u_{r+1},\ r = \overline{r}+1, ..., M-3$,
  – if $u_{i_{\overline{r}}}$ is triple, $u_{i_r} = u_r,\ r = 0, ..., \overline{r},\ u_{i_r} = u_{r+1},\ r = \overline{r}+2, ..., M-3$,
  – if $v_{j_{\overline{s}}}$ is double, $v_{j_s} = v_s,\ s = 0, ..., \overline{s},\ v_{j_s} = v_{s+1},\ s = \overline{s}+1, ..., N-3$,
  – if $v_{j_{\overline{s}}}$ is triple, $v_{j_s} = v_s,\ s = 0, ..., \overline{s},\ v_{j_s} = v_{s+1},\ s = \overline{s}+2, ..., N-3$.

  In order to write $\mathcal{S}$ as a linear combination of $B_{i,j}$'s with coefficients $C_{i,j}$, depending only on $\left\{P_j^{(r)}\right\}_{j=0}^{N-1}$ and $\left\{Q_i^{(s)}\right\}_{i=0}^{M-1}$, we note that from Theorems 3.1, 3.2 and similar cases, we have:

$$C_{i_{\overline{r}}+1,j} = P_j^{(\overline{r})}, \quad \text{if } u_{i_{\overline{r}}} \text{ is double,} \tag{39}$$

$$C_{i,j_{\overline{s}}+1} = Q_i^{(\overline{s})}, \quad \text{if } v_{j_{\overline{s}}} \text{ is double,} \tag{40}$$

$$C_{i_{\overline{r}}+1,j^-} = P_{j^-}^{(\overline{r})}, \ C_{i_{\overline{r}}+2,j^+} = P_{j^+}^{(\overline{r})} \quad \text{if } u_{i_{\overline{r}}} \text{ is triple,} \tag{41}$$

$$C_{i^-,j_{\overline{s}}+1} = Q_{i^-}^{(\overline{s})}, \ C_{i^+,j_{\overline{s}}+2} = Q_{i^+}^{(\overline{s})} \quad \text{if } v_{j_{\overline{s}}} \text{ is triple.} \tag{42}$$

Therefore, since from (39), (40), (41) and (42) we fixed some control points, only the remaining points $C_{i,j}$ for

$$\begin{array}{l} i-1 \neq \overline{r},\ j-1 \neq \overline{s} \text{ in case of double knots } u_{i_{\overline{r}}} \text{ and } v_{j_{\overline{s}}}, \text{ respectively,} \\ i-1 \neq \overline{r},\ i-2 \neq \overline{r},\ j-1 \neq \overline{s}\ j-2 \neq \overline{s} \text{ in case of triple knots } u_{i_{\overline{r}}} \text{ and } v_{j_{\overline{s}}}, \text{ respectively,} \end{array} \tag{43}$$

have to be written in terms of control points of $\{C_r(v)\}$ and $\{D_s(u)\}$.
Therefore in case of $S_{2,2}^{(\overline{\mu}^u,\overline{\mu}^v)}(R_{m,n})$ we have γ unconstrained parameters, one for each subdomain, while for $S_2^{(\overline{\mu}^u,\overline{\mu}^v)}(T_{m,n})$ we have no unconstrained parameters at all.

Remark 3.1 In Fig. 3 we show the above situations. Thick lines denote double knots, $\times$ the points $C_{i,j}$ and $\bullet$ the points $Q_i^{(s)}$, $P_j^{(r)}$. Since $L_u^0 = L_v^0 = 2$, the number of subdomains Ω_r is $\gamma = 9$. Points denoted by $\bigcirc$, one for each Ω_r and conventionally located at its left-down corner, represent the ones from which all others depend. Points corresponding to double knots, i.e. $C_{2,j}$, $C_{4,j}$, $C_{i,4}$, $C_{i,6}$, $j = 1, ..., 7$, $i = 1, ..., 6$, coincide with $P_j^{(1)}$, $P_j^{(2)}$, $Q_i^{(3)}$, $Q_i^{(4)}$ of curves $C_r(v)$ and $D_s(u)$, respectively. Similarly for triple knots, but indices are different, due to the jump. □

Lemma 3.1 If (11) hold, then all constrained points can be written as

$$\begin{aligned} C_{i,j} = \Gamma_{i,j} + & (-1)^{i+j+\rho_1(\overline{r}_\lambda+1)+\rho_2+\phi_1(\overline{s}_\psi+1)+\phi_2} \\ & \cdot \frac{h_i}{h_{1+\rho_1\overline{r}_\lambda+1)+\rho_2}} \frac{k_j}{k_{1+\phi_1(\overline{s}_\psi+1)+\phi_2}} C_{1+\rho_1(\overline{r}_\lambda+1)+\rho_2,1+\phi_1(\overline{s}_\psi+1)+\phi_2}, \end{aligned} \tag{44}$$

$i = 1, ..., M-2$, $j = 1, ..., N-2$ (considering (43)) with

$$\begin{aligned} \Gamma_{i,j} = & \sum_{r=1}^{i-1-\rho_1(\overline{r}_\lambda+1)-\rho_2} (-1)^{r+1} \frac{(h_{i-r} + h_{i-r+1})h_i}{h_{i-r}h_{i-r+1}} P_j^{(i-r-\alpha)} + (-1)^{i+\rho_1(\overline{r}_\lambda+1)+\rho_2} \\ & \cdot \frac{h_i}{h_{1+\rho_1(\overline{r}_\lambda+1)+\rho_2}} \sum_{s=1}^{j-1-\phi_1(\overline{s}_\psi+1)-\phi_2} (-1)^s \frac{(k_{j-s} + k_{j-s+1})k_j}{k_{j-s}k_{j-s+1}} Q_{1+\rho_1(\overline{r}_\lambda+1)+\rho_2}^{(j-s-\beta)} \end{aligned} \tag{45}$$

and

$$\rho_1 = \begin{cases} 0 & \overline{r}_\lambda = 0 \\ 1 & \overline{r}_\lambda \neq 0 \end{cases}, \quad \rho_2 = \begin{cases} 0 & \rho_1 = 0 \text{ or } u_{\overline{r}_\lambda} \text{ double} \\ 1 & \rho_1 = 1 \text{ and } u_{\overline{r}_\lambda} \text{ triple} \end{cases}, \tag{46}$$

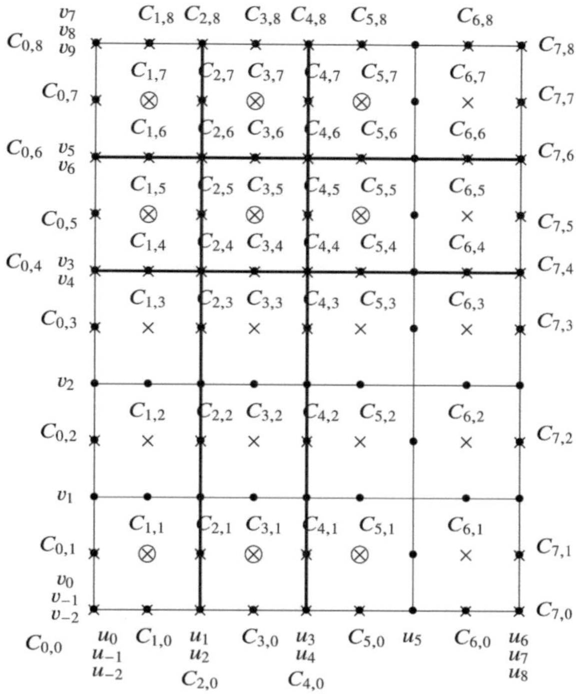

Fig. 3 Example of domain with double knots

$$\phi_1 = \begin{cases} 0 & \bar{s}_\psi = 0 \\ 1 & \bar{s}_\psi \neq 0 \end{cases}, \quad \phi_2 = \begin{cases} 0 & \phi_1 = 0 \text{ or } v_{\bar{s}_\psi} \text{ double} \\ 1 & \phi_1 = 1 \text{ and } v_{\bar{s}_\psi} \text{ triple} \end{cases}, \tag{47}$$

where

- $\bar{r}_\lambda, \bar{s}_\psi$ are the indices of the last multiple knot in U and V before $C_{i,j}$, i.e. $\rho_1(\bar{r}_\lambda + 1) + \rho_2 < i \leq \bar{r}_{\lambda+1}$ and $\phi_1(\bar{s}_\psi + 1) + \phi_2 < j \leq \bar{s}_{\psi+1}$, respectively;
- α, β are the number of multiple knots in U and V before $C_{i,j}$;
- $\sum_{\ell=1}^{0} \cdot = 0$. □

Proof Given a domain Ω, split into subdomains, as in Fig. 4, let us consider the first subdomain where we have MC with all simple knots, so that from Theorem 1 of [16] it results

$$\tau_{s+1} C_{i,s} + \tau'_{s+1} C_{i,s+1} = Q_i^{(s)}, \quad s = 1, ..., \bar{s}_1, \; i = 1, ..., \bar{r}_1. \tag{48}$$

For $i = 1$, applying (48) repeatedly, we can write (see Fig. 5)

$$C_{1,j} = \sum_{s=1}^{j-1} (-1)^{s+1} \frac{Q_1^{(j-s)}}{\tau'_{j-s+1}} \prod_{k=1}^{s-1} \frac{\tau_{j-k+1}}{\tau'_{j-k+1}} + (-1)^{j+1} C_{1,1} \prod_{s=1}^{j-1} \frac{\tau_{j-s+1}}{\tau'_{j-s+1}}, \tag{49}$$

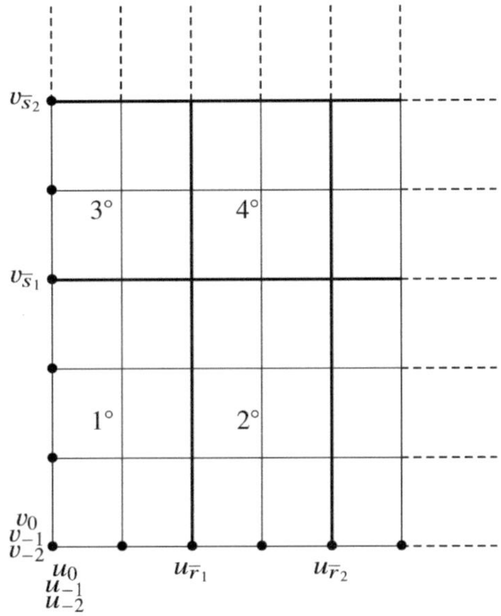

Fig. 4 Example of Ω split into subdomains for Lemma 3.1

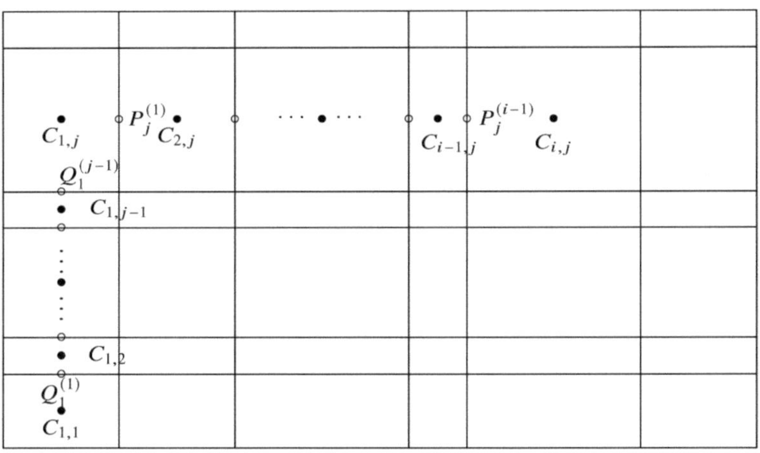

Fig. 5 Path reconstruction to get C_{ij}, starting from C_{11}

with $j = 1, ..., \overline{s}_1$ and $\prod_{l=1}^{0} \cdot = 1$. Moreover from Theorem 1 of [16] it also follows

$$\sigma_{r+1} C_{r,j} + \sigma'_{r+1} C_{r+1,j} = P_j^{(r)}, \quad j = 1, ..., \overline{s}_1, \; r = 1, ..., \overline{r}_1. \tag{50}$$

Then for a fixed j such that $j \le \overline{s}_1$, applying (50) repeatedly, we have

$$C_{i,j} = \sum_{r=1}^{i-1}(-1)^{r+1}\frac{P_j^{(i-r)}}{\sigma'_{i-r+1}}\prod_{k=1}^{r-1}\frac{\sigma_{i-k+1}}{\sigma'_{i-k+1}} + (-1)^{i+1}C_{1,j}\prod_{r=1}^{i-1}\frac{\sigma_{i-r+1}}{\sigma'_{i-r+1}}, \quad i = 1, ..., \overline{r}_1. \tag{51}$$

Therefore, substituting (49) in (51) we obtain

$$C_{i,j} = \Gamma_{i,j} + (-1)^{i+j}C_{1,1}\prod_{s=1}^{j-1}\frac{\tau_{j-s+1}}{\tau'_{j-s+1}}\prod_{r=1}^{i-1}\frac{\sigma_{i-r+1}}{\sigma'_{i-r+1}},$$

with $i = 1, ..., \overline{r}_1$, $j = 1, ..., \overline{s}_1$ and

$$\Gamma_{i,j} = \\ \sum_{r=1}^{i-1}(-1)^{r+1}\frac{P_j^{(i-r)}}{\sigma'_{i-r+1}}\prod_{k=1}^{r-1}\frac{\sigma_{i-k+1}}{\sigma'_{i-k+1}} + (-1)^{i+1}\prod_{r=1}^{i-1}\frac{\sigma_{i-r+1}}{\sigma'_{i-r+1}}\sum_{s=1}^{j-1}(-1)^{s+1}\frac{Q_1^{(j-s)}}{\tau'_{j-s+1}}\prod_{k=1}^{s-1}\frac{\tau_{j-k+1}}{\tau'_{j-k+1}}.$$

Now for the second subdomain, from (14) and (23) it results

$$\tau_{s+1}C_{i,s} + \tau'_{s+1}C_{i,s+1} = Q_i^{(s)}, \quad s = 1, ..., \overline{s}_1, \ i = \rho_1(\overline{r}_1 + 1) + \rho_2 + 1, ..., \overline{r}_2, \tag{52}$$

with ρ_1 and ρ_2 given in (46). For $i = \rho_1(\overline{r}_1 + 1) + \rho_2 + 1$, applying (52) repeatedly,

$$\begin{aligned} C_{\rho_1(\overline{r}_1+1)+\rho_2+1,j} = &\sum_{s=1}^{j-1}(-1)^{s+1}\frac{Q^{(j-s)}_{\rho_1(\overline{r}_1+1)+\rho_2+1}}{\tau'_{j-s+1}}\prod_{k=1}^{s-1}\frac{\tau_{j-k+1}}{\tau'_{j-k+1}} \\ &+(-1)^{j+1}C_{\rho_1(\overline{r}_1+1)+\rho_2+1,1}\prod_{s=1}^{j-1}\frac{\tau_{j-s+1}}{\tau'_{j-s+1}}, \end{aligned} \tag{53}$$

$j = 1, ..., \overline{s}_1$. From (12) and (20) it follows

$$\sigma_{r+1}C_{r,j} + \sigma'_{r+1}C_{r,j} = P_j^{(r-1)}, \quad j = 1, ..., \overline{s}_1, \ r = \rho_1(\overline{r}_1 + 1) + \rho_2 + 1, ..., \overline{r}_2. \tag{54}$$

Then for a fixed j such that $j \le \overline{s}_1$, applying (54) repeatedly,

$$\begin{aligned} C_{\rho_1(\overline{r}_1+1)+\rho_2+t,j} = &\sum_{r=1}^{t-1}(-1)^{r+1}\frac{P_j^{(\rho_1(\overline{r}_1+1)+\rho_2+t-r-1)}}{\sigma'_{\rho_1(\overline{r}_1+1)+\rho_2+t-r+1}}\prod_{k=1}^{r-1}\frac{\sigma_{\rho_1(\overline{r}_1+1)+\rho_2+t-k+1}}{\sigma'_{\rho_1(\overline{r}_1+1)+\rho_2+t-k+1}} \\ +(-1)^{\rho_1(\overline{r}_1+1)+\rho_2+t+\rho_1(\overline{r}_1+1)+\rho_2+1}&C_{\rho_1(\overline{r}_1+1)+\rho_2+1,j}\prod_{r=1}^{t-1}\frac{\sigma_{\rho_1(\overline{r}_1+1)+\rho_2+t-r+1}}{\sigma'_{\rho_1(\overline{r}_1+1)+\rho_2+t-r+1}}, \end{aligned} \tag{55}$$

$t = 2, ..., w$ and w such that $\rho_1(\overline{r}_1 + 1) + \rho_2 + w = \overline{r}_2$. Then, substituting (53) in (55) and setting $i = \rho_1(\overline{r}_1 + 1) + \rho_2 + t$, we obtain

$$C_{i,j} = \Gamma_{i,j} + (-1)^{i+j+\rho_1(\overline{r}_1+1)+\rho_2} C_{\rho_1(\overline{r}_1+1)+\rho_2+1,1} \prod_{s=1}^{j-1} \frac{\tau_{j-s+1}}{\tau'_{j-s+1}} \prod_{r=1}^{i-1-\rho_1(\overline{r}_1+1)-\rho_2} \frac{\sigma_{i-r+1}}{\sigma'_{i-r+1}},$$

with $i = \rho_1(\overline{r}_1 + 1) + \rho_2 + 1, ..., \overline{r}_2$, $j = 1, ..., \overline{s}_1$ and

$$\begin{aligned}\Gamma_{i,j} = & \sum_{r=1}^{i-1-\rho_1(\overline{r}_1+1)-\rho_2} (-1)^{r+1} \frac{P_j^{(i-r-1)}}{\sigma'_{i-r+1}} \prod_{k=1}^{r-1} \frac{\sigma_{i-k+1}}{\sigma'_{i-k+1}} \\ & +(-1)^{i+\rho_1(\overline{r}_1+1)+\rho_2+1} \prod_{r=1}^{i-1-\rho_1(\overline{r}_1+1)-\rho_2} \frac{\sigma_{i-r+1}}{\sigma'_{i-r+1}} \sum_{s=1}^{j-1} (-1)^{s+1} \frac{Q_{\rho_1(\overline{r}_1+1)+\rho_2+1}^{(j-s)}}{\tau'_{j-s+1}} \prod_{k=1}^{s-1} \frac{\tau_{j-k+1}}{\tau'_{j-k+1}}.\end{aligned}$$

Generalizing for $i = \rho_1(\overline{r}_\lambda + 1) + \rho_2 + 1, ..., \overline{r}_{\lambda+1}$, $j = 1, ..., \overline{s}_1$ we have

$$C_{i,j} = \Gamma_{i,j} + (-1)^{i+j+\rho_1(\overline{r}_\lambda+1)+\rho_2} C_{\rho_1(\overline{r}_\lambda+1)+\rho_2+1,1} \prod_{s=1}^{j-1} \frac{\tau_{j-s+1}}{\tau'_{j-s+1}} \prod_{r=1}^{i-1-\rho_1(\overline{r}_\lambda+1)-\rho_2} \frac{\sigma_{i-r+1}}{\sigma'_{i-r+1}},$$

with

$$\begin{aligned}\Gamma_{i,j} = & \sum_{r=1}^{i-1-\rho_1(\overline{r}_\lambda+1)-\rho_2} (-1)^{r+1} \frac{P_j^{(i-r-\alpha)}}{\sigma'_{i-r+1}} \prod_{k=1}^{r-1} \frac{\sigma_{i-k+1}}{\sigma'_{i-k+1}} \\ & +(-1)^{i+\rho_1(\overline{r}_\lambda+1)+\rho_2+1} \prod_{r=1}^{i-1-\rho_1(\overline{r}_\lambda+1)-\rho_2} \frac{\sigma_{i-r+1}}{\sigma'_{i-r+1}} \sum_{s=1}^{j-1} (-1)^{s+1} \frac{Q_{\rho_1(\overline{r}_\lambda+1)+\rho_2+1}^{(j-s)}}{\tau'_{j-s+1}} \prod_{k=1}^{s-1} \frac{\tau_{j-k+1}}{\tau'_{j-k+1}}.\end{aligned}$$

For the third subdomain, from (18) and (32) it results

$$\tau_{s+1} C_{i,s} + \tau'_{s+1} C_{i,s+1} = Q_i^{(s-1)}, \quad s = \phi_1(\overline{s}_1 + 1) + \phi_2 + 1, ..., \overline{s}_2, \ i = 1, ..., \overline{r}_1, \tag{56}$$

with ϕ_1 and ϕ_2 defined in (47). For $i = 1$, applying (56) repeatedly, we get

$$\begin{aligned}C_{1,\phi_1(\overline{s}_1+1)+\phi_2+q} = & \sum_{s=1}^{q-1} (-1)^{s+1} \frac{Q_1^{(\phi_1(\overline{s}_1+1)+\phi_2+q-s-1)}}{\tau'_{\phi_1(\overline{s}_1+1)+\phi_2+q-s+1}} \prod_{k=1}^{s-1} \frac{\tau_{\phi_1(\overline{s}_1+1)+\phi_2+q-k+1}}{\tau'_{\phi_1(\overline{s}_1+1)+\phi_2+q-k+1}} \\ & +(-1)^{\phi_1(\overline{s}_1+1)+\phi_2+q+\phi_1(\overline{s}_1+1)+\phi_2+1} C_{1,\phi_1(\overline{s}_1+1)+\phi_2+1} \prod_{s=1}^{q-1} \frac{\tau_{\phi_1(\overline{s}_1+1)+\phi_2+q-s+1}}{\tau'_{\phi_1(\overline{s}_1+1)+\phi_2+q-s+1}},\end{aligned} \tag{57}$$

with $q = 2, ..., z$, z such that $\phi_1(\overline{s}_1 + 1) + \phi_2 + z = \overline{s}_2$. From (17) and (29) it follows

$$\sigma_{r+1} C_{r,j} + \sigma'_{r+1} C_{r+1,j} = P_j^{(r)}, \quad j = \phi_1(\overline{s}_1 + 1) + \phi_2 + 1, ..., \overline{s} + 2, \ r = 1, ..., \overline{r}_1. \tag{58}$$

Then, considering (51) for a fixed j such that $\phi_1(\overline{s}_1 + 1) + \phi_2 < j \leq \overline{s}_2$, and substituting (57) into it, we obtain

$$C_{i,j} = \Gamma_{i,j} + (-1)^{i+j+\phi_1(\overline{s}_1+1)+\phi_2} C_{1,\phi_1(\overline{s}_1+1)+\phi_2+1} \prod_{s=1}^{j-\phi_1(\overline{s}_1+1)-\phi_2-1} \frac{\tau_{j-s+1}}{\tau'_{j-s+1}} \prod_{r=1}^{i-1} \frac{\sigma_{i-r+1}}{\sigma'_{i-r+1}},$$

with $i = 1, ..., \overline{r}_1$, $j = \phi_1(\overline{s}_1 + 1) + \phi_2 + 1, ..., \overline{s}_2$ and

$$\Gamma_{i,j} = \sum_{r=1}^{i-1} (-1)^{r+1} \frac{P_j^{(i-r)}}{\sigma'_{i-r+1}} \prod_{k=1}^{r-1} \frac{\sigma_{i-k+1}}{\sigma'_{i-k+1}} + (-1)^{i+1} \prod_{r=1}^{i-1} \frac{\sigma_{i-r+1}}{\sigma'_{i-r+1}} \sum_{s=1}^{j-\phi_1(\overline{s}_1+1)-\phi_2-1} (-1)^{s+1} \frac{Q_1^{(j-s-1)}}{\tau'_{j-s+1}} \prod_{k=1}^{s-1} \frac{\tau_{j-k+1}}{\tau'_{j-k+1}}.$$

Generalizing for $i = 1, ..., \overline{r}_1$, $j = \phi_1(\overline{s}_\psi + 1) + \phi_2 + 1, ..., \overline{s}_{\psi+1}$, we get

$$C_{i,j} = \Gamma_{i,j} + (-1)^{i+j+\phi_1(\overline{s}_\psi+1)+\phi_2} C_{1,\phi_1(\overline{s}_\psi+1)+\phi_2+1} \prod_{s=1}^{j-\phi_1(\overline{s}_\psi+1)-\phi_2-1} \frac{\tau_{j-s+1}}{\tau'_{j-s+1}} \prod_{r=1}^{i-1} \frac{\sigma_{i-r+1}}{\sigma'_{i-r+1}},$$

with

$$\Gamma_{i,j} = \sum_{r=1}^{i-1} (-1)^{r+1} \frac{P_j^{(i-r)}}{\sigma'_{i-r+1}} \prod_{k=1}^{r-1} \frac{\sigma_{i-k+1}}{\sigma'_{i-k+1}} + (-1)^{i+1)+\phi_2+1} \prod_{r=1}^{i-1} \frac{\sigma_{i-r+1}}{\sigma'_{i-r+1}} \sum_{s=1}^{j-\phi_1(\overline{s}_\psi+1)-\phi_2-1} (-1)^{s+1} \frac{Q_1^{(j-s-\beta)}}{\tau'_{j-s+1}} \prod_{k=1}^{s-1} \frac{\tau_{j-k+1}}{\tau'_{j-k+1}}.$$

Finally for the fourth subdomain from (18) and (32) it results

$$\tau_{s+1} C_{i,s} + \tau'_{s+1} C_{i,s+1} = Q_i^{(s-1)}, \quad {\scriptstyle s=\phi_1(\overline{s}_1+1)+\phi_2+1,...,\overline{s}_2,\ i=\rho_1(\overline{r}_1+1)+\rho_2+1,...,\overline{r}_2.} \tag{59}$$

For $i = \rho_1(\overline{r}_1 + 1) + \rho_2 + 1$, applying (59) repeatedly, we have

$$\begin{aligned} C_{\rho_1(\overline{r}_1+1)+\rho_2+1,\phi_1(\overline{s}_1+1)+\phi_2+q} &= \sum_{s=1}^{q-1} (-1)^{s+1} \frac{Q_{\rho_1(\overline{r}_1+1)+\rho_2+1}^{(\phi_1(\overline{s}_1+1)+\phi_2+q-s-1)}}{\tau'_{\phi_1(\overline{s}_1+1)+\phi_2+q-s+1}} \\ &\cdot \prod_{k=1}^{\phi_1(\overline{s}_1+1)+\phi_2+q-1} \frac{\tau_{\phi_1(\overline{s}_1+1)+\phi_2+q-k+1}}{\tau'_{\phi_1(\overline{s}_1+1)+\phi_2+q-k+1}} + (-1)^{\phi_1(\overline{s}_1+1)+\phi_2+q+\phi_1(\overline{s}_1+1)+\phi_2+1} \\ &\cdot C_{\rho_1(\overline{r}_1+1)+\rho_2+1,\phi_1(\overline{s}_1+1)+\phi_2+1} \prod_{s=1}^{q-1} \frac{\tau_{\phi_1(\overline{s}_1+1)+\phi_2+q-s+1}}{\tau'_{\phi_1(\overline{s}_1+1)+\phi_2+q-s+1}}, \end{aligned} \tag{60}$$

with $q = 1, ..., z$, z such that $\phi_1(\overline{s}_1 + 1) + \phi_2 + z = \overline{s}_2$. From (12) and (20) it follows

$$\sigma_{r+1} C_{r,j} + \sigma'_{r+1} C_{r+1,j} = P_j^{(r-1)}, \quad {\scriptstyle j=\phi_1(\overline{s}_1+1)+\phi_2+1,...,\overline{s}+2,\ r=\rho_1(\overline{r}_1+1)+\rho_2+1,...,\overline{r}_2.} \tag{61}$$

Then for a fixed j such that $\phi_1(\overline{s}_1+1)+\phi_2 < j \le \overline{s}_2$, applying (61) repeatedly, we have

$$\begin{aligned} C_{\rho_1(\overline{r}_1+1)+\rho_2+t,j} &= \sum_{r=1}^{t-1}(-1)^{r+1}\frac{P_j^{(\rho_1(\overline{r}_1+1)+\rho_2+t-r-1)}}{\sigma'_{\rho_1(\overline{r}_1+1)+\rho_2+t-r+1}}\prod_{k=1}^{r-1}\frac{\sigma_{\rho_1(\overline{r}_1+1)+\rho_2+t-k+1}}{\sigma'_{\rho_1(\overline{r}_1+1)+\rho_2+t-k+1}} \\ &+(-1)^{\rho_1(\overline{r}_1+1)+\rho_2+t+\rho_1(\overline{r}_1+1)+\rho_2+1}C_{\rho_1(\overline{r}_1+1)+\rho_2+1,j}\prod_{r=1}^{i-1}\frac{\sigma_{\rho_1(\overline{r}_1+1)+\rho_2+t-r+1}}{\sigma'_{\rho_1(\overline{r}_1+1)+\rho_2+t-r+1}}, \end{aligned} \tag{62}$$

with $t=1,\ldots,w$ and w such that $\rho_1(\overline{r}_1+1)+\rho_2+w=\overline{r}_2$. Then, substituting (60) in (62) and setting $j=\phi_1(\overline{s}_1+1)+\phi_2+q$, $i=\rho_1(\overline{r}_1+1)+\rho_2+t$, we get

$$\begin{aligned} C_{i,j} &= \Gamma_{i,j}+(-1)^{i+j+\phi_1(\overline{s}_1+1)+\phi_2+\rho_1(\overline{r}_1+1)+\rho_2}C_{\rho_1(\overline{r}_1+1)+\rho_2+1,\phi_1(\overline{s}_1+1)+\phi_2+1} \\ &\cdot \prod_{s=1}^{j-\phi_1(\overline{s}_1+1)+\phi_2-1}\frac{\tau_{j-s+1}}{\tau'_{j-s+1}}\prod_{r=1}^{i-\rho_1(\overline{r}_1+1)+\rho_2-1}\frac{\sigma_{i-r+1}}{\sigma'_{i-r+1}}, \end{aligned}$$

with $i=\rho_1(\overline{r}_1+1)+\rho_2+1,\ldots,\overline{r}_2$, $j=\phi_1(\overline{s}_1+1)+\phi_2+1,\ldots,\overline{s}_2$ and

$$\begin{aligned} \Gamma_{i,j} = &\sum_{r=1}^{i-\rho_1(\overline{r}_1+1)+\rho_2-1}(-1)^{r+1}\frac{P_j^{(i-r-1)}}{\sigma'_{i-r+1}}\prod_{k=1}^{r-1}\frac{\sigma_{i-k+1}}{\sigma'_{i-k+1}}+(-1)^{i+\rho_1(\overline{r}_1+1)+\rho_2+1} \\ &\cdot\prod_{r=1}^{i-\rho_1(\overline{r}_1+1)+\rho_2-1}\frac{\sigma_{i-r+1}}{\sigma'_{i-r+1}}\sum_{s=1}^{j-\phi_1(\overline{s}_1+1)+\phi_2-1}(-1)^{s+1}\frac{Q^{(j-s-1)}_{\rho_1(\overline{r}_1+1)+\rho_2+1}}{\tau'_{j-s+1}}\prod_{k=1}^{s-1}\frac{\tau_{j-k+1}}{\tau'_{j-k+1}}. \end{aligned}$$

Generalizing for $i=\rho_1(\overline{r}_\lambda+1)+\rho_2+1,\ldots,\overline{r}_{\lambda+1}$, $j=\phi_1(\overline{s}_\psi+1)+\phi_2+1,\ldots,\overline{s}_{\psi+1}$, from

$$\frac{\sigma_\ell}{\sigma'_\ell}=\frac{h_\ell}{h_{\ell-1}},\qquad \frac{\tau_\ell}{\tau'_\ell}=\frac{k_\ell}{k_{\ell-1}},$$

then

$$\prod_{r=1}^{i-1-\rho_1(\overline{r}_\lambda+1)-\rho_2}\frac{\sigma_{i-r+1}}{\sigma'_{i-r+1}}=\frac{h_i}{h_{1+\rho_1(\overline{r}_\lambda+1)+\rho_2}},\qquad \prod_{s=1}^{j-1\phi_1(\overline{s}_\psi+1)-\phi_2}\frac{\tau_{j-s+1}}{\tau'_{j-s+1}}=\frac{k_j}{k_{1+\phi_1(\overline{s}_\psi+1)+\phi_2}},$$

hold and the thesis (44)–(45) follows.

□

Remark 3.2 In case of uniform partitions ($h=h_i$, $\forall i$, $k=k_j$, $\forall j$) the $C_{i,j}$'s of Lemma 3.1 become

$$C_{i,j}=\Gamma_{i,j}+(-1)^{i+j+\rho_1(\overline{r}_\lambda+1)+\rho_2+\phi_1(\overline{s}_\psi+1)+\phi_2}C_{1+\rho_1(\overline{r}_\lambda+1)+\rho_2,1+\phi_1(\overline{s}_\psi+1)+\phi_2},$$

for $i=1,\ldots,M-2$ and $j=1,\ldots,N-2$ (considering (43)), with

$$\Gamma_{i,j} = 2\Bigg[\sum_{r=1}^{i-1-\rho_1(\overline{r}_\lambda+1)-\rho_2}(-1)^{r+1}P_j^{(i-r-\alpha)}$$
$$+(-1)^{i+\rho_1(\overline{r}_\lambda+1)+\rho_2}\sum_{s=1}^{j-1-\phi_1(\overline{s}_\psi+1)-\phi_2}(-1)^s Q_{1+\rho_1(\overline{r}_\lambda+1)+\rho_2}^{(j-s-\beta)}\Bigg].$$

At last the interpolating surface can be written as follows.

Theorem 3.4 Given a B-spline curve network as in (1), satisfying the compatibility conditions (2), the surface (8) with isoparametric curves (1) is

$$\mathcal{S}(u,v) = \mathcal{S}_b(u,v) + \mathcal{S}_{\lambda\psi}(u,v) + \mathcal{S}_\Gamma(u,v) + \mathcal{S}_{fp}(u,v),$$

with

$$\begin{aligned}
\mathcal{S}_b(u,v) &= \sum_{j=0}^{N-1}\left(P_j^{(0)}B_{0,j}(u,v) + P_j^{(M-2-L_u^0-L_u^{-1})}B_{M-1,j}(u,v)\right)\\
&+\sum_{i=1}^{M-2}\left(Q_i^{(0)}B_{i,0}(u,v) + Q_i^{(N-2-L_v^0-L_v^{-1})}B_{i,N-1}(u,v)\right),\\
\mathcal{S}_{\lambda\psi}(u,v) &= \sum_{\overline{r}_\lambda\in I_\lambda^1}\sum_{j=1}^{N-2}P_j^{(\overline{r}_\lambda-\alpha)}B_{\overline{r}_\lambda+1,j}(u,v) + \sum_{i=1\ (i-1\notin I_\lambda^1\cup I_\lambda^2,\ i-2\notin I_\lambda^2)}^{M-2}\ \sum_{\overline{s}_\psi\in I_\phi^1}Q_i^{(\overline{s}_\psi-\beta)}B_{i,\overline{s}_\psi+1}(u,v)\\
&+\sum_{\overline{r}_\lambda\in I_\lambda^2}\sum_{j=1}^{N-2}P_{j^-}^{(\overline{r}_\lambda-\alpha)}B_{\overline{r}_\lambda+1,j}(u,v) + \sum_{i=1\ (i-1\notin I_\lambda^1\cup I_\lambda^2,\ i-2\notin I_\lambda^2)}^{M-2}\ \sum_{\overline{s}_\psi\in I_\phi^2}Q_{i^-}^{(\overline{s}_\psi-\beta)}B_{i,\overline{s}_\psi+1}(u,v)\\
&+\sum_{\overline{r}_\lambda\in I_\lambda^2}\sum_{j=1}^{N-2}P_{j^+}^{(\overline{r}_\lambda-\alpha)}B_{\overline{r}_\lambda+2,j}(u,v) + \sum_{i=1\ (i-1\notin I_\lambda^1\cup I_\lambda^2,\ i-2\notin I_\lambda^2)}^{M-2}\ \sum_{\overline{s}_\psi\in I_\phi^2}Q_{i^+}^{(\overline{s}_\psi-\beta)}B_{i,\overline{s}_\psi+2}(u,v),
\end{aligned}$$

$$\begin{aligned}
\mathcal{S}_\Gamma(u,v) &= \sum_{i=1}^{M-2}\sum_{j=1}^{N-2}\Gamma_{i,j}B_{i,j}(u,v), \quad \text{(considering (43))}\\
\mathcal{S}_{fp}(u,v) &= \sum_{\text{in each } [u_{\overline{r}_\lambda},u_{\overline{r}_\lambda+1}]\times[v_{\overline{s}_\psi},v_{\overline{s}_\psi+1}]} C_{1+\rho_1(\overline{r}_\lambda+1)+\rho_2,1+\phi_1(\overline{s}_\psi+1)+\phi_2}\\
&\cdot\Bigg(B_{1+\rho_1(\overline{r}_\lambda+1)+\rho_2,1+\phi_1(\overline{s}_\psi+1)+\phi_2}(u,v)\\
&+\sum_{i,j \text{ in the considered subdomain}}(-1)^{i+j+\rho_1(\overline{r}_\lambda+1)+\rho_2+\phi_1(\overline{s}_\psi+1)+\phi_2}\frac{h_i}{h_{1+\rho_1(\overline{r}_\lambda+1)+\rho_2}}\\
&\cdot\frac{k_j}{k_{1+\phi_1(\overline{s}_\psi+1)+\phi_2}}B_{i,j}(u,v)\Bigg),
\end{aligned}$$

where $\rho_1,\ \rho_2,\ \phi_1,\ \phi_2,\ \alpha,\ \beta$ are defined in Lemma 3.1,

$$\begin{aligned}
I_\lambda^1 &= \{\overline{r}_\lambda \text{ such that } u_{\overline{r}_\lambda} \text{ double}\}, \qquad & I_\lambda^2 &= \{\overline{r}_\lambda \text{ such that } u_{\overline{r}_\lambda} \text{ triple}\},\\
I_\psi^1 &= \{\overline{s}_\psi \text{ such that } v_{\overline{s}_\psi} \text{ double}\}, \qquad & I_\psi^2 &= \{\overline{s}_\psi \text{ such that } v_{\overline{s}_\psi} \text{ triple}\},
\end{aligned}$$

and the points $C_{i,j}$ are given in (44) and depend on the curve network control points $P_j^{(r)}$ and $Q_i^{(s)}$.

□

Proof We write (8) dividing inner and boundary parts:

$$\mathcal{S}(u, v) = \sum_{i=1}^{m+1} \sum_{j=1}^{n+1} C_{i,j} B_{i,j}(u, v)$$
$$+ \sum_{j=0}^{N-1} (C_{0,j} B_{0,j}(u, v) + C_{M-1,j} B_{M-1,j}(u, v)) + \sum_{i=1}^{M-2} (C_{i,0} B_{i,0}(u, v) + C_{i,N-1} B_{i,N-1}(u, v)).$$

Since $R + 1 = M - 2 - L_u^0 - L_u^{-1}$ and $S + 1 = N - 2 - L_v^0 - L_v^{-1}$, substituting the interpolation conditions given in Theorems 3.1, 3.2 and similar cases, we get the boundary part $\mathcal{S}_b(u, v)$.

For the inner part it is necessary to further distinguish. Indeed (39)-(42) let us obtain $\mathcal{S}_{\lambda\psi}(u, v)$, containing just the curve control points corresponding to multiple knots. For all other points with indices in (43), using Lemma 3.1 and since

$$(-1)^{i+j+\rho_1(\overline{r}_\lambda+1)+\rho_2+\phi_1(\overline{s}_\psi+1)+\phi_2} \frac{h_i}{h_{1+\rho_1(\overline{r}_\lambda+1)+\rho_2}} \frac{k_j}{k_{1+\phi_1(\overline{s}_\psi+1)+\phi_2}}$$
$$\cdot C_{1+\rho_1(\overline{r}_\lambda+1)+\rho_2, 1+\phi_1(\overline{s}_\psi+1)+\phi_2}$$

is equal to $C_{1+\rho_1(\overline{r}_\lambda+1)+\rho_2, 1+\phi_1(\overline{s}_\psi+1)+\phi_2}$ for $i = 1 + \rho_1(\overline{r}_\lambda + 1) + \rho_2$ and $j = 1 + \phi_1(\overline{s}_\psi + 1) + \phi_2$, we finally obtain $\mathcal{S}_\Gamma(u, v)$ and $\mathcal{S}_{fp}(u, v)$.

□

Remark 3.3 Since in the MC $\ell = 0$ and $k = 0$, then for $S_{2,2}^{(\overline{\mu}^u, \overline{\mu}^v)}(R_{m,n})$ from Theorem 3.3 there exist γ, defined in (10), unconstrained parameters, i.e. $\overline{N} = \gamma$, while for $S_2^{(\overline{\mu}^u, \overline{\mu}^v)}(T_{m,n})$ there exist no unconstrained parameters, i.e. $\overline{N} = 0$. Indeed, from Theorem 3.4 in case of $S_{2,2}^{(\overline{\mu}^u, \overline{\mu}^v)}(R_{m,n})$ the interpolating surface depends on the choice of γ $C_{i,j}$, each one having preimage at the left-down corner of each subdomain.

For $S_2^{(\overline{\mu}^u, \overline{\mu}^v)}(T_{m,n})$ the surface is unique, as in the case of all simple knots (see Theorem 3.2 of [10]) because $\mathcal{S}$ in (8) is defined by linearly dependent B-splines, i.e. they are not a basis. Indeed, as reported in Theorem 2 of [9], in each Ω_r, $r = 1, ..., \gamma$, a B-spline is redundant and the corresponding $C_{i,j}$ does not influence the shape of the surface. Therefore in Theorem 3.4 the surface is completely defined by the initial data and the points $C_{i,j}$ in $\mathcal{S}_{fp}(u, v)$ can be chosen arbitrarily, without modifying the surface shape.

□

- **In case of NMC**, we observe that the knots in the partitions U and V not associated to a curve of the network are *barriers* identifying $(k + 1)(\ell + 1)$ subdomains in Ω. In each subdomain we have the MC case where the control point in the left-down corner of each subdomain is identified as unconstrained parameter, as in Remark 3.1.

So, taking into account the same logical scheme of the previous configurations and the NMC described in [16], (11) hold and then

$$C_{i,j} = \Gamma_{i,j} + (-1)^{i+j+\rho_1(\overline{r}_\lambda+1)+\rho_2+\rho_3(\overline{r}_\gamma)+\phi_1(\overline{s}_\psi+1)+\phi_2+\phi_3(\overline{s}_\delta)} \frac{h_i}{h_{1+\rho_1(\overline{r}_\lambda+1)+\rho_2+\rho_3(\overline{r}_\gamma)}} \cdot \frac{k_j}{k_{1+\phi_1(\overline{s}_\psi+1)+\phi_2+\phi_3(\overline{s}_\delta)}} C_{1+\rho_1(\overline{r}_\lambda+1)+\rho_2+\rho_3(\overline{r}_\gamma),1+\phi_1(\overline{s}_\psi+1)+\phi_2+\phi_3(\overline{s}_\delta)}, \tag{63}$$

$i = 1, \dots, M-2,\ \ j = 1, \dots, N-2$, neglecting the indices related to unconstrained parameters and to points corresponding to multiple knots, with

$$\begin{aligned}\Gamma_{i,j} = & \sum_{r=1}^{i-1-\rho_1(\overline{r}_\lambda+1)-\rho_2-\rho_3(\overline{r}_\gamma)} (-1)^{r+1} \frac{(h_{i-r}+h_{i-r+1})h_i}{h_{i-r}h_{i-r+1}} P_j^{(i-r-\alpha)} \\ & +(-1)^{i+\rho_1(\overline{r}_\lambda+1)+\rho_2+\rho_3(\overline{r}_\gamma)} \frac{h_i}{h_{1+\rho_1(\overline{r}_\lambda+1)+\rho_2+\rho_3(\overline{r}_\gamma)}} \\ & \cdot \sum_{s=1}^{j-1-\phi_1(\overline{s}_\psi+1)-\phi_2-\phi_3(\overline{s}_\delta)} (-1)^s \frac{(k_{j-s}+k_{j-s+1})k_j}{k_{j-s}k_{j-s+1}} Q_{1+\rho_1(\overline{r}_\lambda+1)+\rho_2}^{(j-s-\beta)}\end{aligned} \tag{64}$$

and

$$\rho_1 = \begin{cases} 0 & \overline{r}_\lambda < \overline{r}_\gamma \\ 1 & \overline{r}_\lambda > \overline{r}_\gamma \end{cases}, \quad \rho_2 = \begin{cases} 0 & \rho_1 = 0 \text{ or } u_{\overline{r}_\lambda} \text{ double} \\ 1 & \rho_1 = 1 \text{ and } u_{\overline{r}_\lambda} \text{ triple} \end{cases}, \quad \rho_3 = \begin{cases} 0 & \overline{r}_\lambda > \overline{r}_\gamma \\ 1 & \overline{r}_\lambda < \overline{r}_\gamma \end{cases},$$

$$\phi_1 = \begin{cases} 0 & \overline{s}_\psi < \overline{s}_\delta \\ 1 & \overline{s}_\psi > \overline{s}_\delta \end{cases}, \quad \phi_2 = \begin{cases} 0 & \phi_1 = 0 \text{ or } v_{\overline{s}_\psi} \text{ double} \\ 1 & \phi_1 = 1 \text{ and } v_{\overline{s}_\psi} \text{ triple} \end{cases}, \quad \phi_3 = \begin{cases} 0 & \overline{s}_\psi > \overline{s}_\delta \\ 1 & \overline{s}_\psi < \overline{s}_\delta \end{cases},$$

where $\overline{r}_\gamma$ and $\overline{s}_\delta$ are the indices of the last barrier before $C_{i,j}$, $\overline{r}_\lambda$, $\overline{s}_\psi$ are the indices of the last multiple knot before $C_{i,j}$, α, β are the number of multiple knots and barriers before $C_{i,j}$.
Finally, concerning the construction of the interpolating surface (8), it is the sum of $\mathcal{S}_b$, $\mathcal{S}_{\lambda\psi}$, $\mathcal{S}_\Gamma$ and $\mathcal{S}_{fp}$, as in Theorem 3.4. In particular, $\mathcal{S}_b$ and $\mathcal{S}_{\lambda\psi}$ are the same of MC of Theorem 3.4, while $\mathcal{S}_\Gamma$ and $\mathcal{S}_{fp}$ are modified taking into account (63) and (64).

Remark 3.4 We underline that in case of simple knots ($\gamma = 1$), then the domain is subdivided only by barriers: in $S_{2,2}^{1,1}(R_{m,n})$ the number of free parameters is $k\ell + k + \ell + 1$, while in $S_2^1(T_{m,n})$ it is $k\ell + k + \ell$. Thanks to the dependence relationship of the B-splines (see [26]), if $C_{1,1} = (c_1, c_2, c_3)$ and we choose the free control points as follows

$$C_{i,j} = \begin{cases} (c_1, c_2, c_3) & \text{if } i+j \text{ even} \\ (-c_1, -c_2, -c_3) & \text{if } i+j \text{ odd} \end{cases},$$

then the triangulated surface is unique. □

To conclude we remark that from the most general case of **NMC with multiple knots** we can obtain all other cases, also described in [10, 16]:

- **MC with simple knots**. Being neither barriers nor multiple knots, i.e. $\overline{r}_\gamma = \overline{r}_\lambda = \overline{s}_\delta = \overline{s}_\psi = 0$, then $\rho_1 = \rho_2 = \rho_3 = \phi_1 = \phi_2 = \phi_3 = 0$ and (63)–(64) coincide with (26)–(27) in Lemma 3.1 of [10];
- **NMC with simple knots**. Being no multiple knots, i.e. $\overline{r}_\lambda = \overline{s}_\psi = 0$, then $\rho_1 = \rho_2 = \phi_1 = \phi_2 = 0$, $\rho_3 = \phi_3 = 1$ and (63)–(64) coincide with (18)-(19) in Lemma 1 of [16], where α, β denote the number of barriers in U, V, respectively;
- **MC with multiple knots**. Being no barriers, i.e. $\overline{r}_\gamma = \overline{s}_\delta = 0$, then $\rho_3 = \phi_3 = 0$, ρ_1, ρ_2, ϕ_1, ϕ_2 are defined as in (46)–(47) in Lemma 3.1 and (63)–(64) coincide with (44)–(45) in Lemma 3.1, where α, β denote the number of multiple knots in U, V, respectively.

4 Numerical Results

In this section we aim to show the performances of the proposed methods for the construction of surfaces interpolating given curve networks. We compare the obtained results in the two spline spaces $S_{2,2}^{(\overline{\mu}^u,\overline{\mu}^v)}(R_{m,n})$ for biquadratic tensor product surfaces and $S_2^{(\overline{\mu}^u,\overline{\mu}^v)}(T_{m,n})$ for quadratic triangulated surfaces, here considered. We have implemented the numerical tests in Matlab, by using a procedure that produce the desired surface given in input the knots and the control points of the curve network, together with the constraints to be satisfied. Such a procedure is based on the results obtained in Theorem 3.4, taking into account (44)–(45) in the MC case and (63)–(64) in the NMC case.

4.1 Test 1

In this example we compare tensor product and triangulated surfaces in case of MC and multiple knots in the parameter domain. In Fig. 6a we show the curve network

$$C_r(v) = \sum_{j=0}^{5} P_j^{(r)} B_j(v|V), \qquad r = 0, 1, 2, 3, \qquad V = \left\{0, 0, 0, \frac{1}{4}, \frac{1}{2}, \frac{3}{4}, 1, 1, 1\right\},$$

$$D_s(u) = \sum_{i=0}^{5} Q_i^{(s)} B_i(u|U), \qquad s = 0, 1, 2, 3, 4, \qquad U = \left\{0, 0, 0, \frac{1}{3}, \frac{2}{3}, \frac{2}{3}, 1, 1, 1\right\},$$

and in Fig. 6(b) the corresponding partition of Ω given by $U \times V$. We remark that $u = \frac{2}{3}$ is a double knot in U: its presence allows us to write the interpolating surfaces depending on the two control points $C_{1,1}$ and $C_{4,1}$ as explained in Remark 3.1.

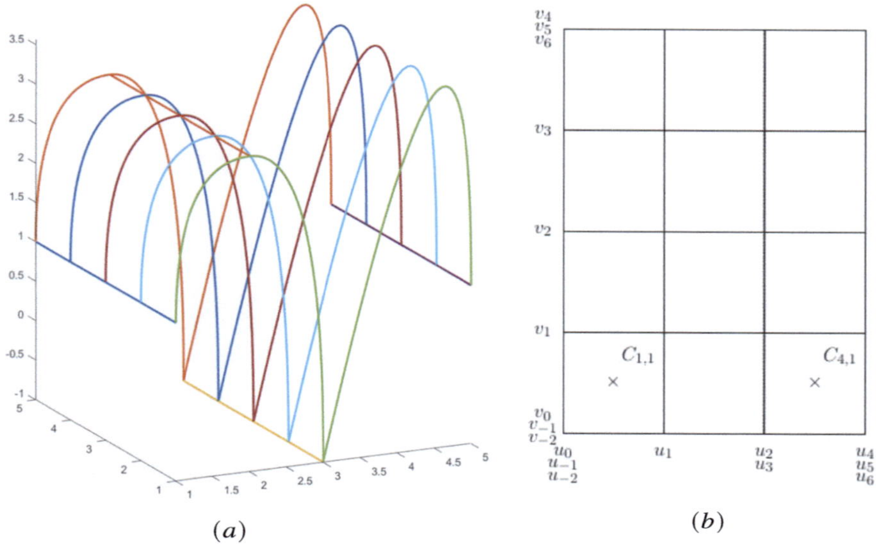

Fig. 6 Test 1—Double knot: **a** Curve network and **b** corresponding parameter domain

We have constructed different tensor product and triangulated surfaces varying the two unconstrained control points in order to show the different behaviour of the two kinds of surfaces. In particular, we can notice that the triangulated surfaces in Figs. 7a, 8a, 9a are all the same surface, due to their uniqueness (see Remark 3.3), while the corresponding control net is varying. Instead, the tensor product surfaces in Figs. 7b, 8b, 9b have different shapes, by adjusting the values of the control points $C_{1,1}$ and $C_{4,1}$.

Now we modify the previous example by considering a curve network with a triple knot (see Fig. 10a):

$$C_r(v) = \sum_{j=0}^{5} P_j^{(r)} B_j(v|V), \qquad r = 0, 1, 2^-, 2^+, 4, \qquad V = \left\{0, 0, 0, \frac{1}{4}, \frac{1}{2}, \frac{3}{4}, 1, 1, 1\right\},$$

$$D_s(u) = \sum_{i=0}^{6} Q_i^{(s)} B_i(u|U), \qquad s = 0, 1, 2, 3, 4, \qquad U = \left\{0, 0, 0, \frac{1}{3}, \frac{2}{3}, \frac{2}{3}, \frac{2}{3}, 1, 1, 1\right\}.$$

Figure 10b shows the corresponding partition of Ω given by $U \times V$. The triple knot $u = \frac{2}{3}$ allows us to write the interpolating surfaces depending on the two control points $C_{1,1}$ and $C_{5,1}$ as explained in Remark 3.1.

As in the previous case of double knot, we have considered different values for the two control points $C_{1,1}$ and $C_{5,1}$ in order to show the different behaviour of the two kinds of surfaces. Also in this case the triangulated surface is unique (see Figs. 11a, 12a, 13a), while the tensor product surfaces in Figs. 11b, 12b, 13b have different shapes, varying the values of the control points. In Fig. 14 the jump at the

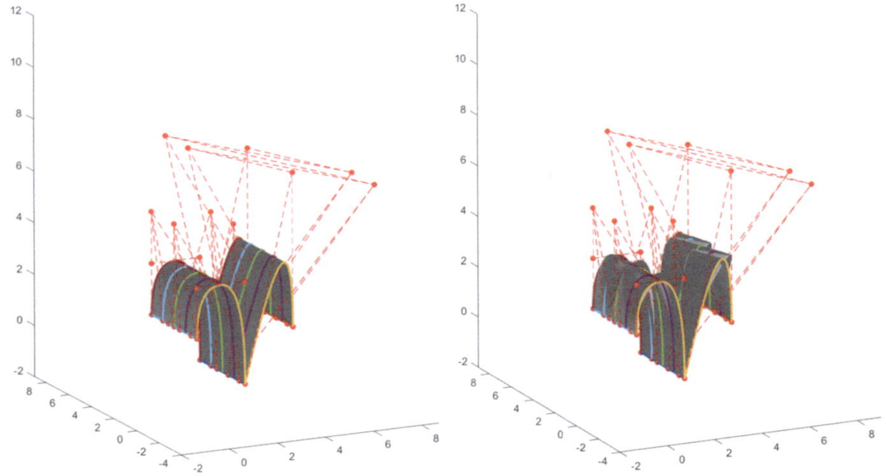

(*a*) (*b*)

Fig. 7 Test 1—Double knot: C^0 **a** triangulated surface and **b** tensor product surface interpolating the curve network in Fig. 6a with $C_{1,1} = (1, 1, 1)$ and $C_{4,1} = (3, 6, 7)$

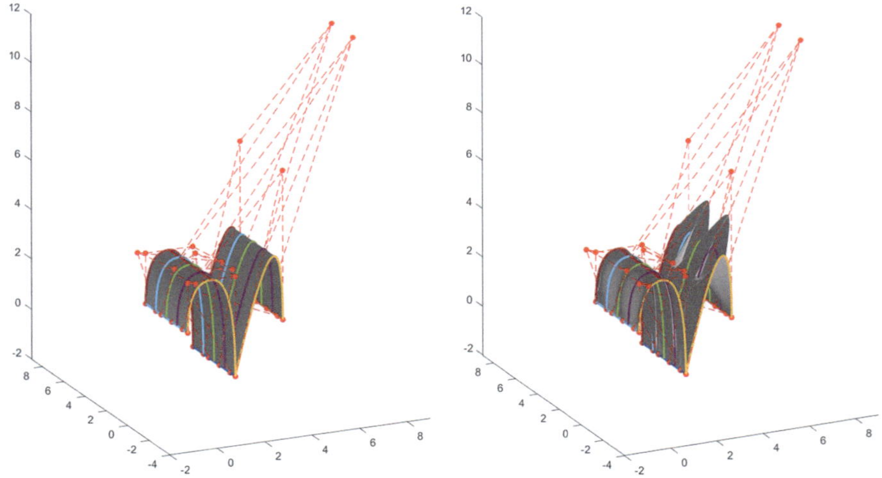

(*a*) (*b*)

Fig. 8 Test 1—Double knot: C^0 **a** triangulated surface and **b** tensor product surface interpolating the curve network in Fig. 6a with $C_{1,1} = (0, -2, 4)$ and $C_{4,1} = (2, 3, 2)$

knot u_2 is highlighted by subdividing the domain into two subdomains and drawing two different surfaces.

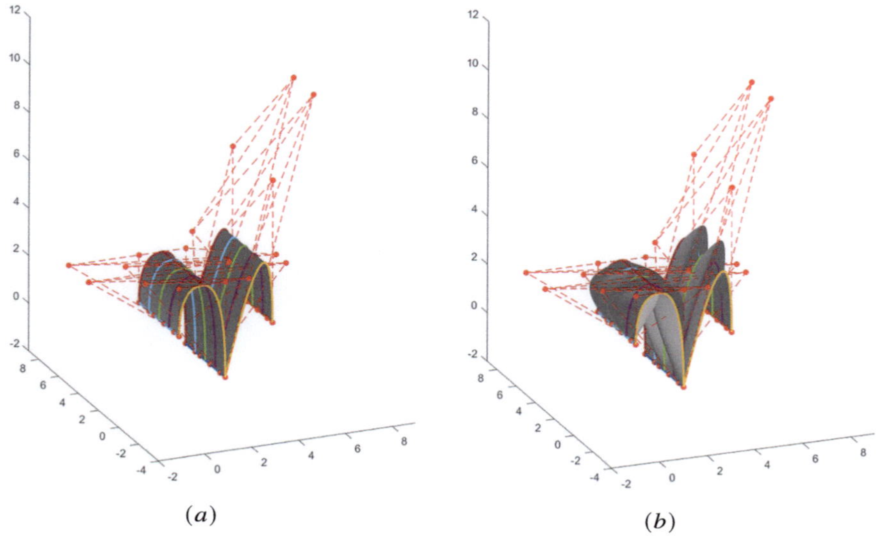

Fig. 9 Test 1—Double knot: C^0 **a** triangulated surface and **b** tensor product surface interpolating the curve network in Fig. 6a with $C_{1,1} = (4, 1, 3)$ and $C_{4,1} = (5, 7, 2)$

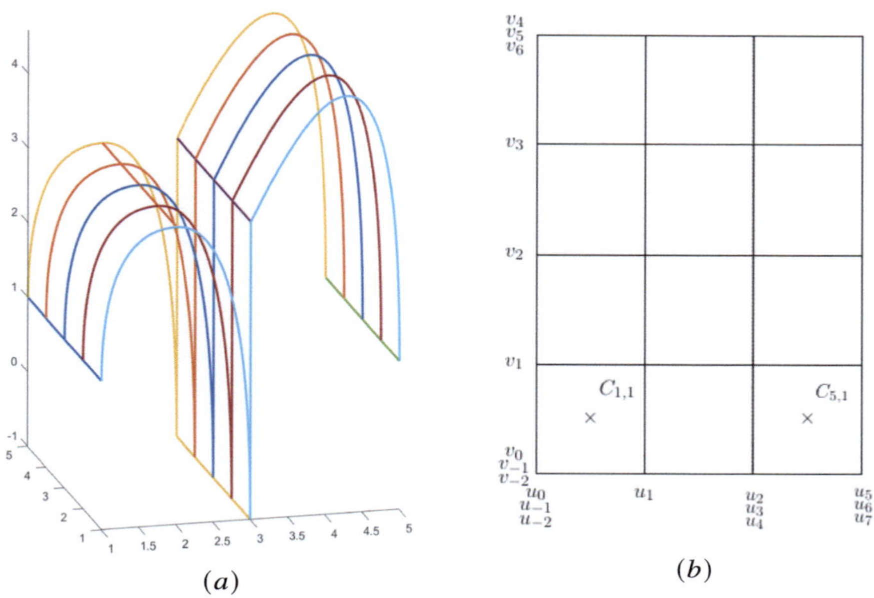

Fig. 10 Test 1—Triple knot: **a** Curve network and **b** corresponding parameter domain

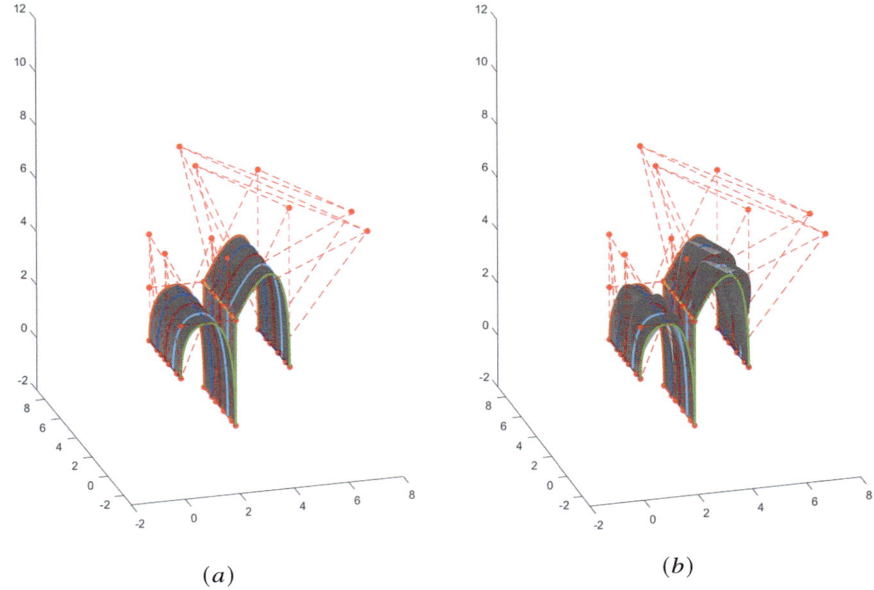

Fig. 11 Test 1—Triple knot: C^{-1} **a** triangulated surface and **b** tensor product surface interpolating the curve network in Fig. 10a with $C_{1,1} = (1, 1, 1)$ and $C_{5,1} = (3, 6, 7)$

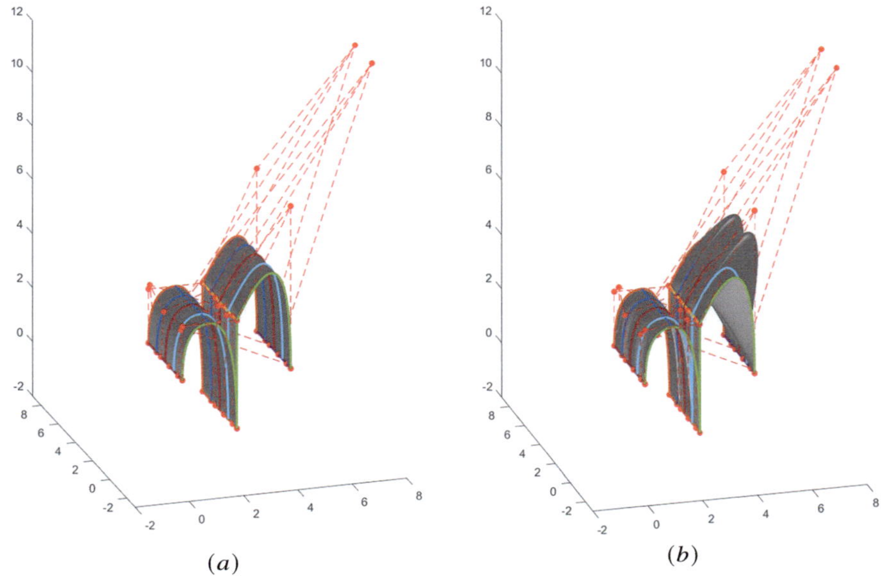

Fig. 12 Test 1—Triple knot: C^{-1} **a** triangulated surface and **b** tensor product surface interpolating the curve network in Fig. 10a with $C_{1,1} = (0, -2, 4)$ and $C_{5,1} = (2, 3, 2)$

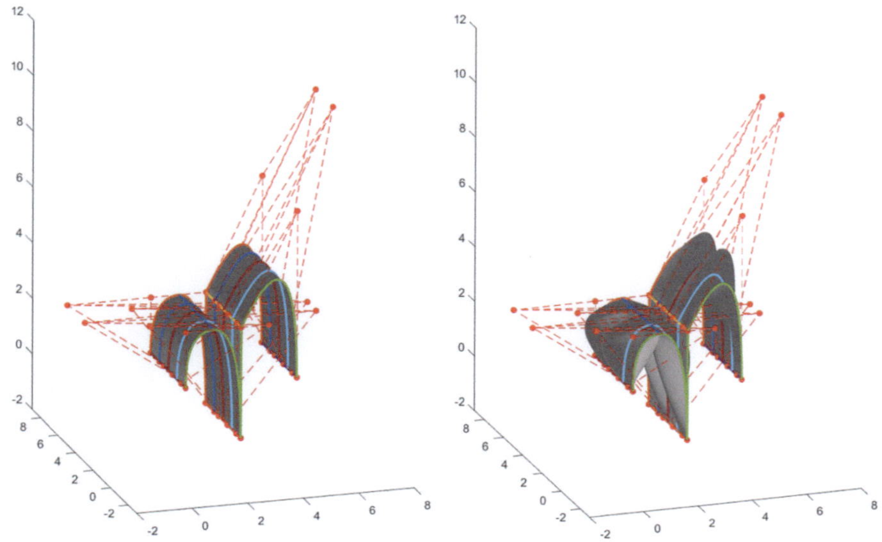

(a) (b)

Fig. 13 Test 1—Triple knot: C^{-1} **a** triangulated surface and **b** tensor product surface interpolating the curve network in Fig. 10a with $C_{1,1} = (4, 1, 3)$ and $C_{5,1} = (5, 7, 2)$

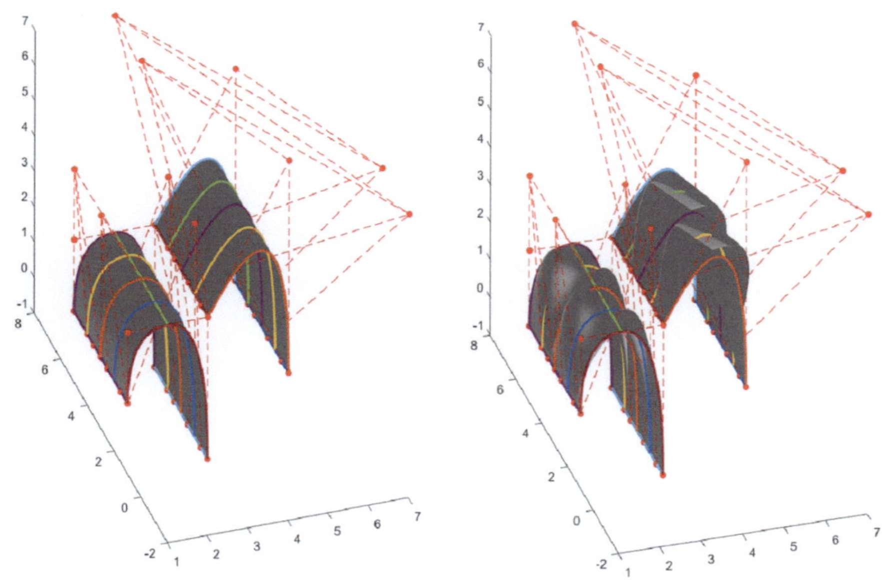

(a) (b)

Fig. 14 Test 1—Triple knot: C^{-1} **a** triangulated and **b** tensor product surfaces of Fig. 11 with the marked jump at u_2

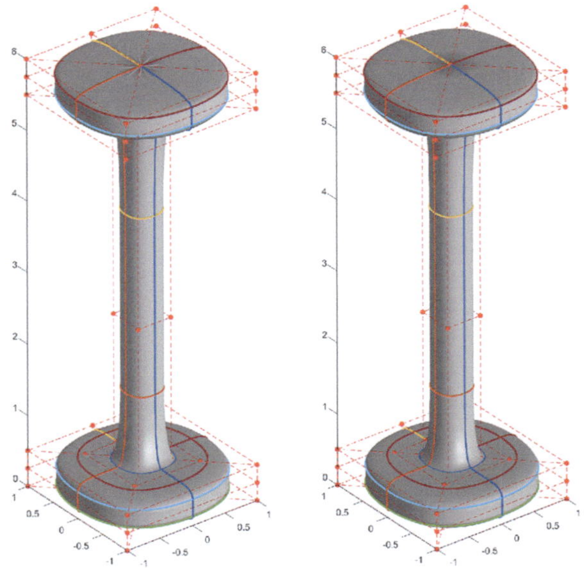

(a) (b)

Fig. 15 Test 2—Simple knots: **a** Triangulated and **b** tensor product surfaces interpolating the bar bell curve network

4.2 Test 2

In this example we want to compare the different results obtained by using simple and double knots. Given a curve network reproducing a bar bell, we interpolate it using only simple knots (see Fig. 15) and inserting double knots in v-direction (see Fig. 16). In Tables 1 and 2 we report knots and control points of the curve network in case of simple and double knots, respectively. The use of double knots allows us to better reconstruct the bar bell shape.

4.3 Test 3

In this example we consider NMC and the space $S_2^{(\overline{\mu}^u,\overline{\mu}^v)}(T_{m,n})$. The aim is to confirm the results of Remark 3.4.

In Fig. 17 we show the curve network together with the corresponding partition of Ω given by $U \times V$. The lines $u = u_3$, $v = v_2$, v_4 are barriers and therefore the domain is subdivided into 6 subdomains and the interpolating surfaces depend on the six control points $C_{1,1}$, $C_{4,1}$, $C_{1,3}$, $C_{4,3}$, $C_{1,5}$, $C_{4,5}$ as explained in Remark 3.1. Here we show the different shape of the triangulated surfaces, varying the unconstrained control points according to Remark 3.4. In particular in Fig. 18a, b, since the rule is not satisfied, the surfaces are different, instead in Fig. 19a, b they are equal.

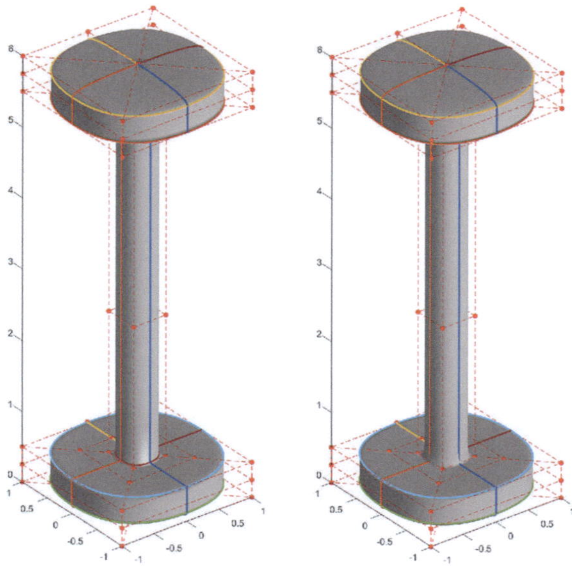

(a) (b)

Fig. 16 Test 2—Double knots: **a** Triangulated and **b** tensor product surfaces interpolating the bar bell curve network

4.4 Test 4

We want to compare the methods here proposed in case of MC and simple knots, with a Gordon-type C^1 biquadratic B-spline surface, [22, Chap. 10] (see also [10]), written as a boolean sum of two skinned surfaces and a tensor product surface interpolating a given set of points. Moreover, regarding the choice of the free parameter appearing in the construction of the tensor product surface, it is possible to choose it such that the surface is "smooth", i.e. the energy of a thin plate described by the surface is minimal, e.g. see [15]. Considering

$$\Theta = \int_a^b \int_c^d [(\mathcal{S}^{uu}(u, v))^2 + 2(\mathcal{S}^{uv}(u, v))^2 + (\mathcal{S}^{vv}(u, v))^2] \mathrm{d}u \mathrm{d}v,$$

approximation of the thin plate energy, with

$$\mathcal{S}^{uu}(u, v) = \frac{\partial^2 \mathcal{S}(u, v)}{\partial u^2}, \ \mathcal{S}^{uv}(u, v) = \frac{\partial^2 \mathcal{S}(u, v)}{\partial u \partial v}, \ \mathcal{S}^{vv}(u, v) = \frac{\partial^2 \mathcal{S}(u, v)}{\partial v^2},$$

then the free parameter C_{11} can be chosen by minimizing Θ with respect to it, obtaining

Table 1 Test 2—Simple knots: control points of the curve network. The knot vectors are $U = \{0, 0, 0, 1/4, 1/2, 3/4, 1, 1, 1\}$ and $V = \{0, 0, 0, 1/11, 2/11, 3/11, 4/11, 5/11, 6/11, 7/11, 8/11, 9/11, 10/11, 1, 1, 1\}$

j	$P_j^{(0)}$	$P_j^{(1)}$	$P_j^{(2)}$	$P_j^{(3)}$	$P_j^{(4)}$	
0	(0, 0, 0)	(0, 0, 0)	(0, 0, 0)	(0, 0, 0)	(0, 0, 0)	
1	(0, 1, 0)	(1, 0, 0)	(0, −1, 0)	(−1, 0, 0)	(0, 1, 0)	
2	(0, 1, 1/4)	(1, 0, 1/4)	(0, −1, 1/4)	(−1, 0, 1/4)	(0, 1, 1/4)	
3	(0, 1, 1/2)	(1, 0, 1/2)	(0, −1, 1/2)	(−1, 0, 1/2)	(0, 1, 1/2)	
4	(0, 1/2, 1/2)	(1/2, 0, 1/2)	(0, −1/2, 1/2)	(−1/2, 0, 1/2)	(0, 1/2, 1/2)	
5	(0, 1/4, 1/2)	(1/4, 0, 1/2)	(0, −1/4, 1/2)	(−1/4, 0, 1/2)	(0, 1/4, 1/2)	
6	(0, 1/4, 5/2)	(1/4, 0, 5/2)	(0, −1/4, 5/2)	(−1/4, 0, 5/2)	(0, 1/4, 5/2)	
7	(0, 1/4, 11/2)	(1/4, 0, 11/2)	(0, −1/4, 11/2)	(−1/4, 0, 11/2)	(0, 1/4, 11/2)	
8	(0, 1/2, 11/2)	(1/2, 0, 11/2)	(0, −1/2, 11/2)	(−1/2, 0, 11/2)	(0, 1/2, 11/2)	
9	(0, 1, 11/2)	(1, 0, 11/2)	(0, −1, 11/2)	(−1, 0, 11/2)	(0, 1, 11/2)	
10	(0, 1, 23/4)	(1, 0, 23/4)	(0, −1, 23/4)	(−1, 0, 23/4)	(0, 1, 23/4)	
11	(0, 1, 6)	(1, 0, 6)	(0, −1, 6)	(−1, 0, 6)	(0, 1, 6)	
12	(0, 0, 6)	(0, 0, 6)	(0, 0, 6)	(0, 0, 6)	(0, 0, 6)	
s	$Q_0^{(s)}$	$Q_1^{(s)}$	$Q_2^{(s)}$	$Q_3^{(s)}$	$Q_4^{(s)}$	$Q_5^{(s)}$
0	(0, 0, 0)	(0, 0, 0)	(0, 0, 0)	(0, 0, 0)	(0, 0, 0)	(0, 0, 0)
1	(0, 1, 1/8)	(1, 1, 1/8)	(1, −1, 1/8)	(−1, −1, 1/8)	(−1, 1, 1/8)	(0, 1, 1/8)
2	(0, 1, 3/8)	(1, 1, 3/8)	(1, −1, 3/8)	(−1, −1, 3/8)	(−1, 1, 3/8)	(0, 1, 3/8)
3	(0, 3/4, 1/2)	(3/4, 3/4, 1/2)	(3/4, −3/4, 1/2)	(−3/4, −3/4, 1/2)	(−3/4, 3/4, 1/2)	(0, 3/4, 1/2)
4	(0, 3/8, 1/2)	(3/8, 3/8, 1/2)	(3/8, −3/8, 1/2)	(−3/8, −3/8, 1/2)	(−3/8, 3/8, 1/2)	(0, 3/8, 1/2)
5	(0, 1/4, 3/2)	(1/4, 1/4, 3/2)	(1/4, −1/4, 3/2)	(−1/4, −1/4, 3/2)	(−1/4, 1/4, 3/2)	(0, 1/4, 3/2)
6	(0, 1/4, 4)	(1/4, 1/4, 4)	(1/4, −1/4, 4)	(−1/4, −1/4, 4)	(−1/4, 1/4, 4)	(0, 1/4, 4)
7	(0, 3/8, 11/2)	(3/8, 3/8, 11/2)	(3/8, −3/8, 11/2)	(−3/8, −3/8, 11/2)	(−3/8, 3/8, 11/2)	(0, 3/8, 11/2)
8	(0, 3/4, 11/2)	(3/4, 3/4, 11/2)	(3/4, −3/4, 11/2)	(−3/4, −3/4, 11/2)	(−3/4, 3/4, 11/2)	(0, 3/4, 11/2)
9	(0, 1, 45/8)	(1, 1, 45/8)	(1, −1, 45/8)	(−1, −1, 45/8)	(−1, 1, 45/8)	(0, 1, 45/8)
10	(0, 1, 47/8)	(1, 1, 47/8)	(1, −1, 47/8)	(−1, −1, 47/8)	(−1, 1, 47/8)	(0, 1, 47/8)
11	(0, 0, 6)	(0, 0, 6)	(0, 0, 6)	(0, 0, 6)	(0, 0, 6)	(0, 0, 6)

$$C_{11}^* = -h_1 k_1 \frac{\int_a^b \int_c^d [(\mathcal{S}_b^{uu} + \mathcal{S}_\Gamma^{uu})\mathcal{S}_{fp}^{uu} + 2(\mathcal{S}_b^{uv} + \mathcal{S}_\Gamma^{uv})\mathcal{S}_{fp}^{uv} + (\mathcal{S}_b^{vv} + \mathcal{S}_\Gamma^{vv})\mathcal{S}_{fp}^{vv}]}{\int_a^b \int_c^d [(\mathcal{S}_{fp}^{uu})^2 + 2(\mathcal{S}_{fp}^{uv})^2 + (\mathcal{S}_{fp}^{vv})^2]}. \tag{65}$$

Such a point is the one minimizing Θ. The integrals in (65) can be exactly evaluated by a composite tensor-product Gauss-Legendre quadrature formula with 3×3 nodes.

In Fig. 20 we show the curve network and the triangulated surface, where $C_{11} = (1, 0, 3)$ has been arbitrarily chosen, while in Fig. 21 we display the biquadratic B-spline surface with $C_{11}^* = (1.11, -0.15, 3)$ obtained by (65) and the Gordon-type biquadratic B-spline surface.

We can remark that the shapes of the generated surfaces are comparable from the graphical point of view.

Moreover, the technique here used for the choice of the free control point C_{11} can be extended in case of NMC and multiple knots, applying it to each subdomain presenting a free parameter in case of both tensor product and triangulated surfaces.

Table 2 Test 2—Double knots: control points of the curve network. The knot vectors are $U = \{0, 0, 0, 1/4, 1/2, 3/4, 1, 1, 1\}$ and $V = \{0, 0, 0, 1/7, 2/7, 2/7, 3/7, 3/7, 4/7, 4/7, 5/7, 5/7, 6/7, 1, 1, 1\}$

j	$P_j^{(0)}$	$P_j^{(1)}$	$P_j^{(2)}$	$P_j^{(3)}$	$P_j^{(4)}$	
0	(0, 0, 0)	(0, 0, 0)	(0, 0, 0)	(0, 0, 0)	(0, 0, 0)	
1	(0, 1, 0)	(1, 0, 0)	(0, −1, 0)	(−1, 0, 0)	(0, 1, 0)	
2	(0, 1, 1/4)	(1, 0, 1/4)	(0, −1, 1/4)	(−1, 0, 1/4)	(0, 1, 1/4)	
3	(0, 1, 1/2)	(1, 0, 1/2)	(0, −1, 1/2)	(−1, 0, 1/2)	(0, 1, 1/2)	
4	(0, 1/2, 1/2)	(1/2, 0, 1/2)	(0, −1/2, 1/2)	(−1/2, 0, 1/2)	(0, 1/2, 1/2)	
5	(0, 1/4, 1/2)	(1/4, 0, 1/2)	(0, −1/4, 1/2)	(−1/4, 0, 1/2)	(0, 1/4, 1/2)	
6	(0, 1/4, 5/2)	(1/4, 0, 5/2)	(0, −1/4, 5/2)	(−1/4, 0, 5/2)	(0, 1/4, 5/2)	
7	(0, 1/4, 11/2)	(1/4, 0, 11/2)	(0, −1/4, 11/2)	(−1/4, 0, 11/2)	(0, 1/4, 11/2)	
8	(0, 1/2, 11/2)	(1/2, 0, 11/2)	(0, −1/2, 11/2)	(−1/2, 0, 11/2)	(0, 1/2, 11/2)	
9	(0, 1, 11/2)	(1, 0, 11/2)	(0, −1, 11/2)	(−1, 0, 11/2)	(0, 1, 11/2)	
10	(0, 1, 23/4)	(1, 0, 23/4)	(0, −1, 23/4)	(−1, 0, 23/4)	(0, 1, 23/4)	
11	(0, 1, 6)	(1, 0, 6)	(0, −1, 6)	(−1, 0, 6)	(0, 1, 6)	
12	(0, 0, 6)	(0, 0, 6)	(0, 0, 6)	(0, 0, 6)	(0, 0, 6)	
s	$Q_0^{(s)}$	$Q_1^{(s)}$	$Q_2^{(s)}$	$Q_3^{(s)}$	$Q_4^{(s)}$	$Q_5^{(s)}$
0	(0, 0, 0)	(0, 0, 0)	(0, 0, 0)	(0, 0, 0)	(0, 0, 0)	(0, 0, 0)
1	(0, 1, 1/8)	(1, 1, 1/8)	(1, −1, 1/8)	(−1, −1, 1/8)	(−1, 1, 1/8)	(0, 1, 1/8)
2	(0, 1, 1/2)	(1, 1, 1/2)	(1, −1, 1/2)	(−1, −1, 1/2)	(−1, 1, 1/2)	(0, 1, 1/2)
3	(0, 1/4, 1/2)	(1/4, 1/4, 1/2)	(1/4, −1/4, 1/2)	(−1/4, −1/4, 1/2)	(−1/4, 1/4, 1/2)	(0, 1/4, 1/2)
4	(0, 1/4, 11/2)	(1/4, 1/4, 11/2)	(1/4, −1/4, 11/2)	(−1/4, −1/4, 11/2)	(−1/4, 1/4, 11/2)	(0, 1/4, 11/2)
5	(0, 1, 11/2)	(1, 1, 11/2)	(1, −1, 11/2)	(−1, −1, 11/2)	(−1, 1, 11/2)	(0, 1, 11/2)
6	(0, 1, 47/8)	(1, 1, 47/8)	(1, −1, 47/8)	(−1, −1, 47/8)	(−1, 1, 47/8)	(0, 1, 47/8)
7	(0, 0, 6)	(0, 0, 6)	(0, 0, 6)	(0, 0, 6)	(0, 0, 6)	(0, 0, 6)

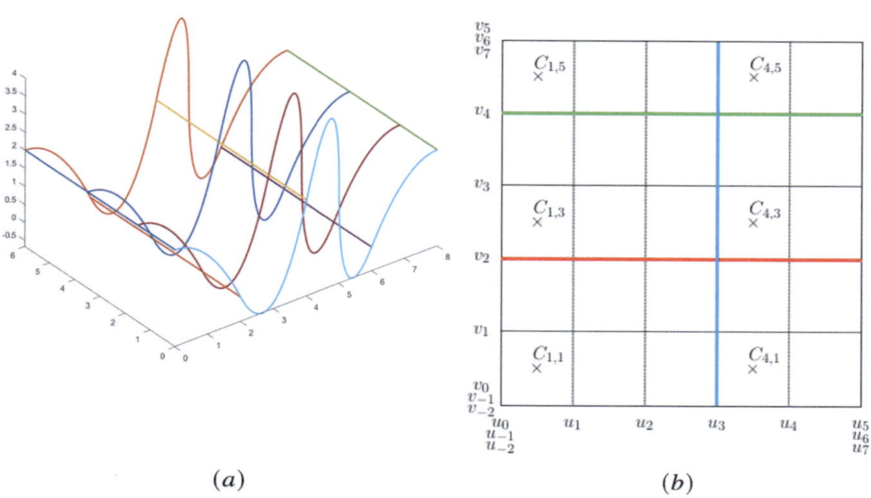

Fig. 17 Test 3: **a** Curve network and **b** corresponding parameter domain

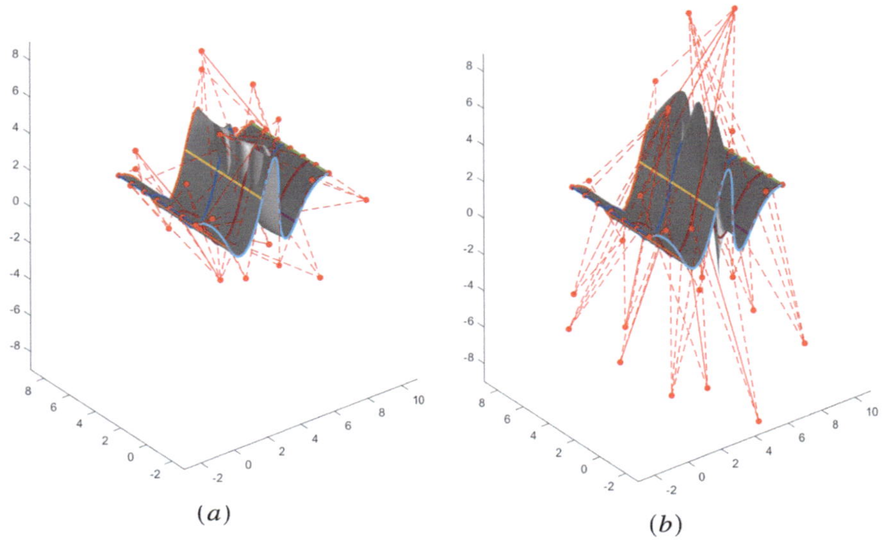

Fig. 18 Test 4: Triangulated surfaces with **a** $C_{1,1} = C_{1,3} = C_{1,5} = C_{4,1} = C_{4,3} = C_{4,5} = (1, 1, 1)$ and **b** $C_{1,1} = C_{1,3} = C_{1,5} = C_{4,1} = C_{4,3} = C_{4,5} = (5, 5, 5)$

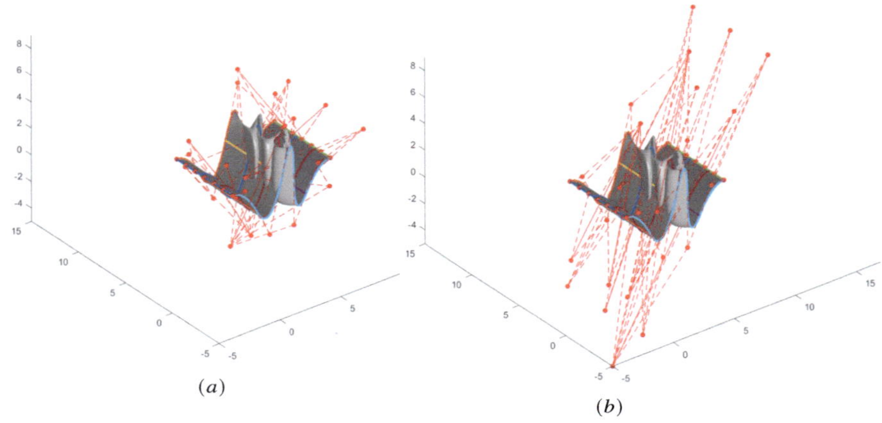

Fig. 19 Test 3: Triangulated surfaces with **a** $C_{1,1} = C_{1,3} = C_{1,5} = (1, 1, 1)$, $C_{4,1} = C_{4,3} = C_{4,5} = (-1, -1, -1)$ and **b** $C_{1,1} = C_{1,3} = C_{1,5} = (5, 5, 5)$, $C_{4,1} = C_{4,3} = C_{4,5} = (-5, -5, -5)$

5 Conclusions

In this paper we have studied the interpolation of a bidirectional B-spline curve network by both biquadratic tensor product B-spline surfaces and quadratic B-spline surfaces on criss-cross triangulations, with the aim of simulating discontinuities in the gradient and even in the surface itself. We have proposed an algorithm for their construction together with graphical results, showing the performance of the method.

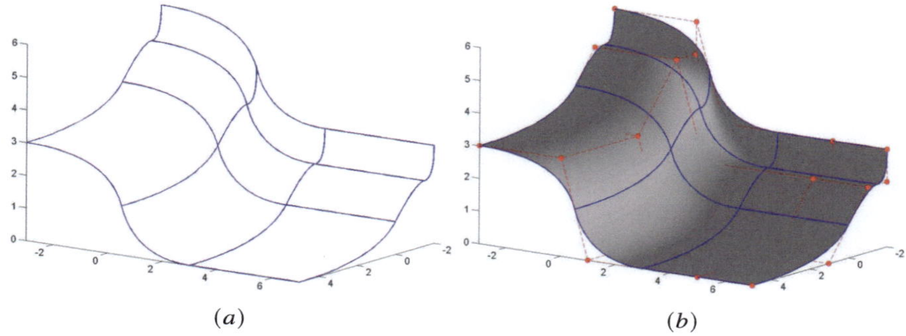

Fig. 20 **a** The curve network and **b** the triangulated surface in $S_2^1(T_{3,3})$ and interpolating the curve network **a**

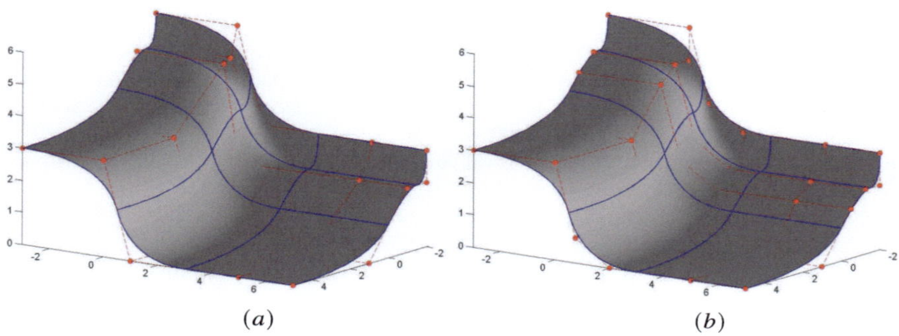

Fig. 21 **a** The tensor product surface defined in $S_{2,2}^{1,1}(R_{3,3})$, and **b** the Gordon type surface defined in $S_{2,2}^{1,1}(R_{8,8})$, interpolating the curve network **a** of Fig. 20

Acknowledgements The authors are members of the INdAM Research group GNCS of Italy.

References

1. Antonelli, M., Beccari, C.V., Casciola, G.: High quality local interpolation by composite parametric surfaces. Comput. Aided Geometric Des. **46**, 103–124 (2016)
2. Barnhill, R.E.: Blending function interpolation: a survey and some new results. In: Collatz, L., Werner, H., Meinardus G. (eds.) Numerische Methoden der Approximations theorie Vol. 3, ISNM Vol. 30, pp. 43–89. Birkhauser Verlag (1976)
3. de Boor, C.: A Practical Guide to Splines, Revised Springer, Berlin (2001)
4. Bos, L.P., Grabenstetter, J.E., Salkauskas, K.: Pseudo-tensor product interpolation and blending with families of univariate schemes. Comput. Aided Geom. Des. **13**, 429–440 (1996)
5. Cho, D.-Y., Lee, K.-Y., Kim, T.-W.: Interpolating G^1 Bézier surfaces over irregular curve networks for ship hull design. CAD Comput. Aided Des. **38**(641–660), 2006 (2006)
6. Chui, C.K.: Multivariate splines. CBMS-NSF Regional Conference Series in Applied Mathematics, vol. 54, SIAM, Philadelphia (1988)

7. Conti, C., Dyn, N.: Blending based Chaikin type subdivision schemes for nets of curves. In: Daehlen, M., Morken, K., Schumaker, L.L. (eds.) Mathematical Methods for Curves and Surfaces, pp. 51–68. Nashboro Press, Brentwood (2005)
8. Dagnino, C., Lamberti, P., Remogna, S.: BB-coefficients of unequally smooth quadratic B-splines on non uniform criss-cross triangulations. Quaderni Scientifici del Dipartimento di Matematica, Università di Torino, n. 24 (2008). http://hdl.handle.net/2318/434
9. Dagnino, C., Lamberti, P., Remogna, S.: B-spline bases for unequally smooth quadratic spline spaces on non-uniform criss-cross triangulations. Numer. Algorithms **61**, 209–222 (2012)
10. Dagnino, C., Lamberti, P., Remogna, S.: Curve network interpolation by C^1 quadratic B-spline surfaces. Comput. Aided Geom. Des. **40**, 26–39 (2015)
11. Delvos, J., Schempp, W.: Boolean Methods in Approximation. Longman Scientific and Technical (1989)
12. Gordon, W.J.: Spline blended interpolation through curve networks. J. Math. Mech. **18**(10), 931–952 (1969)
13. Gordon, W.J.: Blending-function methods of bivariate and multivariate interpolation. SIAM Numer. Anal. **8**, 158–177 (1971)
14. Greshake, S.H., Bronsart, R.: Application of subdivision surfaces in ship hull form modeling. Comput. Aided Des. **100**, 79–92 (2018)
15. Hadenfeld, J.: Local energy fairing of B-spline surfaces. In: Daehlen, M., Lyche, T., Schumaker, L.L. (eds.) Mathematical Methods for Curves and Surfaces: Ulvik, 1994, pp. 203–212. Vanderbilt Univ. Press, Nashville (1995)
16. Lamberti, P., Remogna, S.: Quadratic B-spline surfaces with free parameters for the interpolation of curve networks. Mathematics **10**(4), 543 (2022)
17. Lamnii, A., Lamnii, M., Mraoui, H.: Constructing B-spline representation of quadratic Sibson-Thomson splines. Comput. Aided Geom. Des. **33**, 66–81 (2015)
18. Lamnii, A., Mraoui, H., Sbibih, D., Zidna, A.: A multiresolution method for fitting scattered data on the sphere. BIT Numer. Math. **49**(3), 589–610 (2009)
19. Lin, F., Hewitt, W.T.: Expressing Coons-Gordon surfaces as NURBS. Comput. Aided Des. **26**(2), 145–155 (1994)
20. Lyche, T., Schumaker, L.L.: A Multiresolution Tensor Spline Method for Fitting Functions on the Sphere. SIAM J. Sci. Comp. **22**, 724–746 (2000)
21. Peters, J.: Smooth interpolation of a mesh of curves. Constr. Approx. **7**, 221–246 (1991)
22. Piegl, L., Tiller, W.: The NURBS Book. Springer (1995)
23. Sablonnière, P.: Quadratic spline quasi-interpolants on bounded domains of $\mathbb{R}^d$, $d = 1, 2, 3$. Rend. Sem. Mat. Univ. Pol. Torino **61**(3), 229–246 (2003)
24. Salvi, P., Vaitkus, M., Várady, T.: Constrained modeling of multi-sided patches. Comput. Graph. **114**, 86–95 (2023)
25. Salvi, P., Várady, T.: G^2 surface interpolation over general topology curve networks. Comput. Graph. Forum **33**, 151–160 (2014)
26. Wang, R.H.: Multivariate Spline Functions and Their Applications. Science Press, Beijing/New York, Kluwer Academic Publishers, Dordrecht/Boston/London (2001)

Numerical Solution of Sturm-Liouville-Type Differential Equation Based on Wavelet Bases of Hermite Quadratic Splines

Abdellah Lamnii, Mohamed Lamnii, Chaimae Mouhoub, and Driss Sbibih

Abstract The purpose of this research is to solve the Sturm-Liouville-type differential equation with Dirichlet boundary conditions. We employ a newly constructed locally supported wavelet basis based on Hermite C^1-quadratic splines over a regular grid. The new Hermite quadratic wavelet basis has been carefully adapted to the interval $[0, 1]$ for improved computational simplicity. We validate our approach by presenting numerical examples at the end of this paper.

1 Introduction

Wavelets are versatile mathematical tools with remarkable approximation capabilities [2]. They have multiple applications across various domains, including signal processing and approximation theory. They provide an effective way to analyze and represent signals with varying levels of detail.

In particular, Amaratunga et al. [1] demonstrated that the Galerkin method using wavelets provides an efficient approach to solving one-dimensional partial differential equations, with significant improvements in accuracy. Beylkin, Coifman, and Rokhlin [3] also made a major contribution by developing fast wavelet transform algorithms, which allow for the reduction of computations through matrix compression. Vasilyev and Kevlahan [24] proposed an adaptive multi-level collocation method, leveraging wavelets to efficiently solve elliptic problems with enhanced accuracy, even in computationally intensive settings.

Recent advances, such as the work of Han, Michelle, and Wong [15] on wavelet-based numerical methods, confirm the potential of these tools for complex differential equations, improving both the speed and accuracy as well as the robustness of the solutions. In the context of equations with elliptic operators, the work of Cohen, Dah-

A. Lamnii (✉)
Abdelmalek Essaadi University, ISI, ENSATe, Tetouan, Morocco
e-mail: a_lamnii@yahoo.fr

M. Lamnii · C. Mouhoub · D. Sbibih
Laboratoire LANO, Faculty of Sciences Oujda, University Mohammed First, Oujda, Morocco

© The Author(s), under exclusive license to Springer Nature Switzerland AG 2025
M. J. Ibáñez-Pérez et al. (eds.), *Mathematical And Computational Modelling, Approximation and Simulation*, SEMA SIMAI Springer Series 41,
https://doi.org/10.1007/978-3-032-01114-5_12

men, and DeVore [6] highlighted the convergence rates of adaptive methods based on wavelets, making it possible to address problems with irregular boundary conditions. Furthermore, Dahmen, Proessdorf, and Schneider [10] studied matrix compression for pseudodifferential equations, which enhanced the efficiency of wavelets in applications requiring high computational density. This collective work demonstrates that wavelets are not only effective for numerical resolution but also for optimizing resources and precision in various contexts.

In parallel, Hermite-type wavelets, which combine interpolation and differentiation properties, have proven particularly suited for asymptotic solutions or capturing precise local behaviors. Conti, Romani, and Unser [7] introduced Hermite-like interpolation and subdivision to preserve elliptical shapes, making this technique useful for solutions requiring specific geometric symmetry. Similarly, Cotronei and Moosmüller [8] explored the refinability properties of Hermite B-splines, enabling mask factorization that is useful for precise and memory-efficient calculations.

Other notable works, such as those by Jia, Riemenschneider, and Zhou [16], have developed vector subdivision schemes and multi-wavelets, providing greater flexibility for Hermite applications. The research of Merrien [21] and Dubuc and Merrien ([11, 12]) has contributed to consolidating the use of Hermite subdivision schemes, utilizing Taylor polynomials and developing precise interpolants that enhance the convergence and stability of the solutions. Hermite multi-wavelets, such as those developed by Cotronei, Sissouno, and Moosmüller [9], incorporate polynomial and exponential moments, making these wavelets particularly suitable for differential equations whose solutions exhibit variable and complex behaviors.

In the article [17], Jia et al. developed a pair of wavelets using cubic Hermite splines, which they then applied to solve the Sturm-Liouville equation with Dirichlet boundary conditions. These equations, as pointed out in [23], play an important role in various fields. This work likely illustrates a specific application of wavelets in solving differential equations, highlighting their effectiveness in this context. In this paper, we propose using wavelets based on quadratic splines to solve a similar type of differential equation with the same boundary conditions. For more information on quadratic splines and two-point oscillatory basis splines, the reader is referred to [19].

Our approach employed quadratic splines. This decision was motivated by several critical considerations. First and foremost, by opting for quadratic splines, we achieved similar, if not superior, results while significantly reducing the required computational time. Furthermore, the use of quadratic splines has improved the conditionality of the discretized stiffness matrix in our method. This enhancement is crucial in numerous numerical applications where result reliability heavily depends on numerical stability.

The work is organized as follows: In Sect. 2, we construct Hermite's quadratic spline basis. In Sect. 3, we construct Hermite quadratic composite-spline wavelets.

In Sect. 4, we adapt the wavelets that we have constructed to the interval [0, 1]. In Sect. 5, we focus on the numerical resolution of the Sturm-Liouville equation using wavelets, while taking into account Dirichlet boundary conditions. At the end of this section, several numerical examples are provided.

2 Hermite's Quadratic Spline Base

The construction of Hermite spline interpolants by adding new knots at the initial partition and increasing the number of polynomial pieces is familiar, as in [5, 13, 20, 22]. These methods incorporate the shape parameters to preserve the convexity or monotony of the data. Schumaker in [22] has developed a very economical C^1 quadratic Hermite interpolation scheme with shape constraints using canonical polynomial basis functions. Manni (see [20]) developed a C^2 cubic interpolants with shape constraints by using parametric curves. Recently, Lamnii et al. [19] have constructed splines using the polar forms; the same authors created a Hermite Base based on two points [18]. Bernstein-Bézier representation is one of the fundamental tools for dealing with the building of Hermite's base of quadratic composite-spline, thus we recall some results concerning it.

For a given integer m and $t \in [0, 1]$, the Bernstein polynomials are given by:

$$\mathcal{B}_{j,m} = \binom{m}{j} t^j (1-t)^{m-j}, \quad j = 0, \dots, m$$

Now, let $[a, b]$ be an interval in $\mathbb{R}$, any $x \in \mathbb{R}$ can be expressed as $x = (1-t)a + tb$, where $(1-t, t)$ are the associated barycentric coordinates. When x belongs to $[a, b]$, they are non-negative. A polynomial p with a degree less than or equal to m can be straightforwardly expressed using Bernstein polynomials as follows:

$$p(x) = \sum_{j=0}^{m} b_j \mathcal{B}_{j,m}(t), \tag{1}$$

where $t = \frac{x-a}{b-a}$, and $b_j, \ j = 0, \dots, m$ are the Bézier control points.

In the next, Bernstein basis functions will be used to simplify the development of our new Hermite quadratic splines. More precisely, given the following subdivision $\tau = \{x_0 < x_1 < \cdots < x_n\}$, let $\tau_i = \{x_i < x_{i,1} < x_{i+1}\}$, where $x_{i,1} = (1-\lambda_i)x_i + \lambda_i x_{i+1}$ and $\lambda_i \in (0, 1)$. Assuming we are provided with the data $f_i^{(0)}$ and the derivative $f_{i+1}^{(1)}$ for each vertex x_i in the range $i = 0, \dots, n$, the following result holds.

Proposition 2.1 There exists a unique spline $s_i \in \mathcal{S}_2^1(\tau_i, [x_i, x_{i+1}])$ such that

$$\begin{cases} s_i(x_k) = f_k^{(0)}, \\ s_i^{(1)}(x_k) = f_k^{(1)} \end{cases} k = i, i+1,$$

Proof This spline can be written in B-spline form as:

- For $x \in [x_i, x_{i,1}]$, $s_i(x) = \sum_{j=0}^{2} a_j \, \mathcal{B}_{j,2}(t)$, where $t = \frac{x - x_i}{x_{i,1} - x_i}$.
- For $x \in [x_{i,1}, x_{i+1}]$, $s_i(x) = \sum_{j=0}^{2} b_j \, \mathcal{B}_{j,2}(t)$, where $t = \frac{x - x_{i,1}}{x_{i+1} - x_{i,1}}$.

where a_i, b_i are reel numbers.

To simplify notations, we define $p = s_i|_{[x_i, x_{i,1}]}$ and $q = s_i|_{[x_{i,1}, x_{i+1}]}$. Therefore, to ensure that the connection between the polynomials p and q is of class C^1 at the point $x_{i,1}$, we demonstrate that the interpolation problem yields:

$$s_i(x_i) = p(x_i) = a_0 \text{ hence,} \quad a_0 = f_i^{(0)}.$$

$$s_i'(x_i) = p'(x_i) = \frac{2(a_1 - a_0)}{\lambda_i(x_{i+1} - x_i)} = f_1^{(1)} \text{ hence,} \quad a_1 = f_i^{(0)} + \frac{\lambda_i}{2}(x_{i+1} - x_i) f_i^{(1)}.$$

as well following a similar reasoning we have:

$$b_2 = f_{i+1}^{(0)}, \; b_1 = f_{i+1}^{(0)} + \frac{1 - \lambda_i}{2}(x_i - x_{i+1}) f_{i+1}^{(1)}.$$

The connection of class C^1 at the point $x_{i,1}$ is expressed by:

$$a_2 = b_0 = (1 - \lambda_i) a_1 + \lambda_i b_1.$$

So, the spline s_i is uniquely determined. □

In the following, we will refine the notation to establish a basis for the space $\mathcal{S}_2^1(\tau, [x_0, x_n])$. Specifically, if we designate the coefficients of the spline s_i as c_k, where $k = 0, \ldots, 4$, as illustrated in Fig. 1, we obtain:

$$c_0 = f_i^{(0)}, \quad c_4 = f_{i+1}^{(0)}, \tag{2}$$

$$c_1 = c_0 + \frac{\lambda_i}{2}(x_{i+1} - x_i) f_i^{(1)}, \quad c_3 = c_4 + \frac{1 - \lambda_i}{2}(x_i - x_{i+1}) f_{i+1}^{(1)}, \tag{3}$$

$$c_2 = (1 - \lambda_i) c_1 + \lambda_i c_3. \tag{4}$$

By using the expressions of the coefficients c_k, $k = 0, \ldots, 4$, we can easily calculate the elements of the Hermite's basis $\phi_{i,0}$ and $\phi_{i,1}$ of the space $\mathcal{S}_2^1(\tau, [x_0, x_n])$ satisfying $\phi_{i,j}^{(k)}(x_l) = \delta_{i,l}\delta_{j,k}$, $l = 0, \ldots, n$, $j = 0, 1$.

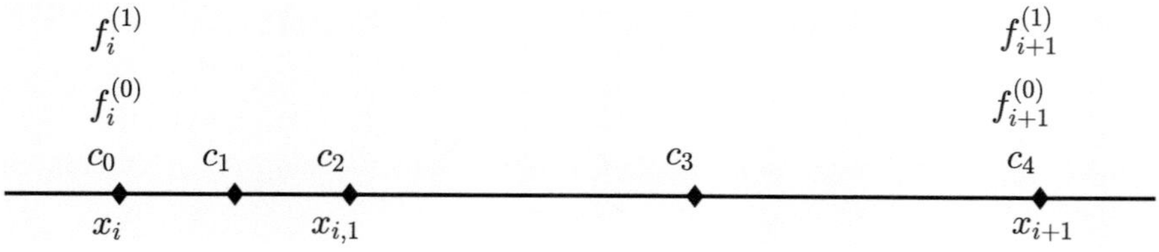

Fig. 1 B-coefficients of spline s_i

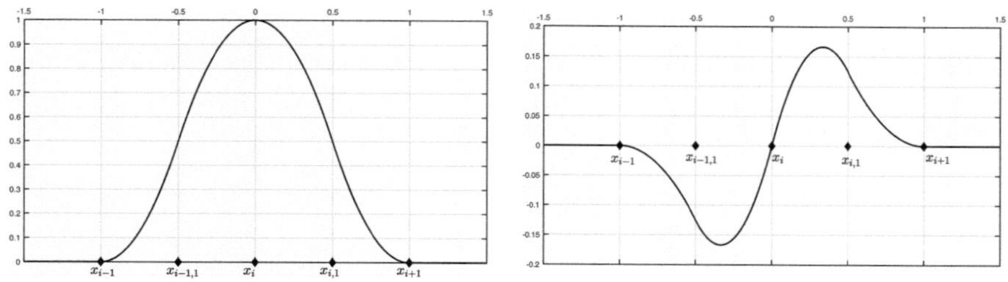

(a) ϕ_0. (b) ϕ_1.

Fig. 2 Quadratic composite-spline Hermite base

Now, considering the case where the interpolation points are in $\mathbb{Z}$ and $\lambda_i = \frac{1}{2},\ i \in \mathbb{Z}$. The Hermite basis of the $\mathcal{S}_2^1(\mathbb{Z}, \mathbb{R})$ space is derived through integer translation of the two functions ϕ_0 and ϕ_1 given by:

$$\phi_0(x) = \begin{cases} 2(x+1)^2, & \text{if } x \in [-1, -\frac{1}{2}] \\ 1-2x^2, & \text{if } x \in [-\frac{1}{2}, \frac{1}{2}] \\ 2(x-1)^2, & \text{if } x \in [\frac{1}{2}, 1]. \end{cases} \quad \text{and } \phi_1(x) = \begin{cases} -\frac{1}{2}(x+1)^2, & \text{if } x \in [-1, -\frac{1}{2}] \\ \frac{3x^2}{2} + x, & \text{if } x \in [-\frac{1}{2}, 0] \\ -\frac{3x^2}{2} + x, & \text{if } x \in [0, \frac{1}{2}] \\ \frac{1}{2}(x-1)^2, & \text{if } x \in [\frac{1}{2}, 1]. \end{cases}$$

The derivatives of these two functions are provided by

$$\phi_0'(x) = \begin{cases} 4(x+1), & \text{if } x \in [-1, -\frac{1}{2}] \\ -4x, & \text{if } x \in [-\frac{1}{2}, \frac{1}{2}] \\ 4(x-1), & \text{if } x \in [\frac{1}{2}, 1]. \end{cases} \quad \text{and } \phi_1'(x) = \begin{cases} -(x+1), & \text{if } x \in [-1, -\frac{1}{2}] \\ 3x+1, & \text{if } x \in [-\frac{1}{2}, 0] \\ 1-3x, & \text{if } x \in [0, \frac{1}{2}] \\ x-1, & \text{if } x \in [\frac{1}{2}, 1]. \end{cases}$$

Figure 2, shows the Hermite composite base elements ϕ_0 and ϕ_1.

3 Hermite Quadratic Composite-Spline Wavelets

In this section, we describe how, similarly to Jia et al. [17], the Hermite composite base elements ϕ_0 and ϕ_1 established in Sect. 2 can be used to develop the corresponding composite spline wavelets.

The following are the main properties of the elements ϕ_0 and ϕ_1:

$$\phi_0(0) = 1, \quad \phi_0'(0) = 0, \quad \phi_1(0) = 0, \quad \phi_1'(0) = 1,$$

Furthermore, as shown in Fig. 2, ϕ_0 exhibits symmetry while ϕ_1 exhibits antisymmetry. Additionally, the B-forms of ϕ_0 and ϕ_1 show that these functions satisfy the scaling equation:

$$\Phi = \sum_{\alpha \in \mathbb{Z}} A(\alpha)\Phi(2. - \alpha), \tag{5}$$

where $\Phi = (\phi_0, \phi_1)^T$, and

$$A(-1) = \frac{1}{8}\begin{pmatrix} 4 & 8 \\ -1 & -2 \end{pmatrix}, \quad A(0) = \frac{1}{2}\begin{pmatrix} 2 & 0 \\ 0 & 1 \end{pmatrix}, \quad A(1) = \frac{1}{8}\begin{pmatrix} 4 & -8 \\ 1 & -2 \end{pmatrix},$$

is the Hermite mask of order 1 reproducing $\mathbb{P}_2$, with $Supp(A) = \{0, \pm 1\}$ in the univariate case (see [14]).

Let S be the shift-invariant space produced by ϕ_0 and ϕ_1. A function v belongs to S only if there are two sequences defined over $\mathbb{Z}$, v_1 and v_2, such that:

$$v(.) = \sum_{i \in \mathbb{Z}} [v_1(i)\phi_0(. - i) + v_2(i)\phi_1(. - i)].$$

Let $L_2(\mathbb{R})$ denote the linear space consisting of all square-integrable real-valued functions on $\mathbb{R}$. We consider the inner product in $L_2(\mathbb{R})$ defined as

$$\langle u, v\rangle = \int_{\mathbb{R}} u(x)v(x)dx, \quad u, v \in L_2(\mathbb{R}).$$

Now, let $\tilde{S}$ denote the space defined by $\tilde{S} := \{v(2\cdot) : v \in S\}$. The refinement of Φ implies that S is a subset of $\tilde{S}$. Our goal is to identify two wavelets, ψ_0 and ψ_1, with one important condition: these wavelets have to satisfy the following property:

$$\langle \psi_0', \phi_m'(. - j)\rangle = \langle \psi_1', \phi_m'(. - j)\rangle = 0, \quad m = 0, 1, \quad \forall j \in \mathbb{Z}. \tag{6}$$

The shifts of these wavelets will generate a wavelets space that we denote by W, this space will be associated in a direct sum with S to produce $\tilde{S}$.

Express the wavelet function ψ as

$$\psi(x) = \sum_{k\in\mathbb{Z}} [v_1(k)\phi_0(2x-k) + v_2(k)\phi_1(2x-k)], \; x \in \mathbb{R}.$$

For any $k \in \mathbb{Z}$, we have $Supp(\phi_m(2x-k)) = [\frac{k-1}{2}, \frac{k+1}{2}]$, $m = 0, 1$.

Therefore, only the splines $\phi_m(2x-k)$, $k = 2j-2, \ldots, 2j+2$ intersect the support of $\phi_m(x-j)$.

Hence, we can express this as:

$$\begin{aligned}\langle\psi', \phi_0'(.-j)\rangle =& \frac{1}{6}\Big(-6v_1(2j-2) + 12v_1(2j) - 6v_1(2j+2) \\ & - v_2(2j-2) + 2v_2(2j-1) - 2v_2(2j+1) + v_2(2j+2)\Big),\end{aligned}$$

and

$$\begin{aligned}\langle\psi', \phi_1'(.-j)\rangle =& \frac{1}{24}\Big(6v_1(2j-2) - 12v_1(2j-1) + 12v_1(2j+1) - 6v_1(2j+2) \\ & + v_2(2j-2) - 4v_2(2j-1) + 6v_2(2j) - 4v_2(2j+1) + v_2(2j+2)\Big).\end{aligned}$$

For $z \in \mathbb{C}\backslash\{0\}$, let:

$$\begin{aligned} q_{11}(z) &= \sum_{j\in\mathbb{Z}} v_1(2j+1)z^{2j+1}, & q_{12}(z) &= \sum_{j\in\mathbb{Z}} v_1(2j)z^{2j} \\ q_{21}(z) &= \sum_{j\in\mathbb{Z}} v_2(2j+1)z^{2j+1}, & q_{22}(z) &= \sum_{j\in\mathbb{Z}} v_2(2j)z^{2j}. \end{aligned}$$

Subsequently, the condition $\langle\psi', \phi_m'(.-j)\rangle = 0$ holds for $m = 1, 2$ and all $j \in \mathbb{Z}$ if and only if

$$B(z)\Big(q_{11}(z), q_{12}(z), q_{21}(z), q_{22}(z)\Big)^T = 0, \quad \forall z \in \mathbb{C}\backslash\{0\},$$

where

$$B(z) = \begin{pmatrix} 0 & -6z^2 + 12 - 6z^{-2} & 2z - 2z^{-1} & -z^2 + z^{-2} \\ -12z^1 + 12z^{-1} & 6z^2 - 6z^{-2} & -4z - 4z^{-1} & z^2 + 6 + z^{-2} \end{pmatrix}$$

As a result, two independent solutions are obtained:

$$\begin{pmatrix} q_{11}(z) \\ q_{12}(z) \\ q_{21}(z) \\ q_{22}(z) \end{pmatrix} = \begin{pmatrix} z^{-1} - z \\ 0 \\ 6z^{-1} + 6z \\ 12 \end{pmatrix} \text{ and } \begin{pmatrix} q_{11}(z) \\ q_{12}(z) \\ q_{21}(z) \\ q_{22}(z) \end{pmatrix} = \begin{pmatrix} -z^{-1} - z \\ 2 \\ -6z^{-1} + 6z \\ 0 \end{pmatrix}.$$

We can deduce the two wavelets, ψ_0 and ψ_1, from these solutions, which are written as:

$$\psi_0(x) = -\phi_0(2x+1) + 2\phi_0(2x) - \phi_0(2x-1) - 6\phi_1(2x+1) + 6\phi_1(2x-1),$$
$$\psi_1(x) = \phi_0(2x+1) - \phi_0(2x-1) + 6\phi_1(2x+1) + 12\phi_1(2x) + 6\phi_1(2x-1).$$

More explicitly, these wavelets are given by

$$\psi_0(x) = \begin{cases} 4(1+x)^2 & \text{if } x \in [-1, -\frac{3}{4}] \\ -14 - 40x - 28x^2 & \text{if } x \in]-\frac{3}{4}, -\frac{1}{2}] \\ 8 + 48x + 60x^2 & \text{if } x \in]-\frac{1}{2}, -\frac{1}{4}] \\ 2 - 36x^2 & \text{if } x \in]-\frac{1}{4}, \frac{1}{4}] \\ 8 - 48x + 60x^2 & \text{if } x \in]\frac{1}{4}, \frac{1}{2}] \\ -14 + 40x - 28x^2 & \text{if } x \in]\frac{1}{2}, \frac{3}{4}] \\ 4(x-1)^2 & \text{if } x \in]\frac{3}{4}, 1] \end{cases}$$

$$\text{and} \quad \psi_1(x) = \begin{cases} -4(1+x)^2 & \text{if } x \in [-1, -\frac{3}{4}] \\ 14 + 40x + 28x^2 & \text{if } x \in]-\frac{3}{4}, -\frac{1}{2}] \\ -10 - 56x - 68x^2 & \text{if } x \in]-\frac{1}{2}, -\frac{1}{4}] \\ 4x(6+23x) & \text{if } x \in]-\frac{1}{4}, 0[\\ 4x(6-23x) & \text{if } x \in [0, \frac{1}{4}] \\ 10 - 56x + 68x^2 & \text{if } x \in]\frac{1}{4}, \frac{1}{2}] \\ -14 + 40x - 28x^2 & \text{if } x \in]\frac{1}{2}, \frac{3}{4}] \\ 4(-1+x)^2 & \text{if } x \in]\frac{3}{4}, 1] \end{cases}$$

with $\psi_0(-x) = \psi_0(x)$ and $\psi_1(-x) = -\psi_1(x)$.

Wavelets ψ_0 and ψ_1 verify equation (6) and, by construction, have the interval $[-1, 1]$ as their support, and their shifts generate the wavelet space W such that $\tilde{S} = S \oplus W$. Figure 3 illustrates that ψ_0 has symmetry across the support interval $[-1, 1]$, while ψ_1 has antisymmetry.

The shift invariance and linear independence of ψ_0' and ψ_1' shifts over the interval $[0, 1]$ ensure that they remain linearly independent on each interval $[k, k+1]$ for every $k \in \mathbb{Z}$. Because of the initial linear independence of the shifts of ψ_0' and ψ_1' over the interval $[0, 1]$, this linear independence remains on the interval $[k, k+1]$ for all $k \in \mathbb{Z}$.

Suppose that b_1 and b_2 are two square-summable sequences defined on $\mathbb{Z}$ and let

$$u(.) := \sum_{j \in \mathbb{Z}} [b_1(j)\psi_0'(. - j) + b_2(j)\psi_1'(. - j)].$$

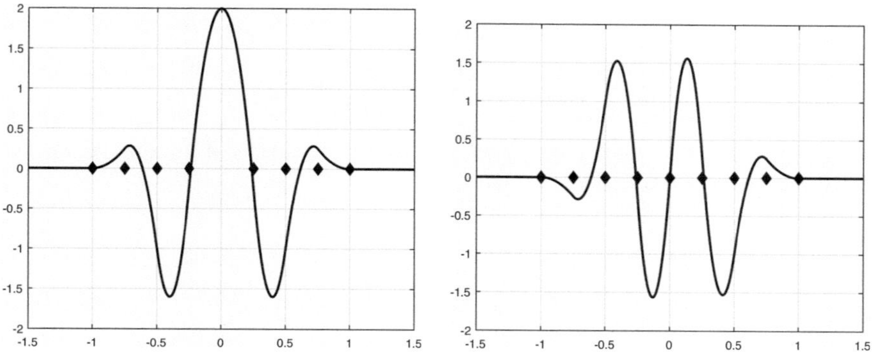

(a) ψ_0. (b) ψ_1.

Fig. 3 Wavelets ψ_0 and ψ_1

The derivatives of ψ_0 and ψ_1, vanishes outside of the interval $[k, k+1]$, and there shifts are linearly independent on the same interval. Which means there exists two positive constants C_1 and C_2 independent of k, b_1 and b_2 such that

$$C_1^2 \sum_{j=k}^{k+1} [|b_1(j)|^2 + |b_2(j)|^2] \leq \int_k^{k+1} |u(x)|^2 dx \leq C_2^2 \sum_{j=k}^{k+1} [|b_1(j)|^2 + |b_2(j)|^2].$$

As a result, for $k \in \mathbb{Z}$, we have

$$2C_1^2 \sum_{j=k}^{k+1} [|b_1(j)|^2 + |b_2(j)|^2] \leq \int_{\mathbb{R}} |u(x)|^2 dx \leq 2C_2^2 \sum_{j=k}^{k+1} [|b_1(j)|^2 + |b_2(j)|^2].$$

Hence the shifts ψ_0' and ψ_1' are stable.

4 Adapting Wavelets to the Interval [0, 1]

This section focuses on establishing a wavelet basis for the space $H_0^1(0, 1)$ using the Hermite quadratic composite-spline wavelets introduced in the preceding section. Consider, for $n \geq 1$, the space V_n consisting of quadratic splines s that belong to $C^1([0, 1])$, satisfying $s(0) = s(1) = 0$, and

$$s|_{\left(\frac{j}{2^{n+1}}, \frac{j+1}{2^{n+1}}\right)} \in \mathbb{P}_2 \quad \text{for} \quad j = 0, \ldots, 2^{n+1} - 1.$$

In what follows, we aim to construct a Riesz wavelet basis for $L_2(0, 1)$. Let

$$\Phi_n := \{\phi_0(2^n - j),\ j = 1, \ldots, 2^n - 1\} \cup \{\phi_1(2^n - j),\ \ j = 1, \ldots, 2^n\}, \qquad (7)$$

and

$$\Psi_n := \{\psi_0(2^n - j), j = 1, \ldots, 2^n - 1\} \cup \{\psi_1(2^n - j), j = 1, \ldots, 2^n\}. \quad (8)$$

Let V_n and W_n denote the linear spaces spanned by the entries of Φ_n and Ψ_n, respectively. Regarding these spaces, we have:

- $dim(V_n) = dim(W_n) = 2^{n+1}$,
- $V_1 \subset V_2 \subset \cdots \subset H_0^1(0, 1)$,
- $\bigcup_{n=1}^{\infty} V_n$ is dense in $H_0^1(0, 1)$,
- $H_0^1(0, 1) = V_1 \oplus \sum_{n=1}^{\infty} W_n$.

For $n = 1, 2, \ldots$ and $x \in [0, 1]$, the normalized wavelets such that $\left\|\psi'_{n,j}\right\|_{L_2(0,1)} = 1$ for all $j = 1, \ldots, 2^{n+1}$ are given by:

$$\psi_{n,j}(x) := \frac{2^{-\frac{n}{2}}}{\sqrt{117.3}}\psi_0(2^n x - \frac{j}{2}) \text{ for } j = 2, 4, \ldots, 2^{n+1} - 2,$$

$$\psi_{n,j}(x) := \frac{2^{-\frac{n}{2}}}{\sqrt{170.6}}\psi_1(2^n x - \frac{j-1}{2}) \text{ for } j = 3, 4, \ldots, 2^{n+1} - 1,$$

and

$$\psi_{n,1}(x) := \frac{2^{-\frac{n}{2}}}{\sqrt{76.8}}\psi_1(2^n x), \ \psi_{n,2^{n+1}}(x) := \frac{2^{-\frac{n}{2}}}{\sqrt{76.8}}\psi_1(2^n x - 2^n).$$

The quadratic composite-spline basis is also given by:

$$\phi_{1,1}(x) := \sqrt{\frac{3}{16}}\phi_0(2x - 1), \qquad \phi_{1,2}(x) := \sqrt{3}\phi_0(2x)$$

$$\phi_{1,3}(x) := \sqrt{\frac{3}{2}}\phi_1(2x - 1), \qquad \phi_{1,4}(x) := \sqrt{3}\phi_1(2x - 2).$$

We recall here a theorem, provided and thoroughly proven by [17], which is easily adaptable to our situation.

Theorem 4.1 The sequence $(\psi'_{n,j})_{n\in\mathbb{N}\setminus\{0\},1\le j\le 2^{n+1}}$ is a Riesz sequence in $L_2(0, 1)$. In other words, there exist two positive constants A and B such that

$$A\left(\sum_{n=1}^{\infty}\sum_{j=1}^{2^{n+1}}|b_{n,j}|^2\right)^{\frac{1}{2}} \le \left\|\sum_{n=1}^{\infty}\sum_{j=1}^{2^{n+1}} b_{n,j}\psi'_{n,j}\right\|_{L_2(0,1)} \le B\left(\sum_{n=1}^{\infty}\sum_{j=1}^{2^{n+1}}|b_{n,j}|^2\right)^{\frac{1}{2}}$$

for every sequence $(b_{n,j})_{n\in\mathbb{N}\setminus\{0\},1\le j\le 2^{n+1}}$. □

Each $\phi_{1,j}$ and $\psi_{n,j}$ is subject to normalization, specifically ensuring that $\|\psi'_{n,j}\|_{L_2(0,1)} = 1$, for $j = 1, \ldots, 2^{n+1}$ and $\|\phi'_{1,j}\|_{L_2(0,1)} = 1$, for $j = 1, \ldots, 4$. The functions $\phi_{1,j}$, $j = 1, 2, 3, 4$ and $\psi_{n,j}$, $n \in \mathbb{N}\backslash\{0\}$, $1 \leq j \leq 2^{n+1}$, collectively span the space $H_0^1(0, 1)$.

In the continuation of this work, and with the aim of enhancing programming convenience and applicative manipulability of the latest $H_0^1(0, 1)$ base, we adopt the following notation:

$$\Gamma_n := \{g_1, \ldots, g_{2^{n+1}}\}, \tag{9}$$

where $g_j = \phi_{1,j}$, $j \in \{1, \ldots, 4, \}$ and $g_{2^{n+1}+j} = \psi_{n,j}$, for $n \in \mathbb{N}^*$ and $j \in 1, \ldots, 2^{n+1}$.

The sequence $(g'_r)_{r\in\mathbb{N}^*}$ also satisfies Theorem 4.1 and can serve as a basis for V_n, $n = 1, 2, \ldots$.

5 Numerical Solution of Sturm-Liouville-Type Differential Equation

The wavelets developed in the previous section are used to solve differential equations in this section. We focus on the Sturm-Liouville equation with the Dirichlet boundary condition. The numerical testing examples are the same as those presented by Jia et al. in [17]. The Galerkin method is the main asset we employ. In this regard, we shall recall certain results and adapt them to our situation. Let us consider the following Sturm-Liouville equation:

$$-\frac{d}{dx}\left(p(x)\frac{du(x)}{dx}\right) + q(x)u(x) = f(x), \ \forall x \in [0, 1], \tag{10}$$

with the Dirichlet Boundary condition $u(0) = u(1) = 0$. The functions p and y are continuous and satisfy the following conditions:

$$p(x) > 0 \quad , q(x) \geq 0, \quad \forall x \in [0, 1].$$

Let:

$$a(u, v) = \int_0^1 p(x)u'(x)v'(x)dx + \int_0^1 q(x)u(x)v(x)dx, u, v \in H_0^1(0, 1).$$

Then the variational form of the above equation with the Dirichlet boundary condition is:

$$a(u_n, \vartheta) = \langle f, \vartheta \rangle, \forall \vartheta \in H_0^1(0, 1). \tag{11}$$

We construct the stiffness matrixes by discretizing the Galerkin approximation problem according to each example for two differential equations of type Sturm-Liouville with the Dirichlet boundary conditions. These examples have the same solution. However, it will be shown that the numerical solution is remarkably different for an empathetic scale. These examples are used in Jia et al. [17] work on cubic spline wavelets, so comparing their results with those given by our method is interesting.

In [4], Bramble et al. proposed the BPX method for preconditioning, which uses a nodal basis composed of piecewise linear functions. However, these functions only provide an approximation of order 2. To improve convergence and achieve orders 3 and 4, it is possible to extend the BPX method by using quadratic and cubic Hermite splines.

The numerical results for the eigenvalues $\lambda_{\max}$ and $\lambda_{\min}$, as well as the conditioning number of the matrix (see the tables below), show that the conditioning number induced by our wavelet basis is lower than that obtained by the BPX method.

In [17], Jia et al. showed that the matrix remains sparse and that its conditioning number is uniformly bounded (independent of n. The numerical calculation confirms that this number is lower than that obtained by the BPX method. Thus, in comparison, the conditioning number of our wavelet basis is better than that obtained by both methods.

5.1 Example 1

The following example is a Dirichlet problem given by:

$$\begin{cases} -u'' = f, & \text{on } [0,1] \\ u(0) = u(1) = 0 \end{cases}$$

where $f(x) = (53.7\pi)^2 \sin(53.7\pi x) + (2.3\pi)^2 \sin(2.3\pi x),\ x \in [0,1]$.
We provide the exact solution of this problem as

$$u(x) = \sin(53.7\pi x) + \sin(2.3\pi x),\ x \in [0,1] \tag{12}$$

The numerical solution of this problem is:

$$u_n = \sum_{k=1}^{2^{n+1}} \mu_k g_k. \tag{13}$$

By using the method suggested in this article, we recall that $g_k,\ k \in \{1, \ldots, 2^{n+1}\}$ denotes an element of the wavelet basis $G_n := \{g_1, \cdots, g_{2^{n+1}}\}$.

We discretize the Galerkin approximation problem as

$$\sum_{k=1}^{2^{n+1}} \langle g_j', g_k' \rangle \mu_k = \langle g_j, f \rangle,\ j = 1, \ldots, 2^{n+1}. \tag{14}$$

$A_n := (\langle g_j', g_k' \rangle)_{1 \le j,k \le 2^{n+1}}$ define the stiffness matrix.

Table 1 Condition number of the stiffness matrix A_n in our method

n	6	7	8	9	10
λ_{max}	1.3535	1.3535	1.3535	1.3535	1.3535
λ_{min}	0.6464	0.6464	0.6464	0.6464	0.6464
κ	2.0938	2.0938	2.0938	2.0938	2.0938

It is a block diagonal, where each block is a banded matrix. We denote by $\lambda_{\max}$ and $\lambda_{\min}$ as the maximal and the minimal eigenvalue of the matrix A_n. Using the Theorem 2 ([17]), Condition number $\kappa = \frac{\lambda_{\max}}{\lambda_{\min}}$ is uniformly bounded (Table 1).

The following tables present the conditioning number of the stiffness matrix for $n = 6, \ldots, 10$, in comparison with the results given in [4, 17].

We define approximation the error depending of each $n = 5, \ldots, 10$ by $error_n = \|u_n - u\|$, while u is the exact solution given at (12) and u_n is the numerical solution giving by (13).

The following table give the computed values of the approximation error and the rate of convergence (Table 4).

5.2 *Example 2*

In this example we explore another aspect of the use of the wavelet basis Γ_n. We propose the same example given by Jia et al. [17].

Let us consider the Dirichlet problem given by:

$$\begin{cases} -u'' + u = f, & \text{on } [0, 1], \\ u(0) = u(1) = 0 \end{cases}$$

where

$$f(x) = ((53.7\pi)^2 + 1)\sin(53.7\pi x) + ((2.3\pi)^2 + 1)\sin(2.3\pi x), \; x \in [0, 1].$$

The exact solution of this example is the same one as the previous example (12).

The Galerkin approximation problem (11) is now discretized as

$$\sum_{k=1}^{2^{n+1}} (\langle g_j', g_k' \rangle + \langle g_j, g_k \rangle)\mu_k = \langle g_j, f \rangle, \; j = 1, \ldots, 2^{n+1}. \tag{15}$$

In this case the stiffness matrix is defined by $A_n := (\langle g_j', g_k' \rangle + \langle g_j, g_k \rangle)_{1 \leq j, k \leq 2^{n+1}}$. As we have already found in the previous example, A_n is a block diagonal with each block as a banded matrix, and Condition number is uniformly bounded .

Table 2 BPX preconditioning for Hermite cubic splines

n	6	7	8	9	10
λ_{max}	3.562	3.632	3.682	3.718	3.743
λ_{min}	0.7693	0.7696	0.7696	0.7696	0.7696
κ	4.630	4.719	4.784	4.831	4.864

Table 3 Condition number of the matrix A_n of the base in [17]

n	6	7	8	9	10
λ_{max}	1.5780	1.5787	1.5789	1.5789	1.5789
λ_{min}	0.4220	0.4213	0.4211	0.4211	0.4211
κ	3.7396	3.7474	3.7494	3.7498	3.7498

Table 4 The error $error_n$ and the convergence rate

n	6	7	8	9	10
$error_n$	0.0276	0.0025	$2.9242\ 10^{-04}$	$3.5957\ 10^{-05}$	$4.4735\ 10^{-05}$
$\log_2(e_{n-1}/e_n)$	2.9178	2.4015	2.1459	2.0959	2.0959

Table 5 Condition number of the stiffness matrix A_n in our method

n	6	7	8	9	10
λ_{max}	1.3841	1.3841	1.3841	1.3841	1.3841
λ_{min}	0.7033	0.7033	0.7033	0.7033	0.7033
κ	1.9678	1.9678	1.9678	1.9678	1.9678

The following table displays Condition number of the stiffness matrix A_n for $n = 6, \ldots, 10$, in comparison with the results from [17] (see Table 3) and the results from [4] (see Table 2):

The approximation error is given on the next table with its convergence rate (Tables 5 and 6).

Table 6 The error $error_n$ and the convergence rate

n	6	7	8	9	10
$error_n$	0.02762	0.0025	$2.9242\ 10^{-04}$	$3.5957\ 10^{-05}$	$4.4735\ 10^{-05}$
$\log_2(e_{n-1}/e_n)$	2.9173	2.4098	2.1482	2.0959	2.0842

References

1. Amaratunga, K., Williams, J.R., Qian, S., Weiss, J.: Wavelet-Galerkin solutions for one-dimensional partial differential equations. Int. J. Numer. Methods Eng. **37**, 2703–2716 (1994)
2. Rioul, O., Vetterli, M.: Wavelets and signal processing. IEEE Signal Process. Mag. **8**(4), 14–38 (1991). https://doi.org/10.1109/79.91217
3. Beylkin, G., Coifman, R., Rokhlin, V.: Fast wavelet transforms and numerical algorithms I. Commun. Pure Appl. Math. **44**, 141–183 (1991)
4. Bramble, J.H., Pasciak, J.E., Xu, J.C.: Parallel multilevel preconditioners. Math. Comput. **55**, 1–22 (1990)
5. Cline, A.K.: Scalar- and planar-valued curve fitting using splines under tension. Commun. ACM **17**(4), 218–220 (1974)
6. Cohen, A., Dahmen, W., DeVore, R.: Adaptive wavelet methods for elliptic operator equations: convergence rates. Math. Comput. **70**(233), 27–75 (2000)
7. Conti, C., Romani, L., Unser, M.: Ellipse-preserving Hermite interpolation and subdivision. J. Math. Anal. Appl. **426**, 211–227 (2015)
8. Cotronei, M., Moosmüller, C.: Hermite B-splines: n-refinability and mask factorization. Mathematics **9** (2021)
9. Cotronei, M., Moosmüller, C., Sauer, T., Sissouno, N.: Level-dependent interpolatory Hermite subdivision schemes and wavelets. Constr. Approx. **50**, 341–366 (2019)
10. Dahmen, W., Proessdorf, S., Schneider, R.: Wavelet approximation methods for pseudodifferential equations II: Matrix compression and fast solutions. Adv. Comput. Math. **1**, 259–335 (1993)
11. Dubuc, S., Merrien, J.-L.: Convergent vector and Hermite subdivision schemes. Constr. Approx. **23**, 1–22 (2006)
12. Dubuc, S., Merrien, J.-L.: Hermite subdivision schemes and Taylor polynomials. Constr. Approx. **29**, 219–245 (2009)
13. Fritsch, F.N., Carlson, R.E.: Monotone piecewise cubic interpolation. SIAM J. Numer. Anal. **17**(2), 238–246 (1980)
14. Han, B., Yu, T.P.-Y., Piper, B.: Multivariate refinable Hermite interpolant. In: Math. Comput. **73**, 1913–1935 (2004)
15. Han, B., Michelle, M.: Derivative-orthogonal Riesz wavelets in Sobolev spaces with applications to differential equations. Appl. Comp. Harmon. Anal. **47**(3), 759–794 (2019)
16. Jia, R.-Q., Riemenschneider, S. D., Zhou, D.-X.: Vector subdivision schemes and multiple wavelets. Math. Comput. **67**(224), 1533–1563 (1998)
17. Jia, R.Q., Liu, S.T.: Wavelet bases of Hermite cubic splines on the interval. Adv. Comput. Math. **25**, 23–39 (2006)
18. Lamnii, A., Lamnii, M., Oumellal, F.: Computation of Hermite interpolation in terms of B-spline basis using polar forms. Math. Comput. **134**, 17–27 (2017)
19. Lamnii, A., Lamnii, M., Oumellal, F.: Two-point osculatory basis and splines. Appl. Comput. **283**, 355–368 (2016)
20. Manni, C.: On shape preserving C^2 Hermite interpolation. BIT Numer. Math. **41**(1), 127–148 (2001)
21. Merrien, J.-L.: A family of Hermite interpolants by bisection algorithms. Numer. Algorithms **2**, 187–200 (1992)
22. Schumaker, L.L.: On shape preserving quadratic spline interpolation. SIAM J. Numer. Anal. **20**(4), 854–864 (1983)
23. Titchmarsh, E.C.: Linear Differential Operators. Oxford University Press (1986)
24. Vasilyev, O.V., Kevlahan, N.K.-R.: An adaptive multilevel wavelet collocation method for elliptic problems. J. Comput. Phys. **206**, 412–431 (2005)

A Deep Network-Based Spline Active Contour Method for Medical Image Segmentation

Rania Sefti, Driss Sbibih, and Rachid Jennane

Abstract Medical image processing is one of the most challenging tasks in computer vision. Many obstacles are encountered when performing automatic lesion segmentation in medical images, such as: shadows, artifacts, image quality, etc. This work aims to automatically segment a variety of medical images, overcoming all such drawbacks. To achieve this goal, we propose a combination of two convolutional neural networks and a deformable model using the spline approximation functions. More specifically, we introduce an automatic parametric active contour method based on an energy functional that combines the auto-encoder model for texture feature analysis and the U-net model for edge feature localization and extraction. The robustness of the proposed deep snake model is evaluated using three different datasets of brain MRI images, breast ultrasound images, and Foot Ulcer Segmentation Challenge 2021. Obtained results confirm that our proposed model improves segmentation performance and is robust to medical image segmentation challenges.

1 Introduction

In recent years, there has been increasing interest in understanding and extracting information from medical images. This helps doctors to improve diagnostic efficiency, reduce misdiagnosis, facilitate organ recognition, etc. Automatic image segmentation, or the extraction of the contours of an object, is a recent technique in digital image processing with various applications; it consists in dividing the image into homogeneous zones based on certain characteristics, such as intensity, texture,

R. Sefti (✉) · D. Sbibih
LANO Laboratory, Faculty of Sciences, University Mohammed First, Oujda, Morocco
e-mail: sefti.rania@ump.ac.ma

D. Sbibih
e-mail: d.sbibih@ump.ac.ma

R. Jennane
IDP Laboratory, UMR CNRS 7013, University of Orleans, Orleans, France
e-mail: rachid.jennane@univ-orleans.fr

© The Author(s), under exclusive license to Springer Nature Switzerland AG 2025
M. J. Ibáñez-Pérez et al. (eds.), *Mathematical And Computational Modelling, Approximation and Simulation*, SEMA SIMAI Springer Series 41,
https://doi.org/10.1007/978-3-032-01114-5_13

or shape. Unfortunately, the quality of medical images can be poor (in-homogeneity of intensity) and they can also have a great variability in the shapes of objects, which often leads to significant difficulties compared to natural images. The segmentation of this type of images is one of the most challenging tasks in computer vision and it has attracted the attention of most researchers from different fields.

Different segmentation methods have been developed in the literature, a classical "contour" approach is implemented by searching for regions of strong gradient, but the obtained contours are often discontinuous. Another approach called Active Contours (AC) or "snake", first introduced by Kass et al. [1], are part of the high-level techniques often used for fitting elastic curves defined in an image domain. This approach has proven to be an efficient method for line and edge detection, where its main advantage is that it automatically provides a closed contour of an object. The snake method has inspired many researchers to develop numerous improvements. These extensions generally differ in the type of representation and the choice of energy expression. For example, in [2] and [3], the authors introduced new balloon and Gradient Vector Flow (GVF) forces in the propagation of the active contour. In [7] and [8], a new image energy term was introduced that uses statistical information to distinguish between different homogeneous regions. To make fitting the optimal snake curve less costly, in [10] Menet et al. proposed the B-spline snake model, or B-snake. This model is based on the polygonal snake model and develops it into a contour structure by approximating curves with fewer control points using B-spline functions, resulting in fast contour generation and reduced computational complexity.

By definition, the edges of an image are the regions of strong gradient where the intensity changes abruptly. However, there are some cases where object differentiation is based on changes in texture or shape, even though the gradient may not show significant fluctuations. Because the B-snake model is an energy based approach using geometric and physical principles, the segmentation process is highly dependent on image content. This limitation makes the model sensitive to abrupt changes in image intensity. As a result, the active contours may not converge on the edges of the object of interest when the image quality is poor, which is very common in medical images.

On the other hand, Convolutional Neural Networks (CNNs), have been successfully applied to many image processing tasks and have reached the State-Of-The-Art (SOTA) in many fields, including medical image segmentation with exceptional performance in diverse applications, including breast ultrasound lesion detection [21], chest CT imaging [5], brain tumor identification [20], and foot ulcer analysis [22], among others. Various models were developed by researchers: they have proposed an encoder-decoder structure that captures input semantic information useful for predicting output, such as Fully Convolution Network (FCN) [11], U-Net [16], and Deeplab [12]. Auto-encoders [17] are a special case of encoder-decoder models where input and output are identical. Ronneberger et al. [16] have proposed an efficient segmentation model for biological microscopy images, namely U-Net. The main advantage of this model is that it allows precise localization by using a symmetric expansion path. The great success of this model has inspired several researchers to introduce similar architectures to achieve some improvements, such as U-Net++ [5],

V-Net [14], Res-UNet [15] and many others. In [4], a semi-automatic segmentation method is presented that combines the AC model with the energy term from the U-net. However, the model is poorly formulated and fails to capture precise boundaries in medical images.

Although U-Net models have shown significant achievements in analysis-based object detection and image feature localization for medical image segmentation, object boundary detection and delineation remains a challenging task even for the most accurate CNN architectures. It is known that a deeper network design can improve the effectiveness of the network and cover a larger receptor field. However, such deeper networks require a significant number of parameters, resulting in increased computational complexity.

This work aims to introduce a new automatic active contour segmentation model based on a deep learning model that focuses on three characteristics of object edges in an image, namely strong gradient, texture and shape. The goal is to obtain improved and robust segmentation results for different types of medical images. In this work, several contributions are proposed to address the previously discussed issues:

- A new constraint energy functional is introduced that forces the snake not to drift toward out-of-contour objects and allows fast convergence to the optimal position.
- A new external energy functional is presented by combining two deep neural networks, namely auto-encoder and U-net models.
- Using the energy expression and incorporating various spline-based theoretical developments, a new expression for the force that attracts the surface snake towards the object boundary is introduced (Eqs. (11) and (12)).

The remainder of this paper is organized as follows: In Sect. 2, we briefly discuss the background of spline theory and the active contour model. In Sect. 3, we describe the general framework of the proposed method. In Sect. 4, we present the experiments and discuss the obtained results. Section 5 concludes the paper.

2 Initiations and Background

B-spline snake models are a representation of the contour structure of a snake using B-spline functions. These functions approximate deformable curves using a small set of control points and spline basis functions. The advantage of this approach is that it can be easily modelled and preserves the regularity of the curve.

Given an integer $k \geqslant 0$, let $t = \{t_i\}_{i=0}^{M+k+1}$ be an increasing sequence of $M + k + 2$ real parameters in the interval [0, 1] such that:

$$0 = t_0 = t_1 = \cdots = t_k < t_{k+1} < \cdots < t_{M+1} = t_{M+2} = \cdots = t_{M+k+1} = 1.$$

The B-spline basis function $B_{i,k}$ of degree k, associated with the knot vector t, is defined recursively by the Cox-de Boor formula [9] for all $s \in [0, 1]$ as follows: For $k = 0$,

$$B_{i,0}(s) = \begin{cases} 1, & \text{if } t_i \leqslant s < t_{i+1} \\ 0 & \text{otherwise,} \end{cases}$$

for all $k \geqslant 1$

$$B_{i,k} = \delta_{i,k}(s) B_{i,k-1} + (1 - \delta_{i+1,k}(s)) B_{i+1,k-1}, \tag{1}$$

where the coefficients $\delta_{i,k}$ are given by

$$\delta_{i,k}(s) = \begin{cases} \dfrac{s - t_i}{t_{i+k} - t_i}, & \text{if } t_i \neq t_{i+k}, \\ 0, & \text{otherwise.} \end{cases}$$

Spline functions are popular for modeling curves and surfaces in various fields. This prevalence can be attributed to their use of piecewise polynomial representation and the advantageous geometric properties they offer. These properties include smoothness, positivity, derivability, and so on. Moreover, another interesting property of the B-spline $B_{i,k}$ is that it has an optimal support given by $[t_i, t_{i+k+1}]$ for all $i = 0, \ldots, M$.

Definition 1 The cubic B-spline curve is composed of smoothly connected curve segments, where each curve segment is represented by a polynomial of degree three. This polynomial is defined as a linear combination of the B-spline functions $\{B_{i,3}(s),\ i = 0, \ldots, M\}$ and uses the control points $\{P_i = (X_i, Y_i),\ i = 0, \ldots, M\}$ as coefficients:

$$\begin{aligned} r(s) &= \sum_{i=0}^{M} P_i B_{i,3}(s) \qquad\qquad s \in [0, 1]. \\ &= (\sum_{i=0}^{M} X_i B_{i,3}(s), \sum_{i=0}^{M} Y_i B_{i,3}(s)) \end{aligned} \tag{2}$$

Since the B-spline basis functions join smoothly, the spline curve segments maintain continuous connections. As a result, even if the control point positions are changed, the continuity of the curve is not affected.

The active contour or "snake" model is a computer vision technique used for image segmentation and object detection. They have a dynamic behavior that evolves from an initial position in the image toward object boundaries through an iterative energy minimization procedure.

Proposition 1 The B-snake model is represented parametrically in the interval [0, 1] by the B-spline curve $r(s) = (x(s), y(s))$. □

Remark 1 To describe the snake by a closed curve, we assume that the set of control points is M-periodic:

$$P_{i+M} = P_i, \quad \forall i \in 0, \ldots, M.$$

Definition 2 The total energy functional of the B-Snake, denoted by $E_{B-snake}$ is defined as

$$E_{B-snake}(r) = E_{int}(r) + E_{ext}(r) + E_{const}(r)$$

where the internal energy E_{int} controls the appearance of the curve, the external energy E_{ext} pulls the curve towards the edges of the object, and the constraint energy E_{const} deals with additional influences that can be chosen based on the needs of the specific problem.

□

Remark 2 1. The choice of the initial position and the total energy term is very critical because it determines the quality of the segmentation.
2. In the case where the curve is sufficiently regular, which is our case because $r(s)$ is of class C^2, the internal energy term E_{int} can be eliminated, as the authors in [7] have demonstrated experimentally.

□

3 Proposed Approach

The snake model evolves in three primary phases: initialization, deformation, and eventual balancing. This section outlines the general framework of the proposed method by discussing each of these phases. We establish the total energy functional for the B-spline snake based on the deep texture network. We then detail the minimization procedure that guides the deformation of the snake to its equilibrium position.

3.1 Preprocessing Step

The preprocessing step plays a critical role in determining the quality of the segmentation results and ensuring the convergence of the snake model. It consists of an initialization step and the expression of the energy functional.

3.1.1 Constraint Energy

The constraint energy is defined by the user depending on the specifics of the problem and guides the snake to the convergence when the image information is too ambiguous. We introduce an energy term that represents the inner area of the snake. This term prevents the snake from deforming toward external or undesired regions, such as noise or artifacts near the curve. It ensures that the curve remains focused on converging to the target boundary without distractions. Furthermore, the term accelerates convergence, enabling faster and more efficient alignment with the desired boundary for optimal curve placement.

We express the constraint energy for $r(s) = (x(s), y(s))$ by:

$$E_{const} = \frac{1}{2}(x(1)y(1) - x(0)y(0)) - \sum_{i,j=0}^{M} X_i Y_j \int_0^1 B'_{i,3}(s) B_{j,3}(s) ds, \quad (3)$$

where, using the support property of the B-spline, we have:

$$(x(0), y(0)) = \left(\sum_{i=0}^{2} X_i B_{i,3}(t_3), \sum_{i=0}^{2} Y_i B_{i,3}(t_3) \right) \quad (4)$$

$$(x(1), y(1)) = \left(\sum_{i=M-2}^{M} X_i B_{i,3}(t_{M+1}), \sum_{i=M-2}^{M} Y_i B_{i,3}(t_{M+1}) \right). \quad (5)$$

Note that it is crucial to recognize the importance of the sign in the above expression, as it indicates the direction of the curve's deformation. If the initial curve is inside the object, simply reversing the sign of the energy will suffice. In fact, the constraint energy E_{const} is designed to minimize the snake's inner area. A positive value encourages the snake to contract toward the object's boundary. Conversely, if the snake is initialized inside the object, the goal is for it to expand outward to reach the boundary. In this case, a negative value is used to drive the snake to enlarge (maximize the area) and move outward until it aligns with the object's contour.

3.1.2 External Energy Functional Based on AE-UNet Learning Model

In this section, our goal is to develop an efficient external energy functional for deforming the snake to object edges that recovers all spatial edge information and neglects image noise by combining two CNN models: Auto-Encoder and Unet, as they provide high quality map features. In addition, Gabor filters are used to highlight textures in the image, and unsupervised discriminant analysis is used to maximize the discrimination of the object from the background. All these techniques are used to make the external energy robust against all the limitations of medical images.

Auto-encoders (AE) are unsupervised learning models that compress input data into a low-dimensional code (encoding) and then reconstruct the input (decoding) [17]. AE is particularly useful when the objects of interest are difficult to detect due to shape variability or poor visibility in the image, making their extraction challenging. This model can eliminate annoying noise and analyze the various features in the image, which can help the snake model achieve convergence. After training the unsupervised auto-encoder model, we extract its learned encoder weights and insert them into the U-Net model to produce a higher quality feature map.

In both the AE and Unet models, we use Gabor texture filters. In image processing, Gabor filters are often used to analyse textures and extract features. They are linear filters that act as orientation-sensitive edge detectors. They are particularly efficient at capturing spatial frequency features at different scales and orientations, including edges. The ability of the models to recognise and represent complicated texture details can be enhanced by adding Gabor filters, which is essential for applications involving complex image features or unclear boundaries.

We then retrieve 64 pretrained deep feature maps from the decoder part before the final 1×1 convolutional layer.

Let $\{f_k\}_{k=2}^{K}$ be the collection of the extracted feature maps of a pre-trained AE-UNet method, we consider the external energy in [4] as

$$E_{ext} = \sum_{k=1}^{K} \omega_k E_k \tag{6}$$

where E_k is an energy term based on $\{f_k\}_{k=3}^{K}$ such that

$$E_k = -\int_0^1 \nabla f_k(r(s))ds, \tag{7}$$

with $f_1 = I(x, y)$ the image intensity at pixel position (x, y), ω_k the weight that distinguishes the object from the background in the pretrained feature maps f_k.

The weights $\omega_k,\ k = 1\ldots, K$ are determined using Fisher Linear Discriminant Analysis (LDA) [18], which is a supervised method for dimensionality reduction and classification. They are determined once, after the initialization of the snake. During the deformation process of the snake, ω_k remains unchanged.

3.1.3 Initialization

The snake model is highly sensitive to the initial contour, which has a great influence on the convergence to the boundaries of the object. The initial position of the snake is generally determined by the user based on a prior knowledge of the analyzed image. Often, the initial position can also be drawn manually for more convenience. To build an automatic segmentation model, we use the predicted segmentation result from the already trained AE-UNet model as the initial position of the snake. In this way, the initial parametric contour can be easily obtained by selecting $M + 1$ control points from the predicted contour, and then constructing the spline curve r by using (2).

3.2 Snake Deformation

Once the evolution of the snake has initiated, we can move on to the next step: minimizing the total energy, which aims to determine the optimal position on the snake's curve.

Theorem 1 The deformation of the B-snake curve to an optimal position of total energy $E_{B-snake}$ is determined by solving the following system:

$$\begin{cases} \mathbf{A}Y + \mathbf{F_X} = 0 \\ \mathbf{A}^T X + \mathbf{F_Y} = 0, \end{cases} \tag{8}$$

where $\mathbf{F_X}$ and $\mathbf{F_Y}$ are the external force vectors deforming the snake, $X = (X_0, \ldots, X_M)^T$, $Y = (Y_0, \ldots, Y_M)^T$ represent the coordinates of the control points, and $\mathbf{A}$ is a matrix of size $(M+1) \times (M+1)$.

□

Proof The process of finding the B-snake curve that minimizes the total energy, $\nabla E_{B-snake} = 0$, is transformed into the search for optimal control points, denoted as $P_i = (X_i, Y_i)$ for $i = 0, \ldots, M$, satisfying the following conditions:

$$\frac{\partial E_{B-snake}}{\partial X_l} = \frac{\partial E_{ext}}{\partial X_l} + \frac{\partial E_{const}}{\partial X_l} = 0, \tag{9}$$

$$\frac{\partial E_{B-snake}}{\partial Y_l} = \frac{\partial E_{ext}}{\partial Y_l} + \frac{\partial E_{const}}{\partial Y_l} = 0. \tag{10}$$

For all $l = 0, \ldots, M$, we have :

$$\begin{aligned} \mathcal{F}_l^X = \frac{\partial E_{ext}}{\partial X_l} &= \sum_{k=1}^{K} \omega_k \frac{\partial E_k}{\partial X_l} \\ &= -\sum_{k=1}^{K} \omega_k \int_0^1 \frac{\partial\left(\nabla f_k(r(s))\right)}{\partial x(s)} \frac{\partial x(s)}{\partial X_l} ds. \end{aligned}$$

As $\dfrac{\partial x(s)}{\partial X_l} = \dfrac{\partial}{\partial X_l}\left(\sum_{i=0}^{M} X_i B_{i,3}(s)\right) = B_{l,3}(s)$, we get:

$$\mathcal{F}_l^X = \int_0^1 F_X(s) B_{l,3}(s) ds, \tag{11}$$

where $F_X(s) = -\sum_{k=0}^{K} \omega_k \dfrac{\partial\left(\nabla f_k(r(s))\right)}{\partial x(s)}$. In the same way, we have :

$$\mathcal{F}_l^Y = \frac{\partial E_{ext}}{\partial Y_l} = \int_0^1 F_Y(s) B_{l,3}(s) ds, \tag{12}$$

where $F_Y(s) = -\sum_{k=0}^{K} \omega_k \frac{\partial (\nabla f_k(r(s)))}{\partial y(s)}$.

The external force function, denoted as $F(s) = (F_X(s), F_Y(s))$, plays a role in deforming the contour by moving the snake toward the object's boundaries.

On the other hand, while the curve is attracted by the external force field, i.e. when $\mathcal{F}_l = (\mathcal{F}_l^X, \mathcal{F}_l^Y) \neq (0, 0)$, the inner snake area is also iteratively minimized, and suppose this process is stopped at the equilibrium position. For all $l = 0, \ldots, M$, by using equation (3) we get:

$$\frac{\partial E_{const}}{\partial X_l} = \begin{cases} -\frac{1}{2} B_{l,3}(t_3) y(0) - \sum_{j=0}^{M} Y_j \mathcal{B}_{l,j} & \forall l = 0, 1, 2; \\ -\sum_{j=0}^{M} Y_j \mathcal{B}_{l,j} & \forall l = 3, \ldots, M-3; \\ \frac{1}{2} B_{l,3}(t_{M+1}) y(1) - \sum_{j=0}^{M} Y_j \mathcal{B}_{l,j} & \forall l = M-2, M-1, M, \end{cases} \tag{13}$$

where $\mathcal{B}_{i,j} = \int_0^1 B'_{i,3}(s) B_{j,3}(s) ds$, and $(x(0), y(0))$ and $(x(1), y(1))$ are given in Eqs. (4) and (5).

Similarly, we get:

$$\frac{\partial E_{const}}{\partial Y_l} = \begin{cases} -\frac{1}{2} B_{l,3}(t_3) x(0) - \sum_{j=0}^{M} X_j \mathcal{B}_{j,l} & \forall l = 0, 1, 2; \\ -\sum_{j=0}^{M} X_j \mathcal{B}_{j,l} & \forall l = 3, \ldots, M-3; \\ \frac{1}{2} B_{l,3}(t_{M+1}) x(1) - \sum_{j=0}^{M} X_j \mathcal{B}_{j,l} & \forall l = M-2, M-1, M. \end{cases} \tag{14}$$

The equations (13) and (14) can be expressed in a matrix form as follows:

$$\frac{\partial E_{const}}{\partial X_l} = \mathbf{A} Y \tag{15}$$

$$\frac{\partial E_{const}}{\partial Y_l} = \mathbf{A}^{\mathbf{T}} X \tag{16}$$

where $\frac{\partial E_{const}}{\partial X} = \left(\frac{\partial E_{const}}{\partial X_0}, \ldots, \frac{\partial E_{const}}{\partial X_M}\right)^T$ and $\frac{\partial E_{const}}{\partial Y} = \left(\frac{\partial E_{const}}{\partial Y_0}, \ldots, \frac{\partial E_{const}}{\partial Y_M}\right)^T$, $X = (X_0, \ldots, X_M)^T$, $Y = (Y_0, \ldots, Y_M)^T$, and $\mathbf{A} = \left(a_{i,j}\right)_{i,j=0}^{M}$, such that for all $j = 0, \ldots, M$

$$a_{i,j} = \begin{cases} \mathcal{B}_{i,j} + B_{j,3}(t_3)B_{i,3}(t_3) & i = 0, 1, 2; \\ \mathcal{B}_{i,j} & i = 3, \ldots, M-3; \\ \mathcal{B}_{i,j} + B_{j,3}(t_{M+1})B_{i,3}(t_{M+1}) & i = M-2, M-1, M. \end{cases} \tag{17}$$

Therefore, after finding the partial derivatives of the total energy with respect to the coordinates of the control points, by substituting the Eqs. (11), (12), (15) and (16) into the equations (9) and (10), we obtain the system (8), where $\mathbf{F_X} = (\mathcal{F}_0^X, \ldots, \mathcal{F}_M^X)^T$ and $\mathbf{F_Y} = (\mathcal{F}_0^Y, \ldots, \mathcal{F}_M^Y)^T$.

□

The deformation of the snake is transformed into an optimization of an energy functional with respect to the control points. By iterative method, a series of curves converging to the contour of the object is produced, then we have a deformation of the snake. The descent method is the most popular method used for the snake model, which has higher convergence speed.

Proposition 2 The B-snake control points are determined using the iterative descent method as follows:

$$X^{k+1} = X^k + \gamma(\mathbf{A}^{\mathbf{T}} X^k + \mathbf{F_Y}), \tag{18}$$

$$Y^{k+1} = Y^k + \gamma(\mathbf{A} Y^k + \mathbf{F_X}), \tag{19}$$

where γ is the step size. □

The B-snake achieves convergence through an iterative process of energy minimization when the external force is balanced. This equilibrium position represents the optimal position of the snake to capture the boundaries of the object in an image.

4 Experiments

In this section, we implement the proposed segmentation method on various types of medical images using three different datasets to demonstrate the effectiveness of our approach. In addition, we provide an evaluation framework to establish a comparison with state-of-the-art techniques. To quantify the performance, we use two evaluation metrics: the mean Intersection over Union (IoU) score and the Dice coefficient [23].

The IoU is commonly used in medical image analysis. This metric quantifies the overlap between a segmentation result, and its corresponding ground truth (GT), and yields values ranging from 0 to 1. The Dice score is a popular metric that measures the similarity of segmented regions to the GT by determining their overlap. Its values also range from 0 to 1.

The development was done on Nvidia TESLA A100 GPUs with 40GB and PyTorch v1.8.1 for data processing. The Adam optimizer was used with a learning rate of 10^{-4} and a batch size of 16. To maintain consistency, all images were resized to dimensions of 256×256. The dataset was partitioned by using 72% for training, 13% for validation, and the remaining 15% for testing.

4.1 Datasets

To evaluate the performance of our model, we used three publicly available medical image datasets: the Figshare Brain dataset [20], the Breast Ultrasound dataset [21], and the Foot Ulcer Segmentation Challenge (FUSeg) dataset [22]. Each of these datasets consists of images and corresponding ground truth labels provided by experts. The Figshare Brain Tumor Identification dataset is a larger and more sophisticated dataset containing 3064 MRI images of the brain with a size of 512×512 pixels, including 930 cases of pituitary tumors, 708 cases of meningiomas, and 1426 cases of gliomas.

The second dataset consists of breast ultrasound images collected in 2012 from different patients at the UDIAT Diagnostic Center, part of the Parc Tauli Corporation in Sabadell, Spain. This dataset contains 163 images with an average image size of 760×570 pixels, and each of the images had abnormalities, including 53 images with cancerous masses and 110 with benign lesions.

Finally, the third dataset of the Foot Ulcer Segmentation Challenge (FUSeg) from MICCAI 2021, which contains 810 images collected from hundreds of patients over a 2-year period and is 512×512 pixels in size. For more details on this dataset, see the challenge.

4.2 Results and Discussion

After the automatic initialization of our snake, we generate the initial control polygon, knowing that the number of control points is fixed at 20 points, and then generate the initial spline curve and extract the ROIs. The deformation process of the snake is started until the equilibrium position where the external force is zero. The step size γ of the iterative process in the Eqs. (18) and (19) is set to 0.1. The introduced constraint energy significantly accelerates the snake's convergence. This is demonstrated by the deformation process's execution times, estimated at under 0.1138 seconds for Brain MRI, 0.1553 seconds for Breast Ultrasound, and 0.1321 seconds for FUSeg. Moreover, the average number of iterations required is fewer than 8 for each dataset. Figure 1 shows the segmentation results of the proposed method for random images of the brain tumor MRI, breast ultrasound, and FUSeg datasets, respectively.

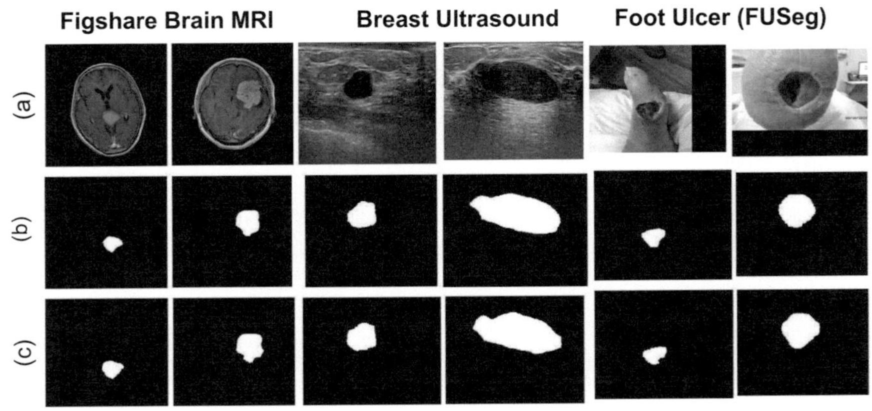

Fig. 1 Lesion segmentation results for three databases: Figshare Brain MRI and breast ultrasound and FUSeg. **a** is the image with the snake curve in blue, **b** is the mask of our method, and **c** is the GT

Table 1 Quantitative comparison to SOTA recent models using the Kvasir-SEG dataset

Method	mIoU score	Mean Dice
U-Net [16]	0.746	0.818
U-Net++ [5]	0.743	0.821
ResU-Net++ [15]	0.793	0.813
SFA [19]	0.611	0.723
Ours	**0.9153**	**0.9206**

The results confirm the effectiveness of our method, as the obtained contours closely match the lesion in the three datasets. Furthermore, the mask of our model is very similar to the GT. This demonstrates the accuracy and robustness of our model in producing segmentation results.

4.3 Evaluation and Comparison to SOTA

To evaluate the performance of our proposed model, a comparison with SOTA models is performed using the Kvasir-SEG database [13]. This database is often used to evaluate the performance of different models.

The Kvasir-SEG dataset contains images of gastrointestinal polyps paired with their corresponding ground truth (GT). These images were carefully annotated by expert physicians and then reviewed by an experienced gastroenterologist. The dataset contains 1000 images, each measuring 352 × 352 pixels.

Table 1 shows the metric results of our model compared to SOTA models. The

Table 2 Quantitative results of the proposed model using the Figshare Brain MRI, the Breast Ultrasound and the FUSeg datasets

Datasets	mIoU	mDice	Accuracy	Error
Figshare brain MRI	0.9068	0.9767	0.9969	0.0032
Breast ultrasound	0.9237	0.9411	0.9936	0.0066
Foot ulcer	0.9301	0.9663	0.9841	0.0167

obtained values of about 91% for the mean IoU score and 92% for the mean Dice are very encouraging. The results demonstrate the effectiveness of the proposed model for medical image segmentation and outperform all existing SOTA models.

4.4 Quantitative Results

To showcase the performance of the proposed model for the three datasets, we highlight in this experiments the quantitative results using various metrics. These metrics include the IoU score, Dice coefficient, Accuracy, and Error. The results are presented in Table 2.

The results clearly indicate that our model achieves excellent performance for all three datasets. This proves the effectiveness of our snake model for segmenting various types of medical images.

5 Conclusion

The proposed model provides high performance in image segmentation despite the difficulties and poor quality of image datasets, such as noise, low contrast, diversity of object shapes, etc. All the techniques we used in this work play an important role in improving the results (constraint energy, AE-Unet, texture filter, unsipervised descreminant analysis, ect). Thus, we can say that this method is a general work that can segment a wide variety of medical image types, despite all the challenges of object detection in medical images. However, despite the encouraging results of our model, segmentation using B-snake models faces inherent limitations in handling complex geometries, particularly in accurately detecting pronounced concavities along object boundaries. These challenges arise from the topological properties of spline curves, which typically only pass through the first and last control points. This limitation can result in the loss of critical information about the object's contour. To address this issue, in our future work we aim to introduce an alignment strategy that better aligns the balanced B-snake curve with the complex regions of the object.

References

1. Kass, M., Witkin, A., Terzopoulos, D.: Snakes: active contour models. Int. J. Comput. Vis. **4**(1), 321–331 (1988)
2. Xu, C., Prince, J.L.: Snakes, shapes, and gradient vector flow. IEEE Trans. Image Process. **7**(3), 359–369 (1998)
3. Blake, A., Curwen, R., Zisserman, A.: A framework for spatiotemporal control in the tracking of visual contours. Int. J. Comput. Vis. **11**(2), 127–145 (1993)
4. Badoual, A., Unser, M., Depeursinge, A.: Texture-driven parametric snakes for semi-automatic image segmentation. Comput. Vis. Image Underst. **188**, 102–793 (2019)
5. Cui, H., Liu, X., Huang, N.: Pulmonary vessel segmentation based on orthogonal fused U-Net++ of chest CT images. In: Proceedings of the International Conference on Medical Image Computing and Computer-Assisted Intervention. Springer, Shenzhen, China, pp. 293–300 (2019)
6. Schumaker, L.: Spline Functions: Basic Theory. Cambridge University Press (2007)
7. Brigger, P., Hoeg, J., Unser, M.: B-spline snakes: a flexible tool for parametric contour detection. IEEE Trans. Image Process. **9**(9), 1484–1496 (2000)
8. Tang, J., Acton, S.T.: Vessel boundary tracking for intravital microscopy via multiscale gradient vector flow snakes. IEEE Trans. Biomed. Eng. **51**(2), 316–324 (2004)
9. De Boor, C.: B(asic)-spline basics. University of Wisconsin-Madison, Mathematics Research Center (1986)
10. Menet, S., Saint-Marc, P., Medioni, G.: Active contour models: overview, implementation and applications. In: IEEE International Conference on Systems, Man, and Cybernetics Conference Proceedings, pp. 194–199 (1990)
11. Long, J., Shelhamer, E., Darrell, T.: Fully convolutional networks for semantic segmentation. In: Proceedings of the IEEE Conference on Computer Vision and Pattern Recognition (CVPR). IEEE, Piscataway, NJ., pp. 3431–3440 (2015)
12. Chen, L.C., Papandreou, G., Schroff, F., Adam, H.: Rethinking atrous convolution for semantic image segmentation, Preprint (2017), arXiv:1706.05587
13. Jha, D., Smedsrud, P.H., Riegler, M.A., Halvorsen, P., de Lange, T., Johansen, D., Johansen, H.D.: Kvasir-seg: A segmented polyp dataset. In: International Conference on Multimedia Modeling. Springer, pp. 451–462 (2020)
14. Milletari, F., Navab, N., Ahmadi, S.A.: V-net: fully convolutional neural networks for volumetric medical image segmentation. In: Conference on 3D Vision (3DV), pp. 565–571 (2016)
15. Xiao, X., Lian, S., Luo, Z., Li, S.: Weighted res-unet for high-quality retina vessel segmentation. In: Conference on Information Technology in Medicine and Education (ITME), pp. 327–331 (2018)
16. Ronneberger, O., Fischer, P., Brox, T.: U-net: Convolutional networks for biomedical image segmentation. In: International Conference on Medical Image Computing and Computer-Assisted Intervention. Springer Cham, pp. 234–241 (2015)
17. Chen, M., Shi, X., Zhang, Y., Wu, D., Guizani, M.: Deep feature learning for medical image analysis with convolutional autoencoder neural network. IEEE Trans. Big Data **7**(4), 750–758 (2017)
18. Fisher, R.A.: The use of multiple measurements in taxonomic problems. Ann. Eugen. **7**(2), 179–188 (1936)
19. Fang, Y., Chen, C., Yuan, Y., Tong, K.Y.: Selective feature aggregation network with area-boundary constraints for polyp segmentation. In: International Conference on Medical Image Computing and Computer Assisted Intervention. Springer, pp. 302–310 (2019)
20. Khan, M.A., Akram, T., Sharif, M., Javed, M.Y., Muhammad, N., Yasmin, M.: An implementation of optimized framework for action classification using multilayers neural network on selected fused features. Pattern Anal. Appl. **22**(4), 1377–1397 (2019)

21. Yap, M.H., Pons, G., Marti, J., Ganau, S., Sentis, M., Zwiggelaar, R., Marti, R.: Automated breast ultrasound lesions detection using convolutional neural networks. IEEE J. Biomed. Health Inform. **22**(4), 1218–1226 (2017)
22. Wang, C., Mahbod, A., Ellinger, I., Galdran, A., Gopalakrishnan, S., Niezgoda, J., Yu, Z.: FUSeg: The Foot Ulcer Segmentation Challenge (2022). arXiv:2201.00414
23. Shelhamer, E., Long, J., Darrell, T.: Fully convolutional networks for semantic segmentation. IEEE Trans. Pattern Anal. Mach. Intell. **39**(4), 640–651 (2017)

MIX
Papier aus verantwortungsvollen Quellen
Paper from responsible sources
FSC® C105338

If you have any concerns about our products,
you can contact us on
ProductSafety@springernature.com

In case Publisher is established outside the EU,
the EU authorized representative is:
Springer Nature Customer Service Center GmbH
Europaplatz 3, 69115 Heidelberg, Germany

Printed by Libri Plureos GmbH
in Hamburg, Germany